Climate Change Handbook

Climate Change Handbook

Edited by Vivian Moritz

SYRAWOOD
PUBLISHING HOUSE

New York

Published by Syrawood Publishing House,
750 Third Avenue, 9th Floor,
New York, NY 10017, USA
www.syrawoodpublishinghouse.com

Climate Change Handbook
Edited by Vivian Moritz

International Standard Book Number: 978-1-64740-126-9 (Hardback)

Cataloging-in-Publication Data

Climate change handbook / edited by Vivian Moritz.
    p. cm.
Includes bibliographical references and index.
ISBN 978-1-64740-126-9
1. Climatic changes. 2. Climatology. 3. Climate change mitigation. I. Moritz, Vivian.
QC903 .C55 2022
551.6--dc23

TABLE OF CONTENTS

                Permissions

                List of Contributors

                Index

PREFACE

This book has been an outcome of determined endeavour from a group of educationists in the field. The primary objective was to involve a broad spectrum of professionals from diverse cultural background involved in the field for developing new researches. The book not only targets students but also scholars pursuing higher research for further enhancement of the theoretical and practical applications of the subject.

The long-term average of the weather is called climate. Changes in the Earth's climate system can produce new weather patterns which may remain the same for more than a few decades. This phenomenon is known as climate change. It takes place when Earth's energy budget is altered by the natural processes inherent to the various parts of the climatic system. The major areas of study within climate change include global warming and its effects. Global warming refers to the long term escalation in the average temperature of the Earth's climate system. It can have a wide range of effects such as regional changes in precipitation, rising sea levels and expansion of deserts. It may also harm food security and diminish crop yields. This book studies and analyses the concepts of climate change and its utmost significance in modern times. It presents studies and researches performed by experts across the globe. This book aims to equip students and experts with the advanced topics and upcoming concepts in this area.

It was an honour to edit such a profound book and also a challenging task to compile and examine all the relevant data for accuracy and originality. I wish to acknowledge the efforts of the contributors for submitting such brilliant and diverse chapters in the field and for endlessly working for the completion of the book. Last, but not the least; I thank my family for being a constant source of support in all my research endeavours.

Editor

The study of potential evapotranspiration in future periods due to climate change in west of Iran

Ahmad Rajabi and Zahra Babakhani
Department of Water Engineering, College of Agriculture, Kermanshah Branch, Islamic Azad University, Kermanshah, Iran

Abstract

Purpose – This study aims to present the climate change effect on potential evapotranspiration (ETP) in future periods.

Design/methodology/approach – Daily minimum and maximum temperature, solar radiation and precipitation weather parameters have been downscaled by global circulation model (GCM) and Lars-WG outputs. Weather data have been estimated according to the Had-CM3 GCM and by A1B, A2 and B1 scenarios in three periods: 2011-2030, 2045-2046 and 2080-2099. To select the more suitable method for ETP estimation, the Hargreaves-Samani (H-S) method and the Priestly–Taylor (P-T) method have been compared with the Penman-Monteith (P-M) method. Regarding the fact that the H-S method has been in better accordance with the P-M method, ETP in future periods has been estimated by this method for different scenarios.

Findings – In all five stations, in all three scenarios and in all three periods, ETP will increase. The highest ETP increase will occur in the A1B scenario and then in the A1 scenario. The lowest increase will occur in the B1 scenario. In the 2020 decade, the highest ETP increase in three scenarios will occur in Khorramabad and then Hamedan. Kermanshah, Sanandaj and Ilam stations come at third to fifth place, respectively, with a close increase in amount. In the 2050 decade, ETP increase percentages in all scenarios are close to each other in all the five stations. In the 2080 decade, ETP increase percentages in all scenarios will be close to each other in four stations, namely, Kermanshah, Sanandaj, Khorramabad and Hamedan, and Ilam station will have a higher increase compared with the other four stations.

Originality/value – Meanwhile, the highest ETP increase will occur in hot months of the year, which are significant with regard to irrigation and water resources.

Keywords Climate change, HadCM3, Lars-WG, Potential evapotranspiration, Statistical downscaling

1. Introduction

Global warming owing to an increase in greenhouse gases leads to climate change all around the earth. Climate change has found a great significance during recent years. Initial studies by intergovernmental panel on climate change (IPCC) indicated changes in different climate parameters, such as temperature, precipitation, snow coverage and sea levels, owing to climate change (IPCC, Climate Change, 2014). The hydrologic cycle has also been affected by this phenomenon (Peng *et al.*, 2013). It is important to study the climate change effect on hydrologic cycle parameters, such as runoff, evapotranspiration (ETP), surface storage and soil moisture, to evaluate conditions of water resources, flood damage amount and hydrologic balance (Xu *et al.*, 2013).

Considering the following factors, population growth, humans' need for food and water resources changes due to climate change, it seems so significant to evaluate the climate change effects on ETP. ETP is needed both in agricultural water requirement planning and hydrologic conceptual models (Allen *et al.*, 1998).

In climate change studies, global circulation models (GCMs) and regional climate models are widely used (IPCC, Climate Change, 2007). GCMs simulate annual or seasonal means of large-scale climate properties. As a matter of fact, their spatial and temporal accuracy do not agree with required hydrologic models' accuracy. By using downscaling methods, we can change GCM outputs in to the data with the scales of a given basin. GCM downscaling statistical methods can predict climate change scenarios in an area better than other methods (Wilby and Dawson, 2013).The present study uses the Lars-WG downscaling model. Lars-WG is a stochastic weather generator which is used to simulate climate data in a given station at present and future conditions affected by climate change. Its data is in the form of daily time series for a set of climate variables, such as precipitation, minimum and maximum temperature and sunshine (Semenov and Barrow, 2007). In the new version of this model, a complete development has been established to provide a powerful model to produce artificial climate data in an extensive climate area. This model has been compared with other stochastic generations with extensive application which use the Markov chain. In the areas with climate variation, it has been specified that if the results are not better than other models, they are at least as good as them (Semenov and Barrow, 2007).

Using the Lars-WG stochastic weather generator, climate change scenarios with high accuracy have been presented to be used in agriculture and hydrology fields (Semenov, 2007). The results have been used to analyse present and future extreme climate events and to study climate change effects on wheat in England. Using the data of 20 stations in different parts of the world with different climates, the ability of the Lars-WG weather generator model to simulate extreme weather events has been studied (Semenov, 2008).

Tiegang *et al.* (2016) studied reference evapotranspiration (ETO) changes in the southwest of China. Results showed that there was a slight downward trend of ETO from 1960 to 2010 and spatially increasing trend from northeast to southeast in an annual time scale. Katerji *et al.* (2016) made a comparison between Allen and Katerji and Perrier formulas for calculating daily ETO. These two formulas have been compared for both the observed period and the future period (2070-2100). Liu *et al.* (2017) investigated the spatiotemporal patterns of evapotranspiration (ET) and primary driving meteorological variables based on a historical and RCP 8.5 scenario daily data set from 40 weather stations over the 3H Plain using linear regression, spline interpolation method, a partial derivative analysis and multivariate regression. Temperature-based methods may be particularly prone to error when extrapolated into the future to assess effects of greenhouse gas-driven warming on potential evapotranspiration (PET). Such biases are derived from the fact that increasing temperature by increasing solar radiation would likely cause a greater increase in PET than would increasing temperature by increasing greenhouse gases, because radiation provides the energy driving ETP (King *et al.*, 2015).

In different research works, the climate change effect on ETP has been evaluated. The climate change effect on hydrologic cycle parameters such as reference ETP was studied in the Guishui River Basin in China (Guo *et al.*, 2014). Reference ETP change in the Haihe River Basin in China owing to climate change was evaluated in the observed period and in the future period (Xu *et al.*, 2013).

During the recent years, many researchers tried to calibrate the Hargreaves model (Gavilan *et al.*, 2006; Tabari and Talaee, 2011; Berti *et al.*, 2014; Shiri *et al.*, 2014, 2015; Marti *et al.*, 2015; Cobaner *et al.*, 2016; Xu *et al.*, 2016; Feng *et al.*, 2017).

The results obtained from most of the mentioned studies indicated that ETP has an increasing trend owing to climate change. This leads to an increase in water requirements of plants and presents worry regarding water resource management.

The main objective of the present study is to provide answers to the following questions:

Q1. Which of the mentioned formulas can simulate the amounts of ETP in the western part of Iran?

Q2. If the more proper ET formula is applied in climate change scenarios for future, what results can be observed?

Changes in ETP in regards to meteorological parameters and agricultural production practice should be given greater attention. It receives more significance when we consider the hydrological process and water management in the coming decades regarding climate change. It is in order to avoid problems such as the overexploitation of groundwater resources.

2. Materials and methods
2.1 Data
The area which is studied includes five provinces in the west of Iran, which is surrounded by mountains and has a semi-arid weather (Figure 1). The required data have been received from the Iran meteorological organization. Regarding the fact that circulation scenarios have been calculated based on period weather parameters since 1960, the stations which possess sufficient data in this period are Kermanshah, Hamedan, Khorramabad, Sanandaj and Ilam synoptic stations. Therefore, the data of these stations have been used to continue the present study (Table I).

2.2 Lars-WG stochastic weather generator
Lars-WG model has been established according to the weather generator series, which has been analysed by Racsko *et al.* (1991). In Lars-WG version 5, minimum and maximum

Figure 1. Location of the stations used in Iran

temperatures for dry and wet days are estimated for each month by a semi-empirical distribution which is calculated by auto-correlation and a monthly cross-correlation coefficient. Applying these changes leads to a significant improvement in the extreme temperature simulation (Semenov and Stratonovitch, 2010).

2.3 Climate change scenarios

For more climate-predicting models, there are different circulation scenarios, including B1, A1B and A1. In the B1 scenario, the assumption is based on an endurable world, rapid change in economic constructions, human rights equality development and a care for protecting our environment. With this assumption, greenhouse gas circulation can be controlled and a pollutant controlling programme for factories and industries will be carried out. In the A1B scenario, the assumption is based on a wealthy world with a rapid economic growth (3 per cent per year), population growth decrease (27 per cent per year), rapid new and effective technology, cultural and economic convergence and a fundamental decrease in regional differences. In the A2 scenario, the assumption is based on the existence of a separate world. Different cultural identities in different parts of the world lead to more differences in the world and a decrease in international cooperation. Local customs and growth increase are emphasized, and there is less emphasis on economic issues (1/65 per cent per year) (IPCC, Climate Change, 2001; IPCC, Climate Change, 2007). In the present study, the Had-CM3 GCM has been used.

2.4 Potential evapotranspiration

There are different methods for estimating daily ETP. The most reliable and common method is the Penman–Monteith (P-M) method. In this method, parameters such as the relative moisture, wind speed and sunshine hours are needed, besides minimum and maximum daily temperature. In this study, regarding the fact that the data cannot be downscaled by Lars-WG, estimation of ETP in future periods is carried out by using methods which make it possible to calculate ETP with present variables. Among different methods, two were selected and used, namely, the Hargreaves-Samani (H-S) method and the Priestly-Taylor (P-T) method. The reason is that these methods possess an almost high accuracy and also need less meteorological data, compare with other methods (Samani, 2000). According to this fact, the P-M method is considered as the reference method. The accuracy of the two methods is evaluated by the P-M method. To compare the results, the values of the root mean square error (RMSE), mean absolute error (MAE) and mean bias error (MBE) statistical indices are used.

The description of the details of ETP methods is available in previous work (Allen et al., 1998; Priestley and Taylor, 1972; Samani, 2000).

Table I. Properties of the desired synoptic stations

Station	Latitude (°N)	Longitude (°E)	Altitude (m a.s.l.)
Hamedan	35° 12'	48° 43'	1,679.7
Khorramabad	33° 26'	48° 17'	1,147.8
Kermanshah	34° 21'	47° 09'	1,318.6
Sanandaj	35° 20'	47° 00'	1,373.4
Ilam	33° 38'	46° 26'	1,337.0

3. Results and discussion

Monthly ETP in the present statistical period available in the mentioned five synoptic stations was calculated by using REF-ET software. Table II presents monthly ETP mean of different methods. Table III demonstrates the difference percentage between monthly ETP by the two methods H-S and P-T with the P-M method (as the reference method) in different stations in the observed period.

As shown, the ETP difference percentage calculated by the H-S method with the P-M method is less than the P-T method.

To make sure about selecting the more suitable method for calculating ETP in future periods, RMSE, MAE and MBE evaluating indices have been used. The results are presented in Table IV in short.

It can be observed that in all stations, the above evaluating indices indicate that ETP amounts of the H-S method are closer to ETP amounts of the P-M method as the reference method in all months. Therefore, this method is used in future periods.

3.1 Generating climate change scenario in future

Daily precipitation, minimum and maximum temperatures and solar radiation data of the mentioned stations have been provided in the observed period by the required format of Lars-WG model, and the model input files have been established according to this. Then, downscaled data of each weather parameter have been simulated for three 20-year periods:

(1)   first period (2020 decade): 2011-2030;

(2)   second period (2050 decade): 2045-2065; and

(3)   third period (2080 decade): 2080-2090.

In the present study, the effects of different scenarios, A1B, A2 and B1, of the GCM model are studied. The GCM which has been used is the Had-CM3 model.

Considering the results of applying the Lars-WG model, the following points are worthy of mention:

* Minimum and maximum temperatures increase considerably in future periods.

* In future periods, minimum and maximum temperature increase peak will be in warm months of the year (July and August) and February.

* The highest solar radiation increase will be in warm months of the year (July and August) and also February.

3.2 ETP estimation in future

After calculating the minimum and maximum temperatures and solar radiation weather parameters in future periods by different scenarios, ETP is estimated by the H-S method. Figures 2 to 6 present daily ETP changes in future periods compared with the observed period in the mentioned stations. As it is obvious, in all five stations, ETP changes in future periods will be almost similar. In all three scenarios and in all three periods, ETP will increase. The 2080 decade will have the highest increase and the 2020 decade will have the lowest increase in ETP. The highest increase in ETP will be in the A1B scenario and then in the A1 scenario, and the lowest increase will be in the B1 scenario. This fact is in accordance with the assumptions of the scenarios. The lowest ETP increase will be in December and January. The highest increase will be in May, July and August. During this period, there will be a lower increase in June.

Table II. Calculated ETP mean by different methods in the observed period (mm/day)

ETP methods	January	February	March	April	May	June	July	August	September	October	November	December	ANN
Kerm													
P-M	1.1	1.7	2.7	3.8	5.1	7.0	7.5	7.1	5.5	3.4	1.8	1.1	4.0
H-S	1.1	1.6	2.6	3.9	5.4	7.4	8.0	7.3	5.5	3.4	1.9	1.2	4.1
P-T	0.8	1.3	2.2	3.2	4.2	5.3	5.3	4.7	3.4	2.1	1.2	0.8	2.9
Ham													
P-M	0.8	1.3	2.4	3.7	4.8	6.7	7.5	7.0	5.1	3.1	1.6	1.0	3.8
H-S	0.8	1.2	2.3	3.7	4.8	6.8	7.3	6.7	5.1	3.1	1.6	0.9	3.7
P-T	0.7	1.1	1.9	3.0	3.9	5.0	5.0	4.4	3.2	1.9	1.0	0.6	2.7
Sanan													
P-M	0.9	1.3	2.4	3.7	4.8	6.4	6.9	6.3	4.9	3.0	1.5	1.0	3.6
H-S	0.9	1.4	2.4	3.8	5.2	7.0	7.6	6.9	5.3	3.2	1.7	1.1	3.9
P-T	0.7	1.2	2.1	3.2	4.1	5.2	5.2	4.5	3.4	2.0	1.1	0.7	2.8
Khora													
P-M	1.1	1.7	2.6	3.5	4.7	6.0	6.2	5.9	4.8	3.2	1.8	1.2	3.6
H-S	1.3	1.9	2.9	4.2	5.8	7.6	8.1	7.4	5.9	3.7	2.1	1.4	4.4
P-T	0.9	1.4	2.2	3.1	4.0	4.8	4.8	4.4	3.3	2.1	1.3	0.9	2.8
Ilam													
P-M	1.2	1.7	2.7	3.9	5.6	7.2	7.5	7.0	5.5	3.5	1.9	1.2	4.1
H-S	1.1	1.5	2.3	3.6	4.9	6.1	6.5	6.0	4.6	3.0	1.7	1.2	3.5
P-T	1.0	1.5	2.3	3.5	4.4	5.3	5.4	4.7	3.5	2.2	1.3	0.9	3.0

Table III. ETP Difference percentage by H-S and P-T methods with P-M method in the observed period

ETP methods	January	February	March	April	May	June	July	August	September	October	November	December
Kerm												
HS	−0.6	−5.9	−4.3	1.9	6.1	5.0	6.4	2.4	0.7	−0.6	5.8	4.5
PT	−23	−21	−20	−16	−17	−24	−29	−34	−37	−39	−35	−31
Ham												
HS	−7.0	−7.0	−6.3	−1.7	0.2	1.0	−3.0	−4.8	−0.6	−2.3	0.5	−3.6
PT	−19	−13	−19	−21	−19	−25	−33	−37	−36	−38	−36	−32
Sana												
HS	6.4	2.8	0.9	3.0	8.4	10.0	8.7	10.0	9.3	5.9	13.7	9.7
PT	−15	−10	−13	−13	−13	−18	−24	−27	−30	−33	−29	−28
Khor												
HS	17.5	10.0	10.2	19.0	22.2	27.7	29.7	27.0	21.1	16.3	20.3	17.7
PT	−17	−16	−15	−11	−15	−19	−22	−25	−31	−33	−28	−25
Ilam												
HS	−5.2	−13	−13	−9.9	−12	−14	−13	−14	−15	−14	−8.0	0.9
PT	−18	−14	−12	−12	−21	−26	−28	−32	−35	−36	−32	−26

Table IV. Evaluating indices ETP estimation methods (mm/day)

Indices	January	February	March	April	May	June	July	August	September	October	November	December	ANN
Kermanshah													
RMSE													
HS	0.2	0.2	0.4	0.4	0.8	1.0	1.2	0.9	0.8	0.5	0.3	0.2	0.6
PT	0.3	0.4	0.6	0.8	1.0	1.9	2.4	2.5	2.2	1.5	0.7	0.4	1.2
MAE													
HS	10	10	11	7.6	10.5	8.5	9.9	8.9	11	10	11	11	10
PT	29	26	25	20	22.3	35	45	53	62	65	54	45	40
MBE													
HS	0.0	0.1	0.1	0.0	−0.4	0.0	0.0	−0.2	0.0	0.0	−0.1	0.0	−0.1
PT	0.3	0.4	0.6	0.6	0.8	1.7	2.3	2.4	2.1	1.4	0.6	0.4	1.1
Hamedan													
RMSE													
HS	0.2	0.2	0.4	0.4	0.9	1.3	1.5	1.3	1.0	0.5	0.3	0.2	0.7
PT	0.3	0.3	0.6	0.9	1.2	1.9	2.7	2.8	2.0	1.3	0.7	0.4	1.3
MAE													
HS	20	18	14	9.3	14.5	13	15	15	14	10	15	16	14
PT	26	20	26	28	27.2	38	54	64	63	66	57	48	43
MBE													
HS	0.1	0.1	0.1	0.1	−0.2	0.0	0.2	0.3	0.0	0.1	0.0	0.0	0.1
PT	0.2	0.2	0.5	0.8	0.8	1.7	2.5	2.6	1.9	1.2	0.6	0.3	1.1
Sanandaj													
RMSE													
HS	0.2	0.2	0.3	0.4	0.8	1.1	1.4	1.2	0.9	0.5	0.3	0.2	0.6
PT	0.2	0.2	0.5	0.7	0.9	1.4	1.9	1.9	1.6	1.1	0.5	0.4	0.9
MAE													
HS	12	10	10	8.7	11.4	11	13	13	12	12	15	14	12
PT	19	16	17	18	19.1	26	35	41	46	51	40	40	31
MBE													
HS	0.0	0.0	0.0	0.0	−0.5	0.0	0.0	−0.6	0.0	0.0	−0.2	0.0	−0.3
PT	0.1	0.1	0.3	0.5	0.6	1.2	1.8	1.7	1.5	1.0	0.4	0.3	0.8

(continued)

Table IV.

Indices	January	February	March	April	May	June	July	August	September	October	November	December	ANN
Khorramabad													
RMSE													
HS	0.3	0.4	0.5	0.9	1.4	2.3	2.6	2.2	1.5	0.9	0.5	0.4	1.2
PT	0.3	0.4	0.6	0.7	1.0	1.5	1.7	1.8	1.8	1.2	0.7	0.5	1.0
MAE													
HS	18	17	15	18	20.3	25	25	23	22	20	20	22	20
PT	25	22	22	18	23.2	27	31	37	49	51	42	36	32
MBE													
HS	0.0	0.0	−0.3	−1	−1.1	−2	−2	−1.6	−1	0.0	−0.4	0.0	−0.8
PT	0.2	0.3	0.4	0.3	0.6	1.1	1.3	1.4	1.4	1.0	0.5	0.3	0.7
Ilam													
RMSE													
HS	0.2	0.3	0.5	0.6	1.0	1.3	1.4	1.3	1.1	0.7	0.3	0.2	0.7
PT	0.3	0.3	0.5	0.6	1.3	2.0	2.3	2.4	2.0	1.4	0.7	0.4	1.2
MAE													
HS	10	16	17	14	16.7	19	19	19	19	21	15	10	16
PT	24	18	16	15	29.2	38	43	52	57	56	48	36	36
MBE													
HS	0.1	0.2	0.4	0.4	0.7	1.0	1.0	1.0	0.8	0.5	0.1	0.0	0.5
PT	0.2	0.2	0.3	0.5	1.1	1.8	2.0	2.2	1.9	1.2	0.6	0.3	1.0

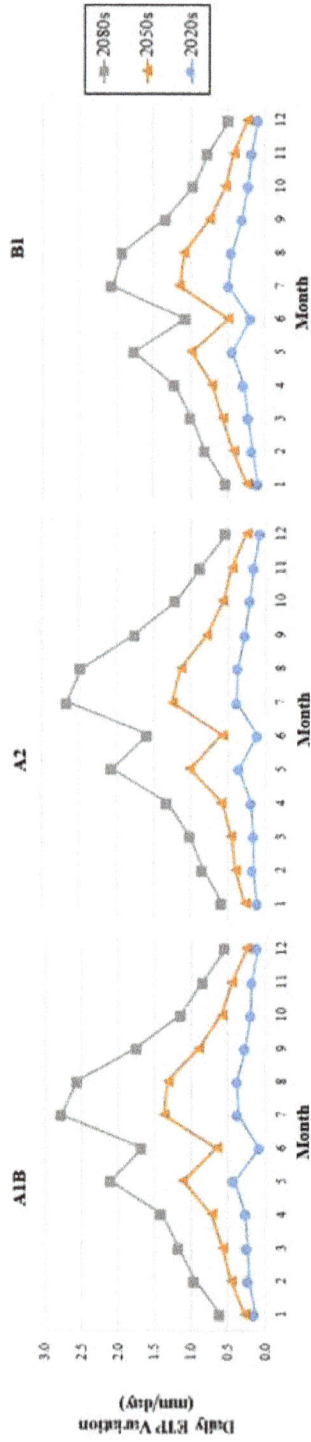

Figure 2. Daily ETP changes compared with the observed period in Kermanshah station

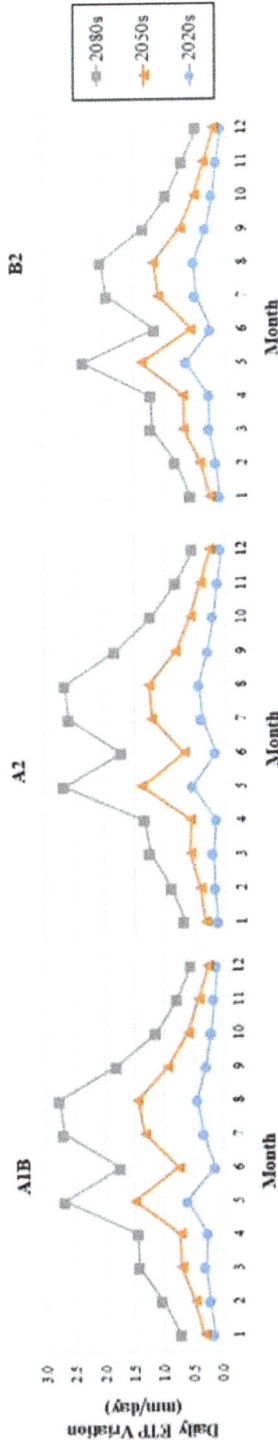

Figure 3. Daily ETP changes compared with the observed period in Hamedan station

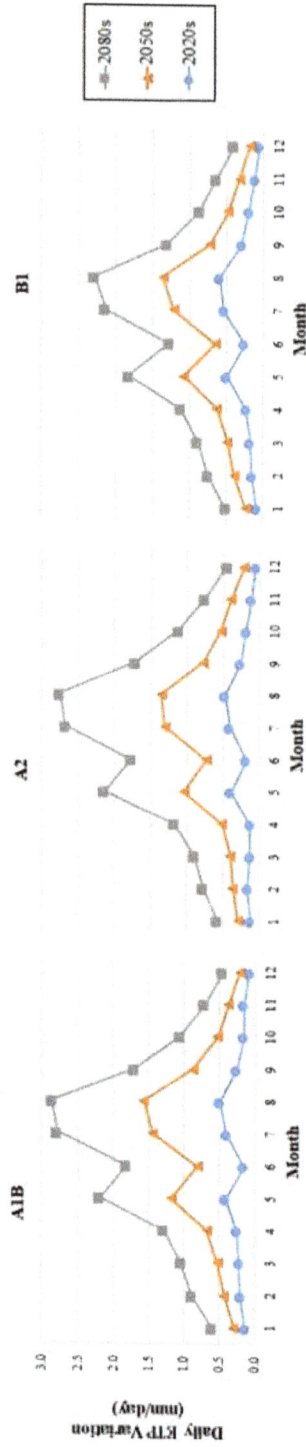

Figure 4. Daily ETP changes compared with the observed period in Sanandaj station

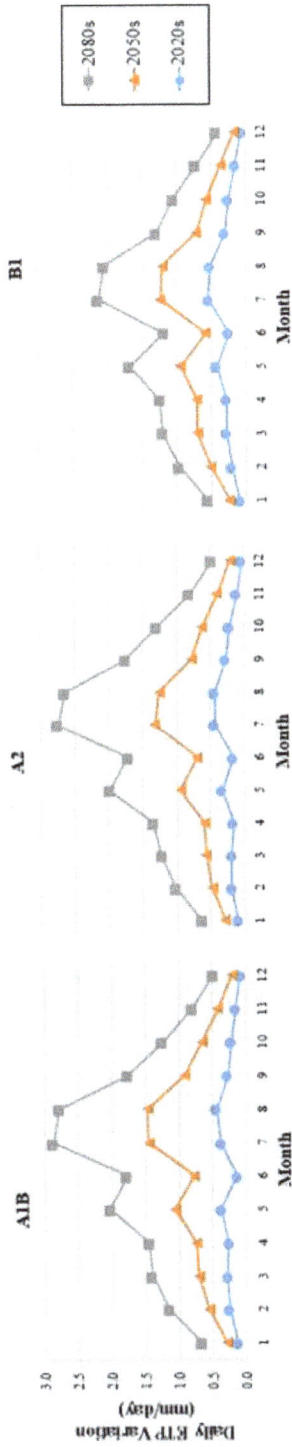

Figure 5. Daily ETP changes compared with the observed period in Khorramabad station

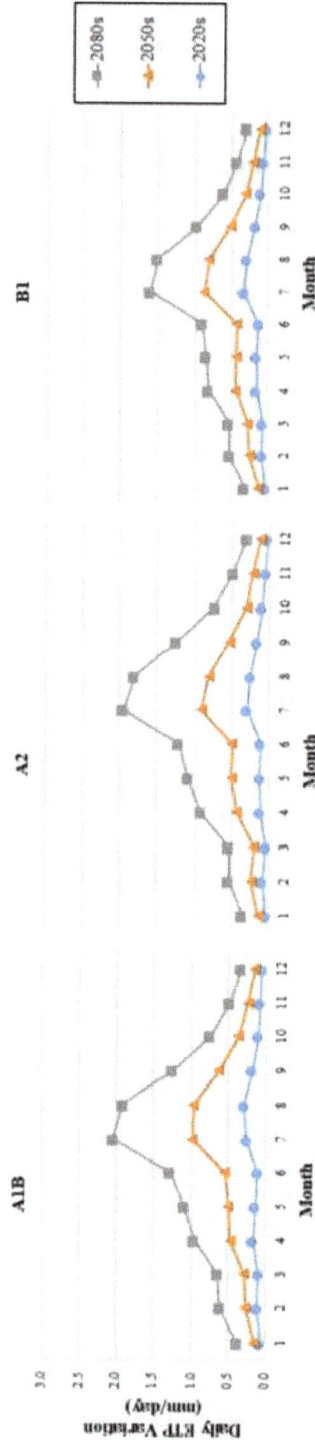

Figure 6. Daily ETP changes compared with the observed period in Ilam station

In 2020s, the highest ETP increase percentage in all three scenarios occurs in the Khorramabad station and then in the Hamedan station; Kermanshah, Sanandaj and Ilam stations come third to fifth, respectively. In 2050s, ETP increase percentage in the scenarios will be close to each other in all five stations in the B1 scenario. It will be about 8.5 per cent in B1, 10.5 per cent in A2 and about 13 per cent in A1B scenario. In the 2080 decade, also, ETP increase percentage will be close to each other in all scenarios in four stations, Kermanshah, Sanandaj, Hamedan and Khorramabad. It will be about 11.5 per cent in B1, about 15.5 per cent in A2 and about 16.5 per cent in A1B scenario. Meanwhile, Ilam station will have a higher increase compared with the other four stations.

3.3 Uncertainty

Most of the uncertainty in climate change studies in hydrology and water resources is owing to GCM projection and regional changes (Berthelot *et al.*, 2005). Anyway, uncertainty is a challenge in hydrologic and weather parameters' prediction accuracy. Besides, a main part of uncertainty is caused by GCMs, circulations scenarios and hydrologic models (Wilby *et al.*, 2006). In A2, B1 and A1B scenarios, which have been used in the present study, there is a considerable uncertainty. The assumptions used in the scenarios may not occur in future.

On the basis of uncertainty analysis results, the contributions of different climatic variables to ET changes at different months and stations can be revealed. In general, solar radiation and daily minimum and maximum temperature are the three major contributors to PET changes in the future periods. However, causes of future PET changes are varied at different stations and months.

The next stage is the uncertainty in predicting methods, analysing climate scenarios and spatial and temporal data accuracy. This uncertainty can affect analysis evaluation accuracy directly. But the consistent trend of climate change predictions and circulation scenarios indicates that the effect of this uncertainty in the prediction results and analysis is not significant (Guo *et al.*, 2014).

The way of selecting the base period is an important factor of uncertainty and will affect the results of analysis. Therefore, the next stage of the present study is the absolute study of uncertainty.

4. Conclusion

It is obvious that in warm months of the year, ETP, which is so important in agriculture, has a higher increase compared to the cold months of the year. This causes worry about providing required water to agricultural products, while there is a decrease in precipitation owing to the climate change. In the present study, ETP has been estimated in the observed period in three future periods by three climate change scenarios, A1B, A2 and B1, by applying the Lars-WG downscaling model and by making use of the H-S equation. Regarding the length of the statistical period of the stations in the west of Iran and the parameters needed by the downscaling model, five stations have been selected for this purpose, namely, Kermanshah, Hamedan, Khorramabad, Sanandaj and Ilam. The results of the study for future periods increase considerably. In future periods, minimum and maximum temperature increase peak will be in warm months of the year (July and August) and in February. ETP increase is sensible in all stations and scenarios, especially in the irrigation season (warm months of the year), in which the highest increase amount can be seen. The results indicate that improved plans and policies in the B1 scenario are effective in the third period and lead to less ETP increase in this period compared with the other two scenarios.

References

Allen, R.G., Pereira, L.S., Raes, D. and Smith, M. (1998), "Crop evapotranspiration: guidelines for computing crop water requirements", *FAO Irrigation and Drainage Paper No. 56*, FAO, Rome, p. 301.

Berthelot, M., Friedlingstein, P., Ciais, P., Dufresne, J.L. and Monfray, P. (2005), "How uncertainties in future climate change predictions translate into future terrestrial carbon fluxes", *Global Change Biology*, Vol. 11 No. 6, pp. 959-970.

Berti, A., Tardivo, G., Chiaudani, A., Rech, F. and Borin, M. (2014), "Assessing reference evapotranspiration by the Hargreaves method in North-Eastern Italy", *Agricultural Water Management*, Vol. 140, pp. 20-25.

Cobaner, M., Citakoglu, H. and Haktanir, T. (2016), "Modifying Hargreaves-Samani equation with meteorological variables for estimation of reference evapotranspiration in turkey", *Hydrology Research*, available at: http://dx.doi.org/10.2166/nh.2016.217

Feng, Y., Jia, Y., Cui, N.B., Zhao, L., Li, C. and Gong, D.Z. (2017), "Calibration of Hargreaves model for reference evapotranspiration estimation in Sichuan Basin of southwest China", *Agricultural Water Management*, Vol. 181, pp. 1-9.

Gavilan, P., Lorite, I.J., Tornero, S. and Berengena, J. (2006), "Regional calibration of Hargreaves equation for estimating reference ET in a semiarid environment", *Agricultural Water Management*, Vol. 81 No. 3, pp. 257-281.

Guo, B., Zhang, J., Gong, H. and Cheng, X. (2014), "Future climate change impacts on the ecohydrology of Guishui river Basin, China", *Ecohydrol Hydrobiol*, Vol. 14 No. 1, pp. 55-67.

IPCC, Climate Change (2001), "The science of climate change", Contribution of Working Group I to the Second Assessment Report of the Intergovernmental Panel on Climate Change, Cambridge University Press, Cambridge, MA.

IPCC, Climate Change (2007), *Fourth Assessment Report of the Intergovernmental Panel on Climate Change*, WMO.

IPCC, Climate Change (2014), "Impacts, adaptation, and vulnerability", *IPCC Working Group II Contribution to AR5*, IPCC, Geneva.

Katerji, N., Rana, G. and Ferrara, R.M. (2016), "Actual evapotranspiration for a reference crop within measured and future changing climate periods in the Mediterranean region", *Theoretical and Applied Climatology*, doi: 10.1007/s00704-016-1826-6.

King, D.A., Bachelet, D.M., Symstad, A.J., Ferschweiler, K. and Hobbins, M. (2015), "Estimation of potential evapotranspiration from extraterrestrial radiation, air temperature and humidity to assess future climate change effects on the vegetation of the northern Great Plains, USA", *Ecological Modelling*, Vol. 297, pp. 86-97.

Liu, Q., Yan, C. and Ju, H. (2017), "Impact of climate change on potential evapotranspiration under a historical and future climate scenario in the Huang-Huai-Hai Plain, China", *Theoretical and Applied Climatology*, doi: 10.1007/s00704-017-2060-6.

Marti, P., Zarzo, M., Vanderlinden, K. and Girona, J. (2015), "Parametric expressions for the adjusted Hargreaves coefficient in Eastern Spain", *Journal of Hydrology*, Vol. 529, pp. 1713-1724.

Peng, H., Jia, Y.W., Qiu, Y.Q. and Niu, C.W. (2013), "Assessing climate change impacts on the ecohydrology of the Jinghe river Basin in the Loess Plateau, China", *Hydrological Sciences Journal*, Vol. 58 No. 3, pp. 651-670.

Priestley, C.H.B. and Taylor, R.J. (1972), "On the assessment of surface heat flux and evapotranspiration using large scale parameters", *Monthly Weather Review*, Vol. 100 No. 2, pp. 81-92.

Racsko, P., Szeidl, L. and Semenov, M. (1991), "A serial approach to local stochastic weather models", *Ecological Modelling*, Vol. 57 Nos 1/2, pp. 27-41.

Samani, Z. (2000), "Estimating solar radiation and evapotranspiration using minimum climatological data", *Journal of Irrigation and Drainage Engineering*, Vol. 126 No. 4, pp. 265-267.

Semenov, M.A. (2007), "Development of high-resolution UKCIP02-based climate change scenarios in the UK", *Agricultural and Forest Meteorology*, Vol. 144 Nos 1/2, pp. 127-138.

Semenov, M.A. (2008), "Simulation of extreme weather events by a stochastic weather generator", *Climate Research*, Vol. 35, pp. 203-212.

Semenov, M.A. and Barrow, E.M. (2007), LARS-WG A Stochastic Weather Generator for Use in Climate Impact Studies, User Manual.

Semenov, M.A. and Stratonovitch, P. (2010), "The use of multi-model ensembles from global climate models for impact assessments of climate change", *Climate Research*, Vol. 41, pp. 1-14.

Shiri, J., Nazemi, A.H., Sadraddini, A.A., Landeras, G., Kisi, O., Fard, A.F. and Marti, P. (2014), "Comparison of heuristic and empirical approaches for estimating reference evapotranspiration from limited inputs in Iran", *Computers and Electronics in Agriculture*, Vol. 108, pp. 230-241.

Shiri, J., Sadraddini, A.A., Nazemi, A.H., Marti, P., Fard, A.F., Kisi, O. and Landeras, G. (2015), "Independent testing for assessing the calibration of the Hargreaves-Samani equation: new heuristic alternatives for Iran", *Computers and Electronics in Agriculture*, Vol. 117, pp. 70-80.

Tabari, H. and Talaee, P.H. (2011), "Local calibration of the Hargreaves and Priestley Taylor equations for estimating reference evapotranspiration in arid and cold climates of Iran based on the Penman-Monteith model", *Journal of Hydrology Engineering*, Vol. 16 No. 10, pp. 837-845.

Tiegang, L., Longguo, L., Jianbin, L., Chao, L. and Wenhua, Z. (2016), "Reference evapotranspiration change and its sensitivity to climate variables in Southwest China", *Theoretical and Applied Climatology*, Vol. 125 Nos 3/4, pp. 499-508.

Wilby, R.L. and Dawson, C.W. (2013), "The statistical DownScaling model: insights from one decade of application", *International Journal of Climatology*, Vol. 33 No. 7, pp. 1707-1719.

Wilby, R.L., Whitehead, P.G., Wade, A.J., Butterfield, D., Davis, R.J. and Watts, G. (2006), "Integrated modelling of climate change impacts on water resources and quality in a lowland catchment: river Kennet", *Journal of Hydrology*, Vol. 330 Nos 1/2, pp. 204-220.

Xu, J.Z., Wang, J.M. and Wei, Q. (2016), "Symbolic regression equations for calculating daily reference evapotranspiration with the same input to Hargreaves-Samani in arid China", *Water Resources Management*, Vol. 30 No. 6, pp. 2055-2073.

Xu, Y.P., Zhang, X.J., Ran, Q.H. and Tian, Y. (2013), "Impact of climate change on hydrology of upper reaches of Qiantang river", *Journal of Hydrology*, Vol. 483, pp. 51-60.

Further reading

Koczot, K.M., Markstrom, S.L. and Hay, L.E. (2011), "Effects of baseline conditions on the simulated hydrologic response to projected climate change", *Earth Interactions*, Vol. 15 No. 27, pp. 1-23.

Xing, W., Wang, W., Shao, Q., Peng, S., Yu, Z., Yong, B. and Taylor, J. (2014), "Changes of reference evapotranspiration in the Haihe River Basin: present observations and future projection from climatic variables through multi-model ensemble", *Global and Planetary Change*, Vol. 115, pp. 1-15.

Corresponding author

Ahmad Rajabi can be contacted at: ahmadrjb@yahoo.com

2

A Ricardian valuation of the impact of climate change on Nigerian cocoa production

William M. Fonta
West African Science Service Centre on Climate Change and Adapted Land Use, Ouagadougou, Burkina Faso

Abbi M. Kedir
Management School, University of Sheffield, Sheffield, UK

Aymar Y. Bossa, Karen M. Greenough and Bamba M. Sylla
West Africa Science Service Center on Climate Change and Adapted Land Use, Ouagadougou, Burkina Faso, and

Elias T. Ayuk
United Nations University Institute for Natural Resources in Africa (UNU-INRA), University of Ghana, Accra, Ghana

Abstract

Purpose – The purpose of this study is to examine the relative importance of climate normals (average long-term temperature and precipitation) in explaining net farm revenue per hectare (NRh) for supplementary irrigated and rainfed cocoa farms in Nigeria.

Design/methodology/approach – NRh was estimated for 280 cocoa farmers sampled across seven Nigerian states. It was regressed on climate, household socio-economic characteristics and other control variables by using a Ricardian analytical framework. Marginal calculations were used to isolate the effects of climate change (CC) on cocoa farm revenues under supplementary irrigated and rainfed conditions. Future impacts of CC were simulated using Six CORDEX regional climate model (RCM) ensemble between 2036-2065 and 2071-2100.

Findings – Results indicate high sensitivity of NRh to Nigerian climate normals depending on whether farms use supplementary irrigation. Average annual temperature increases and precipitation decreases are associated with NRh losses for rainfed farms and gains for supplementary irrigated cocoa farms. Projections of future CC impacts suggest a wide range of NRh outcomes on supplementary irrigated and rainfed farm revenues, demonstrating the importance of irrigation as an effective adaptation strategy in Nigeria.

Originality/value – This paper uses novel data sets for simulating future CC impacts on land values in Nigeria. CORDEX data constitute the most comprehensive RCMs projections available for Africa.

Keywords Nigeria, Climate change, Climate change projections, Net revenue per farm hectare, Ricardian valuation

1. Introduction

Agriculture is a very climate-sensitive sector, and climate-smart agriculture is the way forward to increase agricultural productivity in a sustainable manner. According to the *Fifth Assessment Report* of the Intergovernmental Panel on Climate Change, climate change (CC) will amplify existing stresses on agricultural systems, particularly those in Africa for several reasons (IPCC, 2014). First, about 80 per cent of African agriculture is still mainly rainfed and therefore highly vulnerable to changes in climatic conditions such as droughts, higher temperatures and reduced precipitation levels (World Bank, 2008; Hassan, 2010). Second, African agriculture is mostly extensive and practiced on relatively fragile environments and poor quality soils, with little use of modern inputs and farming methods to cope with CC impacts (Mendelsohn and Dinar, 1999; Mano and Nhemachena, 2006). Third, the presence of multiple stresses such as endemic poverty, weak institutions, inadequate health services, limited access to capital and markets, poor infrastructure and technology and conflicts over natural resources reduce farm households' adaptive capacity to manage the numerous vagaries of CC (IPCC, 2007; Odingo, 2008; Hassan, 2010). Finally, most African governments devote meager financial resources to their agricultural sectors, thereby reducing investment in scientific research needed to better understand and respond to CC impacts (Hassan, 2010). This investment neglect is persistent across many African countries despite the well-recognized poverty-reducing impact of agriculture (Christiaensen *et al.*, 2006).

There is growing concern that CC will be more severe in the west African region because of the problems of huge spatio-temporal rainfall variability, population pressure and limited adaptation and mitigation capacities (Berg *et al.*, 2009; Ibrahim *et al.*, 2014; IPCC, 2014). The impact is expected to be even more severe for a country like Nigeria, where rainfed agriculture is the main source of livelihood for about 50-60 per cent of the population and accounts for over 25.5 per cent of the nation's gross domestic product in real terms (National Bureau of Statistics, 2017). Most of the agricultural value added in Nigeria is mainly from the plantation tree crops sub-sector (i.e. cocoa, rubber, coffee and palm produce) which is highly climate-sensitive. For instance, between 2006 and 2012, a total of about 1.2 million tons of output from plantation tree crops were exported from Nigeria, representing over 40 per cent of the total export value derived from agricultural exports alone (Central Bank of Nigeria, 2014). Cocoa's contribution to the nation's economic development is not in doubt. No other agricultural export commodity has earned more foreign exchange than cocoa. The sub-sector offers substantial employment, both directly and indirectly, and supplies significant volumes of raw materials to cocoa-producing states, as well as revenue to their respective provincial governments. Because of its importance, the Federal Government of Nigeria, concerned with diversifying the nation's export base, has identified cocoa as the most important export tree crop.

However, there are growing concerns that CC may be impacting negatively on cocoa production in Nigeria. The first real evidence of CC impacts on cocoa production was during the 1972/73 drought, when production declined from about 216,000 metric tons to less than 150,000 metric tons (Central Bank of Nigeria, 2014). Thereafter, the production trend has consistently declined despite the agricultural conditionality of the structural adjustment program that was introduced immediately after the major drought occurrence. Some of the key questions that still remain unanswered and of great concern to policy makers are follows:

Q1. What proportion of change in cocoa production is due to the impacts of CC?

Q2. What are the economic implications of CC on cocoa production in Nigeria?

Q3. How do cocoa farmers in Nigeria adapt to changing climatic conditions?

Q4. As climate variables worsen, what is the likely future for this important plantation tree crop in the country?

The academic literature on CC impact on Nigerian cocoa agriculture in general is still very scanty, despite the importance of cocoa as a major export earner. In fact, a review of the extant literature suggests that there are few peer-reviewed documented studies for Nigeria. These include the works of Lawal and Emaku (2007), Omolaja *et al.* (2009), Ajewole and Iyanda (2010), Ajayi *et al.* (2010), Fonta *et al.* (2011), Nwachukwu *et al.* (2012) and Oyekale (2017). Lawal and Emaku (2007) find that while temperature is positively correlated with cocoa yield, rainfall and relative humidity are negatively correlated with cocoa yield. Omolaja *et al.* (2009) find that increasing precipitation and favorable temperature promote the flowering intensity of cocoa in Nigeria. For Ajewole and Iyanda (2010), cocoa yield is less sensitive to decreasing rainfall but more sensitive to increasing temperatures. In line with these findings, Ajayi *et al.* (2010) find cocoa yield also less sensitive to annual rainfall levels, compared to temperature increases. Contrastingly, Nwachukwu *et al.* (2012) finds cocoa productivity to be more rainfall-sensitive. Similar findings were obtained by Fonta *et al.* (2011) and Oyekale (2017). In both papers, net cocoa revenue per household farm hectare and yields was more sensitive to increasing precipitations.

Given the dearth of empirical literature on the impacts of CC on Nigerian agriculture, this study seeks to contribute to the academic literature by investigating the economic impacts of CC on Nigerian cocoa agriculture. It intends to do so by using a climate-land value analytical framework (i.e. a theoretical Ricardian cross-section model). The framework is applied to seven different Nigerian cocoa-producing states to:

- assess the potential economic impacts of CC on Nigeria cocoa agriculture under supplementary irrigated and rainfed farming conditions;
- evaluate the importance of irrigation (supplementary) as an alternative pathway to CC mitigation in Nigeria; and
- examine the likely future impacts of CC on cocoa farming in Nigeria.

The rest of the paper is structured as follows. In Section 2, the authors present the Ricardian model (RM) developed for the analysis; Section 3 describes the data and methodology, followed by a description of empirical results in projections with climate scenarios in Section 4. Section 5 concludes the paper with potential policy implications of the findings and suggests policy response options or recommendations to help mitigate the effects of CC on Nigerian tree crop agriculture.

2. Analytical framework

In his pioneering work on the theory of "economic rents" in 1817, David Ricardo (1817) observed that land values reflect land productivity at a site under a perfectly free competition market structure without monopoly. This invariably implies that any factor of production (e.g. capital, labor, technical progress, conducive weather or their combination) that influences land productivity will be reflected in sale value of land or land rent. As climate affects land productivity through cultivation, it is intuitive to assume that the value of land (rents) contains information about the value of climate as one attribute of land productivity. This is the theoretical basis of the Ricardian approach pioneered by Mendelsohn *et al.* (1994), which is adopted to predict the potential economic damage to US agriculture from CC. Specifically, Mendelsohn *et al.* (1994) showed that by regressing land value (rents) on climate, household covariates and other environmental control variables, it is possible to measure the marginal contribution of each variable to land rents as capitalized in land value. However, in specific cases like in most developing countries where land values (rents) are difficult to compute because of the absence of well-functioning land markets, Dinar *et al.* (1998) and Mendelsohn *et al.* (2001, 2009) suggest the use of net revenue per farm hectare (NRh). Hence, this principle is usually captured through a net farm revenue equation (Mendelsohn *et al.*, 2009) of the form:

$$\text{NRh} = \sum P_i Q_i(X, C, S, Z) - \sum P_x X \qquad (1)$$

where, NRh represents the net annual profit from the production crop Q_i, P_i is the market price for crop i; Q_i is the output of crop i; X is a vector of purchased inputs other than land; C is a vector of climatic variables such as temperature and precipitation; S is a vector of control variables such as soil types, population density, farmland altitude and irrigation; Z is a set of socio-economic variables; and P_x is a vector of input prices. Under this framework, a farmer is assumed to maximize NRh by choosing input (X) subject to C, S and Z. Thus, the RM is a reduced form model that examines how exogenous variables such as C, S and Z affect farmland values or, in this case, NRh. Note that the product of price and quantity gives revenue, and all the variables in the parentheses are critical factors affecting the output or quantity of the crop under consideration (i.e. cocoa). As it is intuitively portrayed in equation (1) above, the difference between total revenue (first term on the right-hand side of the equation) and total costs (second term on the right-hand side of the equation) gives us profit.

Because values far above or below preferred climatic conditions reduce crop productivity, the relationship between NRh and the climatic variables (C) is non-linear and postulated to be inverted U-shaped (Reinsborough, 2003; Seo *et al.*, 2005; Kurukulasuriya and Mendelsohn, 2008; Deressa and Hassan, 2009; Mendelsohn *et al.*, 2009). To control for this potential non-linear relationship in the empirical specification of the NRh function, the authors adopt a quadratic functional form on the CC indicator variable, C:

$$\text{LnNRh}_i = \alpha_0 + \alpha_1 C + \alpha_2 C^2 + \alpha_3 S + \alpha_4 Z + \varepsilon, \qquad (2)$$

where ε represents the error term which is normally distributed with zero mean and constant variance and α_i represents the regression coefficients to be estimated, including the intercept term. The marginal impact of a single climate variable (c_i) on NRh evaluated at the mean of that variable is given as:

$$E\left[\frac{d\text{NRh}}{dc_i}\right] = \alpha_{1,i} + 2 * \alpha_{2,i} * E[c_i]. \qquad (3)$$

One major advantage of using NRh is that it accounts for the direct impacts of climate on yields of different crops, as well as the indirect substitution of different inputs, the introduction of different activities and other possible adaptations by farmers to different climatic shocks (Mendelsohn and Dinar, 1999). However, the RM has been extensively criticized on several grounds, first because crops evaluated under the method are not subject to controlled experiments across farms as with the production function model, the agro-economic model or the agro-ecological zone model (Hassan, 2010). To control for this concern, the authors included other important variables such as soil quality and market access in the model, as suggested by Mendelsohn and Dinar (1999). Second, it fails to account for future change in technology, policies and institutions (Mendelsohn *et al.*, 2009). Third, it assumes that input and output prices remain constant, introducing a potential bias in the analysis (Mendelsohn *et al.*, 2009), and fourth, it does not account for the effect of factors consistent across space, e.g. carbon dioxide concentrations being beneficial to crops (Seo *et al.*, 2005; Mendelsohn *et al.*, 2009; Hassan, 2010). Finally, the model may overestimate potential damage caused by CC to farmland values when irrigation is not factored into account (Reinsborough, 2003; Seo *et al.*, 2005).

This study recognizes the various problems highlighted above and attempts to address them to generate reliable results. However, one also should not expect drastic price changes due to CC. Equally, to make the analysis less prone to potential empirical concerns, one can include carbon dioxide exogenously in the analysis, new technology and irrigation as important control variables.

3. Data and econometric procedure
3.1 Data description
Temperature and precipitation data are key to climate-land value analysis. Because CC involves long-term trends, the use of monthly climate normals is usually preferable (Reinsborough, 2003). To determine seasonal temperatures and precipitations, this study used January-December monthly means from 1981 to 2010 from 32 weather stations around the country from Nigeria's Meteorological Agency. Data on soil types for the seven cocoa-producing states in Nigeria were obtained from the Food and Agricultural Organization (FAO). The FAO soil statistics include information about the major and minor soils in each location, as well as the slope and texture. In all, there exists four types of soil in the respective states, and all of them were used in the analysis. Farm-level data on NRh and its determinants were collected from 280 cocoa farmers spread over the various agro-ecological zones of the seven selected states – Cross River, Abia, Edo, Ondo, Ekiti, Oyo and Ogun (Figure 1). These states represent the major cocoa-producing regions in the country with significant variations in temperatures and precipitations driven primarily by elevation. Four enumeration areas (EAs) were used in each state, for a total of 28 EAs. From each EA, ten farmers were randomly selected from farmers with a cocoa production record of more than 20 years (i.e. long enough to have experienced the CC effects). To collect data, enumerators with extensive fieldwork experience were recruited through Nigeria's National Bureau of Statistics.

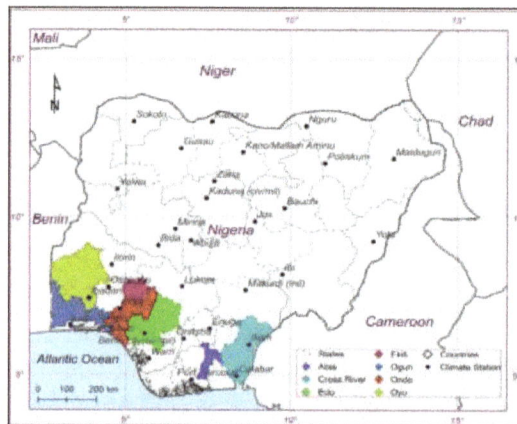

Figure 1. Cocoa-producing states used for the study

The questionnaire contained seven sections that gathered information on household demographic characteristics, the employment status of the household head, land tenure and household labor composition for farming activities, including labor costs. In the fourth section, responses from more in-depth questions concerning farming activities were collected, such as primary crops grown, harvested and sold in the past twelve months; average yields obtained in a normal year; water source for farming, including the specific irrigation system used for farming; animal ownership in terms of the total number of livestock, poultry or other farm animals kept, sold, lost and consumed during the last growing season; and various costs associated with seeds, fertilizer and pesticides purchased, as well as those related to the use of farm machinery (light, heavy and animal power) and the cost of farm buildings. The last sections of the questionnaire helped to collect data on access to sources of information on farming activities and costs involved in obtaining information; estimates of the household's total income (for both farming and non-farming activities), taxes paid and subsidies received and farmers' perceptions of short- and long-term CC effects and their adaptation strategies in response to these changes.

Based on the answers to these questions, it was possible to calculate, for each of the 280 sampled households, the net farm revenue per cocoa hectare (NRh), defined as the difference between gross revenue per farm hectare and the per hectare costs of production, which includes seeds, fertilizer, pesticide, insecticide, herbicide, farm labor, depreciation on machineries and other farming costs.

3.2 Econometric procedure

In the empirical estimation procedure, NRh was regressed on climate, household socioeconomic characteristics and other environmental and control variables. As none of the explanatory variables (e.g. climate variables, household socio-economic condition variables and soil type variables) are endogenous, a log-linear ordinary least squares technique was used to generate the coefficient estimates. Therefore, the data at hand and the nature of the variables considered do not necessitate the use of instrumental variables (IVs) or other econometric models that are designed to account for potential endogeneity for identification of estimated coefficients. Following Reinsborough (2003) and Seo et al. (2005), after several trials and iterations of different definitions of seasonal climate, three seasons were used to define our climate variables that correspond to the three predominant seasons experienced in Nigeria, that is, the dry season (average precipitation and temperature data for October-March), the rainy season (average precipitation and temperature data for April-September) and the Harmattan period[1] (average precipitation and temperature data for December-February).

As indicated earlier, both linear and squared terms of the temperature and precipitation variables were included in the empirical specification of equation (2) to capture the optimum temperature and precipitation levels of cocoa and account for a potential non-linear relationship between CC and net farm revenue. When the coefficient of the quadratic term is positive with a negative linear term, the climate response function is U-shaped, and when the linear term is positive and the quadratic term is negative, the function is hill-shaped or bell-shaped (Reinsborough, 2003). However, note that because seasonal climate variables are often used, the process is somehow complex and likely to result in a mixture of positive and negative quadratic coefficients across seasons (Mendelsohn et al., 2009).

In addition, the climate variables were specified in their interactive forms to check whether climate effects on NRh are dependent on the season under consideration. Hence, the analysis includes interacting dry season temperature with precipitation, rainy season precipitation with temperature and Harmattan temperature with precipitation. With regards to other control variables, urban/rural characteristics, farmland altitude and irrigation, among others, have been shown to significantly affect land value or net farm revenue other than CC alone (Reinsborough, 2003; Seo *et al.*, 2005; Ouedraogo *et al.*, 2006; Kurukulasuriya and Mendelsohn, 2006; Mendelsohn *et al.*, 2009). For instance, to account for demographic and other important factors in the empirical specification of the climate response function [equation (2)], population density was incorporated to capture urban/rural characteristics, main source of water (irrigation or not) and farm altitude.

Furthermore, two sets of climatic scenarios were used to examine the likely future impacts of CC on Nigerian cocoa agriculture. The first set, comprising two representative concentration pathways, i.e. RCP4.5 and RCP8.5 (Moss *et al.*, 2010), and two future time slices (2036-2065 and 2071-2100), was generated from a multimodel ensemble of regional climate models (RCMs) run under the recent coordinated regional climate downscaling experiment (CORDEX) initiative (Giorgi *et al.*, 2009). The CORDEX RCMs downscale many global climate models (GCMs) from the Coupled Model Intercomparison Project, Phase 5 (Taylor *et al.*, 2012; Sylla *et al.*, 2016). The RCMs are run over the whole of Africa at 50 km resolution. To date, the CORDEX data constitute the most comprehensive RCMs projections available for the continent. The authors are not aware of any African application that has utilized CORDEX projections.

Figure 2 shows absolute precipitation and temperature changes (Future minus Historical; in mm d-1 and Kelvin, respectively) for each of the CORDEX RCMs averaged over the whole of Nigeria and for both RCP4.5 and RCP8.5 during the two future time slices (2036-2065 and 2071-2100). In general, an increase in precipitation is projected in all cases, except for the Climate Limited-area Modelling Community (CCLM), which produces a substantial decrease for the RCP8.5 during 2071-2100. The highest precipitation increase of more than 1 mm d-1 is simulated by RegCM4 of the International Centre for Theoretical Physics, Italy. Although the multimodel ensemble projects more rainfall, it is clear that the various CORDEX RCMs produce different magnitude and sign of the precipitation changes over Nigeria. However, the temperature change (i.e. Figure 2) shows a consistent warming across all RCMs with the highest change (more than 4°C) projected during 2071-2100 for the RCP8.5.

Because the CORDEX ensemble data (Table I), which show increases in all variables (rainfall and temperature) regardless of the scenarios (RCP4.5 and RCP8.5), balance out the decreases in precipitation predicted by the regional climate model CCLM, the authors not only use CORDEX but also base the second set of scenarios on hypotheses of 2.5°C increase in temperature and 5 per cent decrease in precipitation from the historical observations (1981-2010). Similar hypotheses (i.e. +2.5 to 5°C and 7 to 14per cent decrease in precipitation) were tested earlier for African croplands by Kurukulasuriya and Mendelsohn (2006) and Mano and Nhemachena (2006). The projected changes from the different climate scenarios generated are summarized in Table I.

4. Results and discussion

4.1 Sample statistics

The summary statistics of the key variables used in the analysis are reported in the Appendix. On average, the NRh was Nigerian Naira 42,558.9 (hereafter NGN) for all farms, NGN45,339.12 for supplementary irrigated farms and NGN31,386.7 for rainfed farms. Mean temperatures and precipitations corresponding to the three seasons were 32.4°C and 47.7 mm for the hot dry season, 25.7°C and 169 mm for the heavy rainy season and 19.6°C and 12.7 mm for the Harmattan season. The soil type on which the farmers operate is a function of geographical location. The soil types are:

(a)

(b)

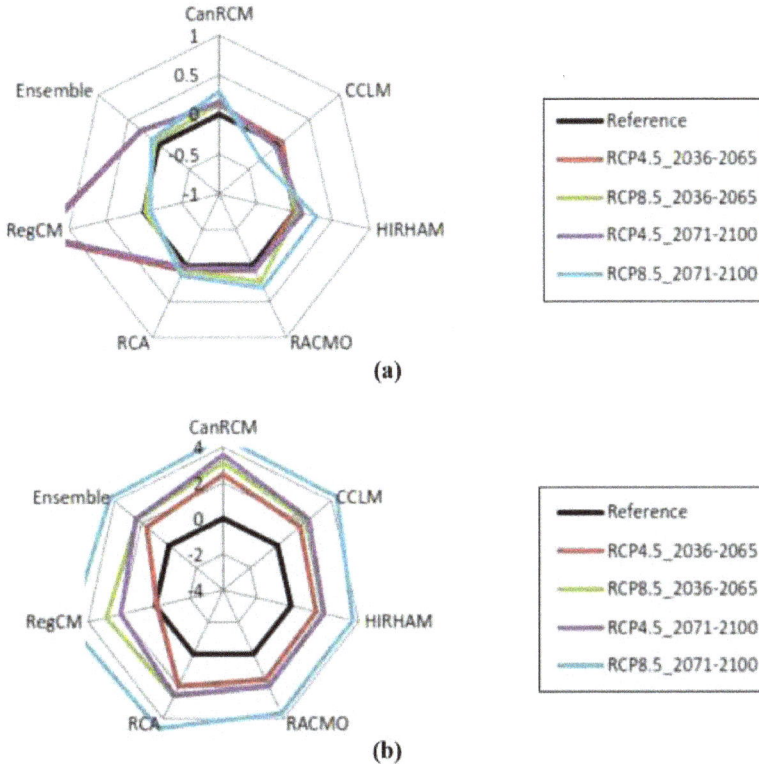

Notes: *Note that the set of RCMs used are from: (i) the CANRCM of the Canadian Centre for Climate Modelling and Analysis (CCCMA); (ii) the CCLM of the Climate Limited-area Modelling Community (CLM-C); (iii) the RACMO of the Royal Netherlands Meteorological Institute (KNMI); (iv) the RCA of the Swedish Meteorological and Hydrological Institute (SMHI); (v) the RegCM of the International Centre for Theoretical Physics (ICTP), Italy; and, (vi) the HIRHAM of the Danish Meteorological Institute (DMI)

Figure 2. Absolute temperature and precipitation changes (in °C and mm d^{-1}) for both the climate scenarios RCP4.5 and RCP8.5 and for two future time windows 2036- 2065 and 2071-2100

- *La*: These are ferralitic soils of the coastal plain sand and escarpment; dominant color is yellowish brown (not differentiated).
- *Jc*: These are ferruginous tropical soils on crystalline acid rocks.
- *Li*: These are ferrallite soils, and the dominant color is red on loose sandy sediments.
- *LVf*: These are ferric luvisol soils, which are seen as reddish sandy clay loam.

More than 42.9 per cent of the sampled farmers cultivated on La soil type, about 28.7 per cent of them farmed on the Jc soil type and about 28.4 per cent of the farmers were split between the Li and LVf soil types.

The average area devoted to cocoa cultivation was about 2.4 hectares from a total average household farm size of 6.5 hectares, suggesting that a non-negligible proportion (about 31 per cent) of the total available farming land is dedicated to cocoa. Men-headed

Table I. Monthly seasonal temperature and precipitation levels

	Scenario 1 (1)					Scenario 2 (2)		
(a) *CORDEX (temperature projections)*						2.5% precipitation decrease from observations (see Table II)		
	Historical	RCP4.5	RCP8.5	RCP4.5	RCP8.5	Tb + 2.5°C (all farms)	Tb + 2.5°C (rainfed farms)	Tb + 2.5°C (irrigated farms)
Period	(1981-2010)	(2036-2065)	(2036-2065)	(2071-2100)	(2071-2100)	(2050-2100)	(2050-2100)	(2050-2100)
Dec-Feb	22.58	24.19	24.93	24.98	26.88	22.1	27.5	28.4
Apr-Sep	25.2	26.8	27.53	27.59	29.45	28.2	27.6	27
Oct-Mar	24.46	26.12	26.84	26.89	28.85	34.9	31.1	30.5
(b) *CORDEX (precipitation projections)*						5% precipitation decrease from observations (see Table II)		
	Historical	RCP4.5	RCP8.5	RCP4.5	RCP8.5	Pb: 5% (all farms)	Pb: 5% (rainfed farms)	Pb: 5% (irrigated farms)
Period	(1981-2010)	(2036-2065)	(2036-2065)	(2071-2100)	(2071-2100)	(2050-2100)	(2050-2100)	(2050-2100)
Dec-Feb	2.79	10.5	3.24	10.53	3.42	12.065	20.14	17.1
Apr-Sep	174.12	183.78	177.66	183.09	178.92	160.55	132.24	143.165
Oct-Mar	30.51	38.19	30.78	38.22	32.19	44.365	35.435	32.015

Notes: N.B., where Tb: baseline temperature; Pb: baseline precipitation; (a) monthly seasonal temperature (in °C) as generated from CORDEX and hypotheses of +2.5°C from observations (baseline period of 1981-2010); (b) monthly seasonal precipitation (in mm/month) as generated from CORDEX and hypotheses of −5% precipitation decrease from observations (baseline period of 1981-2010).

households dominated the sample (95 per cent), with an average age for the entire sample of about 55 years, with 22 years of farming experience and about 9 years of schooling (i.e. junior secondary school). On average, nearly three household members worked in cocoa farming as their primary occupation from a total household size of about seven members. Only about 20 per cent of sampled farmers reported using irrigated farming (i.e. supplementary irrigation); the rest relied only on rain-fed agriculture. Less than 33 per cent of the farmers acknowledged having access to any form of credit facilities; on the other hand, about 14 per cent reported receiving farm subsidies. Yearly average use of fertilizer came to about 776 kg for the whole farm, and more than 93 per cent of farmers reported using pesticides. Average farm visit time from an agricultural extension worker was estimated to be less than 2 h a week; however, more than 66 per cent of sampled farmers reported having received advice from an agricultural extension worker in the last 12 months preceding the survey. More than 72 per cent of farmers reported using multiple farmlands for cocoa farming, and about 74 per cent reported practicing mixed farming. Finally, more than 52 per cent sold their cocoa produce in urban areas, with a mean market distance of about 90 km.

4.2 Ricardian estimates

Table II presents three log linear regressions or models on determinants of NRh for Nigerian cocoa farms. In Column two (Model 1), NRh is regressed on climate variables alone. Though this model fits well, as many of the climatic variables including their squared terms are statistically significant, it does not account for the effects of household socio-economic characteristics on NRh, an important category of independent variables reflecting farmlands' potential for alternative uses (Reinsborough 2003). Notwithstanding, Model 1 highlights the direction and magnitude of climate impacts on NRh. The implications of these findings are taken up later.

In the second regression or Model 2 (i.e. Column 3 of Table II), net farm revenue is regressed on climate and climate interaction effect variables conditional on seasons (dry, rainy and *Harmattan*). Note that many of the climatic variables which were statistically significant in the first model are dropped because of multicollinearity. Equally, it was also observed that many of the squared terms for temperature and precipitation remained statistically significant as in the first model and unchanged in sign and magnitude. However, the squared term of rainy season precipitation is no longer significant, though negative, as in Model 1. This implies that there is an optimal level of precipitation during the rainy season that, above and beyond, NRh decreases. Of the temperature–precipitation interaction effect variables, rainy season temperature–precipitation interaction has a very large significant negative effect on net farm revenue, whereas *Harmattan* temperature–precipitation interaction has a significant positive effect on net farm revenue. The former suggests that during the rainy season, net farm revenue decreases for hotter and wetter cocoa farms; while during *Harmattan*, net farm revenue increases for hotter and wetter cocoa farms.

Column 4 or Model 3 corresponds to the climate response function [equation (2)], where NRh is regressed on climate variables (C), their squared terms (C^2), soil types and three important control variables (S), including household socio-economic characteristics (Z). This model, which comprehensively includes as many determining variables as possible, provides the best fit, with results one would expect, and is therefore used for all subsequent climate calculations. The adjusted R^2 for the model has a higher goodness of fit value of 0.46. The effects of some of the significant variables from the climate "only" Model 1 are unchanged in sign and quite similar in magnitude. Dry season temperatures and

Table II. Loglinear specification results of Ricardian model for all farms

Variables	Climate only (2) Coef.	t-stat.	Climate with CC interaction[a] (3) Coef.	t-stat.	Full model without interaction[b] (4) Coef.	t-stat.
Dry season temp.	−0.917	−0.27			−1.193	−2.78
Dry season temp.²	−0.023	−9.10	−0.021	−8.27	−0.058	−5.64
Rainy season temp.	−6.568	−3.52			−0.079	−0.71
Rainy season temp.²	−0.0045	−1.27	−0.0045	−1.27	−0.0032	−0.96
Harmattan temp.	−8.805	−7.72	−8.805	−7.72	0.534	1.68
Harmattan temp.²	1.877	7.79	1.877	7.79	0.075	10.47
Dry season precip.	−0.086	−3.25			−0.113	−3.46
Dry season precip.²	0.00075	2.41	−0.00023	−5.76	−0.0010	−4.49
Rainy season precip.	−0.0123	−0.40			0.088	0.22
Rainy season precip.²	−0.00023	−5.40	−0.00005	−0.58	−0.0009	−4.80
Harmattan precip.	−0.0035	−0.12			−0.210	−0.56
Harmattan precip.²	0.00019	5.65	0.0017	6.32	0.0061	4.82
Dry season temp. × precip.						
Rainy season precip. × temp.			−0.0027	−3.19		
Harmattan temp. × precip.			0.0015	2.23		
La_soil_type					−3.005	−7.35
Jc_soil_type					−2.432	−6.93
Li_soil_type					−2.576	−7.10
LVf					−2.416	−6.74
Farm altitude					0.014	1.82
Irrigation					0.191	3.24
Population_density					−0.030	−3.24
Access_credit					0.199	4.05
Education_head					−0.034	−1.64
Farm_experience					−0.0037	−0.60
Crop_area					0.0045	0.43
Household_ size					−0.018	−1.02
Market_distance (km)					−0.446	−2.85
No of visit by ext. worker					0.335	1.24
Total farm area (hectare)					−0.067	−1.53
Constant	12.9	7.64	13.1	7.72	26.4	10.4
F-statistics	12.1		12.1		13.6	
Adjusted R^2	0.29		0.27		0.46	
Observations	280		280		280	

Notes: [a]Showing estimated results for model with only climatic variables used as the independent variables. [b]Results for climate response function specified in equation (2); all figures in italic signifies statistical significance of variables at 1% and 5% levels of confidence

precipitations are significant, with large negative effects on NRh as in Model 1. The squared terms for dry and rainy season remained significant and unchanged in sign as in Model 1, indicating that above optimal levels of temperature–precipitation NRh decreases. The four soil types all have a large significant negative effect on net revenue per cocoa farm hectare as expected and consistent with most previous Ricardian findings in Africa. This is perhaps so because African farmers often make extensive use of chemical fertilizers and pesticides, including agricultural practices that enhances soil degradation and worsens soil infertility. This partly explains why net farm revenue is often negatively correlated with poor quality soils in Africa (Kurukulasuriya and Mendelsohn, 2006; Mano and Nhemachena, 2006; Mendelsohn *et al.*, 2009).

Of control and socio-economic variables, irrigation, population density, access to credit and distance to urban markets are significant as expected. Farmers with greater access to credit facilities have higher NRh, a predictable empirical finding considering the capital-intensive nature of cocoa farming. This has an important policy implication with regard to the financial resource allocation by the banking sector for agriculture. Similarly, farmers closest to urban markets have higher NRh, probably because of market proximity and lower transportation cost. Hence, distance to urban markets has a statistically significant negative coefficient. Irrigation has a beneficial effect on NRh, suggesting its potential use as an adaptive intervention to mitigate the impact of CC. On the other hand, population density was found to be detrimental to NRh, as its significant negative coefficient shows.

Column 2 of Table III reports the marginal impacts of climate on NRh of the full model (i.e. Model 3) reported in Table II (i.e. Column 4). The marginal analysis simply shows the effect of an infinitesimal change in temperature and precipitation on cocoa farming in Nigeria. As observed (i.e. Row 6 of Column 2 in Table III), the annual temperature impact is close to NGN−5,698.4 or about US$38.0 per degree celsius evaluated at the mean of the sample. However, the most harmful temperature effects are associated with the dry and rainy season. Conversely, for the precipitation impacts, the marginal calculations also revealed that annual precipitation impact is NGN40 or about US$0.30 increase in net revenue per mm/month, and the most harmful precipitation effect is largely due to dry season precipitation. The marginal impact analysis equally reveals that while temperature increase is strictly detrimental to cocoa farming in Nigeria, infinitesimal precipitation increase is strictly beneficial to cocoa farms in general. This may be linked to the use of supplementary irrigation by cocoa farmers. The implication of this can be profound for Nigeria and other African and developing countries, and we discuss it in the subsequent paragraphs.

Table IV presents a comparison of cocoa farms with supplementary irrigation versus rainfed farms. In the irrigated farm model, the inclusion of the quadratic terms complicated the estimation as many climatic variables were dropped because of collinearity. Similarly, inclusion of the temperature–precipitation interaction effect and control variables produced unrealistic confidence intervals for the estimates. By omitting the C^2 and S variables, more sensible and precise estimates for irrigation and rainfed farms were obtained; these are presented in Column 2 and Column 3, respectively. Many significant differences are

Table III. Marginal impacts of climate on cocoa NRh (Naira)

Variables	Full model Marginal (2)	Irrigated Marginal (3)	Rainfed Marginal (4)
Temperature			
Dry season	−9,302.7	−349.9	−13,528.7
Rainy season	−7,165.2	12,403.8	−16,118.7
Harmattan season	−627.2	5,187.2	−1,547.6
Annual	−5,698.4	5,747.0	−10,398.3
Precipitation			
Dry season	−218.8	192.96	−368.92
Rainy season	88.8	−742.51	−11.16
Harmattan season	250.1	1,124.34	766.94
Annual	40.0	191.60	128.95

Note: N.B. Marginals of each climate variable calculated at the mean of the sample

Table IV. Loglinear specification of rainfed and supplementary irrigated farm samples

Variables	Irrigated model (2) Coef.	t-stat.	Rainfed model (3) Coef.	t-stat.
Dry season temp.	−0.0035	−0.18	−0.902	−0.62
Dry season temp.^{2}			−0.135	*−3.06*
Rainy season temp.	0.124	*3.53*	−0.601	−1.40
Rainy season temp.^{2}			−0.012	*−3.67*
Harmattan temp.	0.519	*2.30*	−0.315	−0.50
Harmattan temp.^{2}			0.034	*8.51*
Dry season precip.	0.0019	0.32	0.104	*3.23*
Dry season precip.^{2}			−0.00005	−1.17
Rainy season precip.	−0.0074	−1.75	0.191	0.02
Rainy season precip.^{2}			−0.00003	*−2.06*
Harmattan precip.	0.0112	1.82	−0.808	−0.65
Harmattan precip.^{2}			0.0290	0.02
Rainy season precip. × temp.				
Harmattan temp. × precip.			0.00060	*2.00*
La			2.927	*9.62*
Jc	−0.106	*−2.47*	−0.239	*−5.75*
Li	2.927	*9.62*		
LVf				
Farm altitude			−0.025	*−9.23*
Access_credit	0.083	*3.82*	0.180	*3.10*
Farm_experience	−0.011	*−2.63*	−0.032	−1.91
Cropland	−0.0020	−0.69	−0.213	*−3.12*
Household_size	0.0036	0.75	−0.0150	−1.08
Market_distance (Km)	−0.051	−1.89	−0.045	*−2.85*
Constant	16.4	*2.20*	21.4	*8.56*
F-statistics	37.2		17.1	
Adjusted R^2	0.87		0.56	
Observations	56		224	

Note: All figures in italic signifies statistical significance of variables at 1% and 5% levels of confidence

obtained between the estimates of the two models. The irrigated model, for example, shows that high temperatures in the rainy and *Harmattan* seasons are best for NRh, whereas dry season precipitations are best for rainfed farms. Irrigated NRh is significantly affected by Li soil types and rainfed NRh is significantly affected by La, whereas both farm types are affected by Jc soil types. Of the socio-economic variables, number of years of farming cocoa is more influential for irrigated farm revenues than rainfed farms, while distance to urban markets significantly influences rainfed NRh more than irrigated NRh, as does farm altitude.

The marginal calculations of Columns 3 and 4 of Table III show the marginal impacts of climate on the irrigated and rainfed regression models based on the results of Table IV. The annual temperature marginals for rainfed farms are much larger than those for supplementary irrigated farms, with an impact of NGN5,747 per degree celsius, in constrast to NGN10,393.3 per degree celsius for rainfed farms. Harmful temperature effects are associated with rainfed farms in all the three seasons, while in the irrigated model, they occur largely in the dry season, with beneficiary effects in the rainy and *Harmattan* seasons. Similarly, annual changes associated with increased precipitation impacts show net gains of NGN191.6 per mm per month for irrigated farms and NGN128.9 per mm per month for rainfed farms. In general, irrigated farms are more sensitive to increased precipitation during the rainy season, while rainfed farms are most vulnerable to decreased precipitation during both dry and rainy seasons.

4.3 Projections with climate scenarios

There are many contentious issues surrounding the use of estimated Ricardian results to predict how temperature and precipitation will affect future net farmland revenues (Reinsborough, 2003, p. 31-32). First, estimating beyond the range of observed data may be problematic, especially when the estimated relationships are not exactly as expected. For instance, one expects negative square terms of the climatic variables, assuming cocoa has optimum temperature and precipitation levels. However, while the squared terms of dry and rainy season temperature and precipitation variables exhibit a hill-shaped relationship, *Harmattan* temperature and precipitation squared exhibit a U-shaped relationship, possibly making an accurate forecasting of future CC effects on the cocoa farms' NRh problematic. Second, variables such as soil types and farmers' socio-economic characteristics, which are assumed to be constant with temperature and precipitation changes, will certainly be affected in reality. For instance, increased rainfall and sunshine affect soil moisture content and hence plant growth and productivity (Deressa and Hassan, 2009). However, as the changes to be estimated are quite moderate, using the results of the full model (i.e. Table II, Column 3) and those of Table IV to predict future climate impacts on NRh for all farms, irrigated and rainfed, should be not be too problematic. The aim of the projection is not to examine how NRh actually changes, but it is simply to isolate the statistically significant effect of CC on NRh, assuming all other conditions (e.g. price changes investment, population and the use of technology) are held constant (Mendelsohn *et al.*, 2009).

The first set of CC scenarios (i.e. Column 1 of Table I), generated from the CORDEX Ensemble of six RCMs, were fitted into the results of Table II's full model (Column 4) and those of Table IV (Columns 2 and 3) to examine how future CCs would affect NRh under both irrigated and rainfed conditions. The simulated results are reported in Tables V and VI. Annual and seasonal impacts vary a great deal between the different farms. In the RCP4.5 scenario, the calculations reveal marked variations in NRh across the different farms. Both scenarios predict drastic dry season declines in NRh between 2036-2065 and 2071-2100 for rainfed farms. Seasonal losses range from NGN−58.0 to −83 for all farms and from NGN−110.9 to 148.8 for rainfed farms. A striking prediction from the CORDEX Ensemble suggests that irrigated farm revenues will increase across all seasons, with the most beneficial effects expected during the *Harmattan* season. However, in terms of annual revenue losses, only rainfed farms are expected to record future NRh decline, with expected losses of NGN−25.2 and NGN−29.1 for 2036-2065 and 2071-2100, respectively. The projections suggest that the increase in benefits to farmers can be gained incrementally via adaptive interventions such as irrigation.

In the second scenario (Table VI), first, a 2.5°C increase in temperature only is associated with seasonal net farm revenues per hectare loss of NGN−126.2 ($−0.84) for all farms and NGN−153.9 ($−1.03) for rainfed farms during the period (2050-2100). These losses are expected also during the dry season as initially predicted by the six CORDEX RCMs ensemble. Second, reducing rainfall by 5 per cent is equally associated with net farm revenues loss of NGN−108.5 ($−0.72) for all farms and NGN−151.2 ($−1.0) for rainfed farms during the dry season per hectare of farmland. Similar seasonal losses are associated when a simultaneous 2.5°C increase in temperature and a 5 per cent reduction in rainfall are considered. The combined effect is a reduction in NRh of about NGN−125.8 ($−0.84) for all farms and NGN−173.9 ($−1.2) for rainfed farms during this same period. Irrigated farms

Table V. Impacts of CORDEX ensemble (six RCMs) scenarios on cocoa NRh

Seasonal variables	Historical (1981-2010)	CORDEX scenarios			
		RCP4.5 (2036-2065)	RCP4.5 (2071-2100)	RCP8.5 (2036-2065)	RCP8.5 (2071-2100)
All farms					
Dec-Feb (Harmattan)	76.16	81.67	85.00	85.71	94.30
Apr-Sep (rainy season)	10.41	7.76	7.73	9.03	8.23
Oct-Mar (dry season)	−58.01	−68.53	−72.92	−71.28	−83.44
Annual	*9.52*	*6.97*	*6.60*	*7.82*	*6.36*
Rainfed farms					
Dec-Feb (Harmattan)	29.59	28.39	29.45	32.36	35.07
Apr-Sep (rainy season)	30.98	30.76	29.65	28.75	26.51
Oct-Mar (dry season)	−110.94	−134.72	−140.97	−129.93	−148.77
Annual	*−16.79*	*−25.19*	*−27.29*	*−22.94*	*−29.06*
Supplementary irrigated farms					
Dec-Feb (Harmattan)	28.15	29.07	29.48	29.37	30.39
Apr-Sep (rainy season)	18.24	18.36	18.47	18.50	18.73
Oct-Mar (dry season)	16.37	16.38	16.38	16.36	16.36
Annual	*20.92*	*21.27*	*21.44*	*21.41*	*21.83*

Table VI. Impacts of +2.5°C and −5% changes in temperature and precipitation on cocoa NRh

Seasonal variables	Baseline Tb and Pb (1981-2010)	IPCC Scenarios based		
		Tb + 2.5°C and Pb (2050-2100)	Tb and Pb −5% Pb (2050-2100)	Tb + 2.5°C and Pb − 5% Pb (2050-2100)
All farms				
Dec-Feb (Harmattan)	64.00	73.15	64.03	73.19
Apr-Sep (rainy season)	11.42	10.79	13.19	12.56
Oct-Mar (dry season)	−108.94	−126.22	−108.46	−125.75
Annual	*−11.17*	*−14.09*	*−10.42*	*−13.33*
Rainfed farms				
Dec-Feb (Harmattan)	30.68	34.35	30.26	33.94
Apr-Sep (rainy season)	24.76	21.68	23.49	20.40
Oct-Mar (dry season)	−153.93	−176.34	−151.19	−173.59
Annual	*−32.83*	*−40.10*	*−32.48*	*−39.75*
Supplementary irrigated farms				
Dec-Feb (Harmattan)	30.04	31.34	30.03	31.33
Apr-Sep (rainy season)	18.32	18.63	18.38	18.69
Oct-Mar (dry season)	16.37	16.36	16.36	16.35
Annual	*21.58*	*22.11*	*21.59*	*22.12*

Notes: N.B: Tb: baseline temperature; and Pb: baseline precipitation

are, in general, less sensitive to any of these changes, as was predicted by the six CORDEX RCMs ensemble. However, note that while the six CORDEX RCMs ensemble predicts that future net revenues decline only for rainfed farms, the second-case climate scenarios considered in the paper predict that annual net revenues decline for all farms and rainfed farms.

5. Conclusions

Using 20 years of data, this paper assesses the economic implications of CC (i.e. average long-term temperature and precipitation changes) on NRh in Nigerian cocoa farms under rainfed and irrigated (supplementary) conditions. The results indicate a high sensitivity of NRh to climate normals, dependent on whether cocoa farms are irrigated. In general, rainfed farms' revenues were more sensitive to marginal changes in both temperature and precipitation than irrigated farms. For instance, the annual temperature marginals for irrigated cocoa farms is about NGN5,747 ($38.3) per degree celsius compared to NGN−10,393.3 ($−69.3) per degree celsius for rainfed farms. Similarly, the annual changes associated with precipitation change are a net gain of NGN191.6 ($1.3) per mm per month for irrigated farms compared to NGN128.9 ($0.86) per mm per month for rainfed farms. Furthermore, the marginal analysis reveals that irrigated farms are sensitive to decreased precipitation only during the rainy season, whereas rainfed farms are most vulnerable in both dry and rainy seasons.

Two sets of future climate impact projections are included in the analysis. The first set of scenarios, based on the CORDEX ensemble, suggests a wide range of outcomes on NRh for all farms between 2036-2065 and 2071-2100. Dry season losses range from NGN−110.9 ($−0.74) to NGN−148.8 ($−1.0) for rainfed farms and from NGN−58.0 ($−0.39) to NGN−83 ($−0.55) for all farms. Irrigated farms are generally less sensitive to the different CORDEX scenarios. The results suggest the need to disaggregate farms by type for a reliable CC impact analysis so that policy can be directed where impacts are felt strongly.

The second hypothesized scenario set also predicts drastic declines in NRh per hectare for rainfed farms and all farms between 2050 and 2100. The combined effect of a simultaneous 2.5°C temperature increase and a 5 per cent rainfall reduction are associated with reduction in NRh of NGN−173.9 ($−1.2) for rainfed farms and about NGN−125.8 ($−0.84) for all farms, and again, irrigated farms are generally less sensitive to any changes. It is not surprising to find similar impacts for rainfed and all farms as there are more rainfed farms (224) in the overall sample farms compared to irrigated farms (56). Hence, the statistical results of all farms are mainly driven by the results that pertain to rainfed farms. A study on larger number of sampled irrigated farms will be beneficial in future.

The results clearly demonstrate the importance of supplementary irrigation as an adaptation strategy to reduce harmful CC effects on Nigerian cocoa agriculture. Serious neglect by the national and sub-national Nigerian governments of irrigation farming systems may be partly responsible for the country's declining trend in cocoa productivity. As a concrete future policy direction, this demands for more investment (not less investment) in irrigation to support local farmers' management of CC uncertainty. The regression results suggest that access to credit/working capital and proximity to markets by improving the road infrastructure that minimize the distance to final destination of agricultural output create the right developmental incentives for cocoa farmers to generate better farm revenues and improve their welfare. In irrigated farms, the experience of farmers was found to be beneficial. In addition to policy measures that increase investment in irrigation, extension services can be used as a vehicle to share experiences among farmers on best practices of cocoa production and environmental management. This may include pilot initiatives such as weather index-based crop insurance that can support farmers in mitigating adverse and heterogeneous effects of CC and natural catastrophes encountered during farming activities (Barnett, 2014; Barnett and Mahul, 2007; Kunreuther, 1996; FAO, 2011; IFAD, 2011; World Bank, 2010). The results clearly demonstrated the detrimental impact of distance to markets. Therefore, the obvious policy implication is the urgent need to provide conducive market infrastructure by improving access to market nodes such as urban areas. This enhances farmers' productivity, the welfare of their respective households and the dynamism of the cocoa sub-sector in Nigeria, with potential implications for farmers in other African and other developing countries across the globe.

In conclusion, the authors want to highlight the fact that they are not arguing that irrigation is the only adaptation strategy among other alternative adaptation strategies (e.g. developing drought resistant crops) that can be deployed to cushion the impact of CC. However, the authors believe that irrigation in the Nigerian context has advantages, is manageable in terms of technological capability of the farmers and is relatively cost-effective route to adapt to CCs compared to, for instance, breeding new crops that can withstand climate stresses such as water shortage and drought. Generating and using new breed of crops is based on long-term research and development investments. However, high yielding varieties are not readily available for farmers who cannot wait for many years or decades of field experiments which are required to validate the new crops. In this context, irrigation provides a practically feasible alternative to the harsh CC that Nigerian farmers find themselves in.

Acknowledgments

The fieldwork for the study was generously funded by the African Economic Research Consortium (AERC) through Research Grant No.: RT10509. The insightful comments from three anonymous reviewers and the deputy editor of the journal are gratefully acknowledged. The authors specially thank the German Federal Ministry of Education and Research (BMBF) for the financial support through the West African Science Service Center on CC and Adapted Land Use (WASCAL), where the work was carried out.

Note

1. The Harmattan season is peculiar to the West African subcontinent. It usually occurs between the month of December and early March and signifies the beginning of the dry season. It is often characterized by harsh weather conditions such as cold, dry and dust-laden wind, very low humidity and relatively lower temperatures.

References

Ajayi, I.R., Afolabi, M.O., Ogunbodede, E.F. and Sunday, A.G. (2010), "Modelling rainfall as a constraining factor for cocoa yield in Ondo State", *American Journal of Scientific and Industrial Research*, Vol. 1 No. 2, pp. 127-134.

Ajewole, D.O. and Iyanda, S. (2010), "Effect of climate change on cocoa yield: a case of cocoa research institute (CRIN) farm, Oluyole local government Ibadan, Oyo state, Nigeria", *Journal of Sustainable Development in Africa*, Vol. 12 No. 1, pp. 350-358.

Barnett, B. (2014), "Multi-peril crop insurance: successes and challenges", *Agricultural Finance Review*, Vol. 74 No. 2, pp. 200-216.

Barnett, B. and Mahul, O. (2007), "Weather index insurance for agriculture and rural areas in lower income countries", *American Journal of Agricultural Economics*, Vol. 89 No. 5, pp. 1241-1247.

Berg, A., Quirion, P. and Sultan, B. (2009), "Weather-index drought insurance in Burkina-Faso: assessment of its potential interest to farmers", *Weather, Climate, and Society*, Vol. 1 No. 1, pp. 71-84.

Central Bank of Nigeria (2014), *Annual Statistical Report*, Abuja.

Christiaensen, L., Lionel, D. and Jesper, K. (2006), "The role of agriculture in poverty reduction – an empirical perspective", World Bank Policy Research Working Paper No. 4013, World Bank, Washington, DC.

Deressa, T. and Hassan, R. (2009), "Economic impact of climate change on crop production in Ethiopia: evidence from cross-section", *Journal of African Economies*, Vol. 18 No. 4, pp. 529-554.

Dinar, A., Mendelsohn, R., Evenson, R., Parikh, J., Sanghi, A., Kumar, K., McKinsey, J. and Lonergan, S. (1998), "Measuring the impact of climate change on Indian agriculture", World Bank Technical Paper No. 402, Washington, DC.

FAO (2011), *Agricultural Insurance in Asia and the Pacific Region, FAO (Food and Agricultural Organization), RAP Publication 2011/12.*

Fonta, W.M., Ichoku, E.H. and Urama, E.N. (2011), "Climate change and plantation agriculture: a Ricardian analysis of farmlands in Nigeria", *Journal of Economics and Sustainable Development*, Vol. 2 No. 4, pp. 63-75.

Giorgi, F., Jones, C. and Asrar, G. (2009), "Addressing climate information needs at the regional level: the CORDEX framework", WMO Bulletin.

Hassan, R. (2010), "Implications of climate change for agricultural sector performance in Africa: policy challenges and research agenda", *Journal of African Economies*, Vol. 19 No. 2, pp. ii77-ii105.

Ibrahim, B., Karambiri, H., Polcher, J., Yacouba, H. and Ribstein, P. (2014), "Changes in rainfall regime over Burkina Faso simulated by 5 regional climate models", *Climate Dynamics*, Vol. 42 Nos 5/6, pp. 1363-1381.

IFAD (2011), "Weather index-based insurance in agricultural development: a technical guide", available at: www.ifad.org/documents/10180/2a2cf0b9-3ff9-4875-90ab-3f37c2218a90 (accessed 10 January 2016).

IPCC (2007), *Impacts, Adaptations, and Vulnerability, Fourth Assessment Report*, IPCC (Intergovernmental Panel on Climate Change), Cambridge University Press, Cambridge, MA.

IPCC (2014), "Climate change 2014: synthesis report and summary for policymakers", available at: www.ipcc.ch/report/ar5/syr/ (accessed 20 February 2016).

Kunreuther, H. (1996), "Mitigating disaster losses through insurance", *Journal of Risk and Uncertainty*, Vol. 12 Nos 2/3, pp. 171-187.

Kurukulasuriya, P. and Mendelsohn, R. (2006), "A Ricardian analysis of the impact of climate change on African cropland", CEEPA Discussion Paper No. 8, CEEPA (Center for Environmental Economics and Policy in Africa), University of Pretoria.

Kurukulasuriya, P. and Mendelsohn, R. (2008), "A Ricardian analysis of the impact of climate change on African crop land", *African Journal of Agriculture and Resource Economics*, Vol. 2 No. 1, pp. 1-23.

Lawal, J.O. and Emaku, L.A. (2007), "Evaluation of the effect of climatic changes on cocoa production in Nigeria: cocoa research institute of Nigeria (CRIN) as a case study", *African Crop Science Conference Proceedings*, Vol. 8, pp. 423-426.

Mano, R. and Nhemachena, C. (2006), "Assessment of the economic impacts of climate change on agriculture in Zimbabwe: a Ricardian approach", CEEPA Discussion Paper No. 11, CEEPA (Center for Environmental Economics and Policy in Africa), University of Pretoria.

Mendelsohn, R. and Dinar, A. (1999), "Climate change, agriculture, and developing countries: does adaptation matter?", *World Bank Research Observer*, Vol. 14 No. 2, pp. 277-293.

Mendelsohn, R., Arellano-Gonzalez, J. and Christensen, P. (2009), "A Ricardian analysis of mexican farms", *Environment and Development Economics*, Vol. 15 No. 2009, pp. 153-172.

Mendelsohn, R., Dinar, A. and Sanghi, A. (2001), "The effect of development on the climate sensitivity of agriculture", *Environment and Development Economics*, Vol. 6 No. 1, pp. 85-101.

Mendelsohn, R., Nordhaus, W. and Shaw, D. (1994), "The impact of global warming on agriculture: a Ricardian analysis", *American Economic Review*, Vol. 84, pp. 753-771.

Moss, R.H., Edmonds, J.A., Hibbard, K.A., Manning, M.R., Rose, S.K., van Vuuren, D.P., Carter, T.R., Emori, S., Kainuma, M., Kram, T., Meehl, G.A., Mitchell, J.F., Nakicenovic, N., Riahi, K., Smith, S.J., Stouffer, R.J., Thomson, A.M., Weyant, J.P. and Wilbanks, T.J. (2010), "The next generation of scenarios for climate change research and assessment", *Nature*, Vol. 463 No. 7282, pp. 747-756.

National Bureau of Statistics (2017), *Nigerian Gross Domestic Product Report: Q42016*, National Bureau of Statistics, Abuja.

Nwachukwu, I.N., Ezeh, C.I. and Emerole, C.O. (2012), "Effect of climate change on cocoa productivity in Nigeria", *African Crop Science Journal*, Vol. 20 No. s2, pp. 487-491.

Odingo, R.S. (2008), "Climate change and economic development: issues, challenges and opportunity for Africa in the decades ahead", AERC Senior Policy Seminar Paper No. 10, pp. 1-22.

Omolaja, S.S., Aikpokpodion, P., Adedeji, S. and Vwioko, D.E. (2009), "Rainfall and temperature effects on flowering and pollen productions in cocoa", *African Crop Science Journal*, Vol. 17 No. 1, pp. 41-48.

Ouedraogo, M., Some, L. and Dembele, Y. (2006), "Economic impact assessment of climate change on agriculture in Burkina Faso: a Ricardian approach", CEEPA Discussion Paper No 24, CEEPA (Center for Environmental Economics and Policy in Africa), University of Pretoria.

Oyekale, S.A. (2017), "Impact of climate change on cocoa agriculture and technical efficiency of cocoa farmers in South-West Nigeria", *Journal of Human Ecology*, Vol. 40 No. 2, pp. 143-148.

Reinsborough, M.J. (2003), "A Ricardian model of climate change in Canada", *Canadian Journal of Economics/Revue Canadienne d'Economique*, Vol. 36 No. 1, pp. 21-40.

Ricardo, D. (1817), *The Principles of Political Economy and Taxation*, John Murray, London.

Seo, N., Mendelsohn, R. and Munasinghe, M. (2005), "Climate change and agriculture in Sri Lanka: a Ricardian valuation", *Environment and Development Economics*, Vol. 10 No. 5, pp. 581-596.

Sylla, M.B., Elguindi, N., Giorgi, F. and Wisser, D. (2016), "Projected robust shift of climate zones over west africa in response to anthropogenic climate change for the late 21st century", *Climatic Change*, Vol. 134 Nos 1/2, pp. 241-253.

Taylor, K.E., Stouffer, R.J. and Meehl, G.A. (2012), "An overview of CMIP5 and the experiment design", *Bulletin of the American Meteorological Society*, Vol. 93 No. 4, pp. 485-498.

World Bank (2008), *World Development Report 2008*, Washington, DC.

World Bank (2010), "Agricultural insurance", Primer Series on Insurance, No. 12, Washington DC.

Further reading

Ajetomobi, O.J., Abiodun, A. and Hassan, R. (2011), "Impacts of climate change on rice agriculture in Nigeria", *Tropical and Subtropical Agroecosystems*, Vol. 14 No. 2011, pp. 613-622.

Deressa, T.T. (2006), "Measuring the economic impact of climate change on Ethiopian agriculture: Ricardian approach", CEEPA Discussion Paper No. 25, CEEPA (Center for Environmental Economics and Policy in Africa), University of Pretoria, SA.

Druyan, L.M. (2011), "Studies of 21st-century precipitation trends over west africa", *International Journal of Climatology*, Vol. 31 No. 10, pp. 1415-1424.

Eid, M.H., El-Marsafawy, S.M. and Ouda, S.A. (2006), "Assessing the economic impacts of climate change on agriculture in Egypt: a Ricardian approach", CEEPA Discussion Paper No. 16, CEEPA (Center for Environmental Economics and Policy in Africa), University of Pretoria, SA.

Gbetibouo, G. and Hassan, R. (2005), "Measuring the economic impact of climate change on major South African field crops: a Ricardian approach", *Global and Planetary Change*, Vol. 47 Nos 2/4, pp. 143-152.

Hassan, R. and Nhemachena, C. (2008), "Determinants of climate adaptation strategies of African farmers: multinomial choice analysis", *African Journal of Agricultural and Resource Economics*, Vol. 2 No. 1, pp. 83-104.

Jain, S. (2006), "An empirical economic assessment of impacts of climate change on agriculture in Zambia", CEEPA Discussion Paper No. 27, CEEPA (Center for Environmental Economics and Policy in Africa), University of Pretoria.

Jones, C., Giorgi, F. and Asrar, G. (2011), "The coordinated regional downscaling experiment (CORDEX): an international downscaling link to CMIP5", *CLIVAR Exchanges*, Vol. 56 No. 16, pp. 34-40.

Mariara, K.J. and Karanja, F.K. (2006), "The economic impact of climate change on Kenyan crop agriculture: a Ricardian approach", CEEPA Discussion Paper No. 12, CEEPA (Center for Environmental Economics and Policy in Africa), University of Pretoria.

Molua, E.L. and Lambi, C.M. (2006), "The economic impact of climate change on agriculture in Cameroon", CEEPA Discussion Paper No. 33, CEEPA (Center for Environmental Economics and Policy in Africa), University of Pretoria.

Onyekuru, A.N. and Marchant, R. (2016), "Assessing the economic impact of climate change on forest resource use in Nigeria: a Ricardian approach", *Agricultural and Forest Meteorology*, Vol. 220, pp. 10-20.

Sene, M.I., Diop, M. and Dieng, A. (2006), "Impacts of climate change on the revenues and adaptation of farmers in Senegal", CEEPA Discussion Paper No. 20, (Center for Environmental Economics and Policy in Africa), University of Pretoria.

Seo, N. and Mendelsohn, R. (2006), "The impact of climate change on livestock management in Africa: a structural Ricardian analysis", CEEPA Discussion Paper No. 23, CEEPA (Center for Environmental Economics and Policy in Africa), University of Pretoria.

World Bank (2014), *World Development Indicator*, World Bank, Washington, DC.

Corresponding author
William M. Fonta can be contacted at: fonta.w@wascal.org

Appendix

Table AI. Summary statistics of the sampled cocoa farms

Variable definition and measurement	All farms		Rainfed		Irrigated	
	Mean	SD	Mean	SD	Mean	SD
Socioeconomic variables						
NRh (in Naira)	42,558.9 ($283.7)	46,774.8 ($311.8)	31,386.7 ($209.2)	17387.2 ($115.9)	45,339.1 ($302.3)	51,202.7($341.4)
Age of household head (years)	55.3	12.72	56.7	13.3	52.4	11.7
Education (years)	9.1	4.1	6.5	5.01	12.5	10.6
Farming experience (cocoa)	22.8	10.8	23	3.95	21.9	5.0
Household size (number of persons)	7.5	3.8	5.3	2.19	6.2	2.7
Agricultural variables						
Total farm area (hectares)	6.5	4.8	5.3	2.2	7.9	5.3
Visit from extension worker (number)	2.5	2.6	2.6	2.9	2.2	2.4
Farm labor (cocoa farming)	2.9	2.1	3.6	2.8	2.3	2.2
Crop area (cocoa hectares)	2.4	1.0	1.4	1.4	1.6	1.0
Distance to urban market (km)	90.5	142.7	86.4	78.5	82.9	62.6
Aggregate measures (proportions)						
% of farms headed by male	95.0		90.0		97.0	
% of farmers with access to credit	33.6		39.0		37.0	
% of farms that received subsidy	14.0		5.5		45.5	
% of farms using suppl. irrigation	19.9		0		100	
% of farmers keeping livestock	46.7		40.0		37.0	
% of farms using pesticides	93.0		95.0		97.0	
% of farms over 5 hectares	11.5		7.4		22.6	
% of farms with extension contacts	66.0		56.9		55.6	
% of farms on La soils	42.9		52.2		19.5	
% of farms on Jc soils	28.7		21.3		46.2	
% of farms on Li soils	14.2		14.2		25.6	
% of farms on LVf soils	14.2		12.3		8.8	
Climate variables						
Monthly dry season temp. (°C)	32.4	4.82	28.6	1.12	28.9	1.05
Monthly rainy season temp. (°C)	25.7	1.13	25.1	0.91	24.5	0.82
Monthly Harmattan temp. (°C)	19.6	6.5	25.0	3.0	25.9	3.41

(continued)

Table AI.

Variable definition and measurement	All farms Mean	SD	Rainfed Mean	SD	Irrigated Mean	SD
Monthly dry season precip. (mm)	46.7	33.9	37.3	23.8	33.7	21.15
Monthly rainy season precip. (mm)	169.0	15.9	139.2	28.4	150.7	22.9
Monthly Harmattan precip. (mm)	12.7	21.4	21.2	13.1	18	14.5
Control variables						
Population density	251.9	19.93	251.7	19.9	252.3	20.4
Irrigation	19.9	0.40				
Altitude (m)	1,238.8	139.3	1,217.3	146.3	1,332.8	110.2
Number of observations	280		224		56	

3

Domestic water demand forecasting in the Yellow River basin under changing environment

Xiao-jun Wang, Jian-yun Zhang, Shamsuddin Shahid, Lang Yu,
Chen Xie, Bing-xuan Wang and Xu Zhang
(Author affiliations can be found at the end of the article)

Abstract

Purpose – The purpose of this paper is to develop a statistical-based model to forecast future domestic water demand in the context of climate change, population growth and technological development in Yellow River.

Design/methodology/approach – The model is developed through the analysis of the effects of climate variables and population on domestic water use in eight sub-basins of the Yellow River. The model is then used to forecast water demand under different environment change scenarios.

Findings – The model projected an increase in domestic water demand in the Yellow River basin in the range of 67.85×10^8 to $62.20 \times 10^8\,\mathrm{m}^3$ in year 2020 and between 73.32×10^8 and $89.27 \times 10^8\,\mathrm{m}^3$ in year 2030. The general circulation model Beijing Normal University-Earth System Model (BNU-ESM) predicted the highest increase in water demand in both 2020 and 2030, while Centre National de Recherches Meteorologiques Climate Model v.5 (CNRM-CM5) and Model for Interdisciplinary Research on Climate- Earth System (MIROC-ESM) projected the lowest increase in demand in 2020 and 2030, respectively. The fastest growth in water demand is found in the region where water demand is already very high, which may cause serious water shortage and conflicts among water users.

Originality/value – The simple regression-based domestic water demand model proposed in the study can be used for rapid evaluation of possible changes in domestic water demand due to environmental changes to aid in adaptation and mitigation planning.

Keywords Climate change, Statistical model, Water demand forecasting,
Water resources management, Yellow river

1. Introduction

Climate change, rapid social economic development and population growth have imposed significant challenges to sustainable development all around the world (Nigel and Ben, 2014; Wang *et al.*, 2014a, 2014b; Shahid *et al.*, 2014, 2016). As a unique and increasingly scarce commodity, water is among the most significantly affected resources by these changes. Ensuring continuous and adequate water supply to growing population and developing economy is one of the major concerns for policy makers and scientists (Butler and Memon, 2006; Kim *et al.*, 2014). It is much more challenging in arid and semi-arid river basins in Australia, Middle East, Africa and Northern China due to limited availability of water resources (Howe *et al.*, 2005; Butler and Memon, 2006; Wang *et al.*, 2012a). Understanding possible future changes in water demand is essential for the development of effective climate change adaptation policies (Shahid *et al.*, 2017).

Water demand is influenced by various factors such as population growth, technological development, climate change, social-economic (Wang *et al.*, 2016a). The complex interactions of different influential factors have made the forecasting of water demand a difficult task (Vörösmarty *et al.*, 2000; Downing *et al.*, 2003; Alvisi *et al.*, 2003; Babel *et al.*, 2007; Khatri and Vairavamoorthy, 2009; Wang *et al.*, 2015). A number of studies have been carried out to predict water demand (Combalicer *et al.*, 2010; Dursun, 2010; Jakimavičius and Kriaučiūnienė, 2013; Wang *et al.*, 2016b). The approaches generally used for forecasting water demand can be classified into three broad classes, namely, time series, econometric and end-use forecasting (Zhou *et al.*, 2000; Alvisi *et al.*, 2003; Altukaynak *et al.*, 2005; Bougadis *et al.*, 2005; Khatri and Vairavamoorthy, 2009; Caiado, 2010; Wang *et al.*, 2017). Besides that, artificial neural network, support vector regression, etc. have been used in a few cases (Ghiassi *et al.*, 2008; Adamowski, 2008). Unfortunately, yet only a few of the traditional forecasting models can adequately manage the dynamics of a water supply system because of the limitations in modeling structures. Potential challenges also arise, as these traditional methods require long-term, continuous historical records of water demand and other variables (Simonovic, 2002; Qi and Chang, 2011).

The impact of climate change on water demand is a major issue in recent years. Most of the recent studies on climate change impacts on water demand are based on the historical relationship between water demand and climate variables such as rainfall, air temperature, sunshine duration, relative humidity and wind speed (Protopapas *et al.*, 2000; Downing *et al.*, 2003; Karamouz *et al.*, 2011). It is widely accepted that water demand will increase due to the increased evapotranspiration caused by higher temperature. However, how to take consideration of climate variables and population growth together for forecasting future domestic water demand is still a challenge.

The Yellow River, originated in the Qinghai-Tibetan plateau of Qinghai province, flows across eight other provinces and autonomous regions of China before it ends into the Bohai Sea (Liu and Zhang, 2002; Wang *et al.*, 2012b). The river plays a key role not only in the country's economic development, but also in the historic and cultural identity of the Chinese people. With approximately 9 per cent of China's population and 17 per cent of its agricultural area depending on the Yellow River basin, it is designated as "the cradle of Chinese Civilization" (Liu and Zhang, 2002; Giordano *et al.*, 2004). Population and economy of the River Basin have grown rapidly in the recent years. This has caused a significant increase in water demand for agriculture, industry and households in the Basin. Consequently, the gap between water supply and demand has widened in recent years (Giordano *et al.*, 2004; Barnett *et al.*, 2006; Gonçalves *et al.*, 2007). Some of the main tributaries of the Yellow River, such as Kuye River and Tuwei River, showed a decreasing trend of runoff, even drying up (Wang *et al.*, 2014b). Increase in demand of water for domestic use due to climate change and population growth may aggravate the existing situation of water shortage in the Basin. Forecasting possible changes in residential water demand is very important for planning and management of water resources of the Basin.

The objective of the present paper is to develop a statistical model to forecast domestic water demand in the context of climate change, population growth and technological development. The proposed model is developed through the analysis of the effects of climate and population on domestic water use and then applied in the Yellow River basin to forecast water demand under different environment change scenarios.

2. Materials and methods

2.1 Data and sources

2.1.1 Water consumption and population growth data. Population growth, domestic water use and economic development data of the Yellow River basin for the period 1980-2012 were collected from China Water Resources Bulletins (MWR, 2000-2012). The recent data of domestic water use in eight sub-basins of the Yellow River, namely, Uperstream of Longyangxia (UL), Longyangxia-Lanzhou (LL), Lanzhou-Hekouzhen (LH), Hekouzhen-Longmen (HL), Longmen-Sanmenxia (LS), Sanmenxia-Huayuankou (SH), Downstream of Huayuankou (DH) and Interior Area (IA), were collected from the Yellow River Water Resources Bulletins (YRCC, 1998-2012). Table I summarizes the basic statistics of collected data for different sub-basins of the Yellow River.

2.1.2 Meteorological data. Daily temperature data of 61 stations for the period 1961-2012 were obtained from National Climate Center of China (NCCC). Climate projections by a suit of seven general circulation models (GCMs), namely, Beijing Climate Center-Climate System Model version 1 (BCC-CSM1-1), BNU-ESM, CNRM-CM5, Goddard Institute for Space Studies Model E version 2 with Russell ocean model (GISS-E2-R), MIROC-ESM, Max Planck Institute-Earth System Model-Low Resolution (MPI-ESM-LR) and Meteorological Research Institute-Coupled Global Climate Model version 3 (MRI-CGCM3), under Representative Concentration Pathways (RCPs) scenario RCP4.5 were also collected from NCCC.

2.2 Domestic water demand forecast modeling

Domestic water demand includes water needs for all residential purposes such as in-house water use for drinking, preparing food, bathing, washing clothes and dishes and flushing toilets, etc. as well as outdoor water needs for gardening, lawn watering, etc. (Alvisi *et al.*, 2003; Garcia *et al.*, 2004; Blokker *et al.*, 2010; Wang *et al.*, 2017). According to the water resources planning technical specifications of China, domestic water demand and population growth can be linked through the following equations:

$$LW^{/} = LuW^{/} + LrW^{/}$$

$$= Pu^{/} \times LQu^{/} \times \frac{365}{1000} + Pr^{/} \times LQr^{/} \times \frac{365}{1000} \qquad (1)$$

Table I. Climate and population of different sub-basins of the Yellow River

Sub-basin	Area (10^4 km^2)	Precipitation (mm) (1956-2000)	Population (10^4 p) in 2010 Urban	Rural	Urbanization rate (%) in 2010	Population density (P/km^2) in 2010
Yellow River	79.5	445.8	4543.27	6824.96	40.0	143
UL	13.1	478.3	14.28	50.95	21.9	5
LL	9.1	478.9	327.12	590.29	35.7	101
LH	16.3	261.9	850.36	755.62	52.9	99
HL	11.2	433.5	265.02	605.98	30.4	78
LS	19.2	540.6	2066.34	3053.14	40.4	268
SH	4.2	659.5	529.66	810.61	39.5	319
DH	2.2	647.8	463.68	928.22	33.3	633
IA	4.2	271.9	26.81	30.15	47.1	14

where LW' is the domestic water demand (10^4 m^3), which can be estimated by multiplying the projected population (10^4 p) with the projected per capita water consumption (l/p.d). The urban water demand, LuW' is estimated by multiplying the urban population, Pu' (10^4 p), with the per capita water consumption by urban population, LQu' (l/p.d); the rural water demand, LrW', is estimated by multiplying the rural population, Pr' (10^4 p), with the per capita water consumption by rural population, LQr' (l/p.d) (Wang *et al.*, 2011).

Equation (1) does not consider climate change for forecasting water demand. Therefore, the equation is modified in this study to incorporate the impact of climate change on water demand, ΔLW^t, as follows:

$$\Delta LW^t = d_T \cdot \Delta T \cdot LW^t \tag{2}$$

where:

$$d_T = \frac{\Delta LW}{\Delta T} \tag{3}$$

where d_T is the climatic elasticity of domestic water demand; ΔLW is the change in domestic water demand (10^4 m^3); and ΔT is the change in temperature (°C).

3. Domestic water demand under changing environment
3.1 Population growth and domestic water use
The historical changes and future projection of population in different sub-basins of the Yellow River basin are shown in Figure 1. The highest population in the Yellow River basin is in LS sub-basin, while the lowest population in UL sub-basin. Figure 1 shows that the population in all the sub-basins has increased rapidly over the past years. Population in the Yellow River basin was 8,177 × 10^4 in 1980 and the urbanization rate was only 17.8 per cent. The highest rate of urbanization was in LH (33.5 per cent). According to the future plan of River Basin management authority, the total population and urbanization in the Basin will reach 12,658 × 10^4 and 50.4 per cent, respectively, in 2020 and 13,094 × 10^4 and 58.8 per cent, respectively, in 2030.

The growing population has caused a drastic increase in domestic water demand in the Basin. The domestic water consumption in the Yellow River basin has increased from 11.72 × 10^8 m^3 in 1980 to 22.66 × 10^8 m^3 in 2000. The per capita water demand has also increased due to urbanization and changes in socioeconomic condition of the people. The changes in per capita domestic water demand in different sub-basins are shown in Figure 2. The figure shows

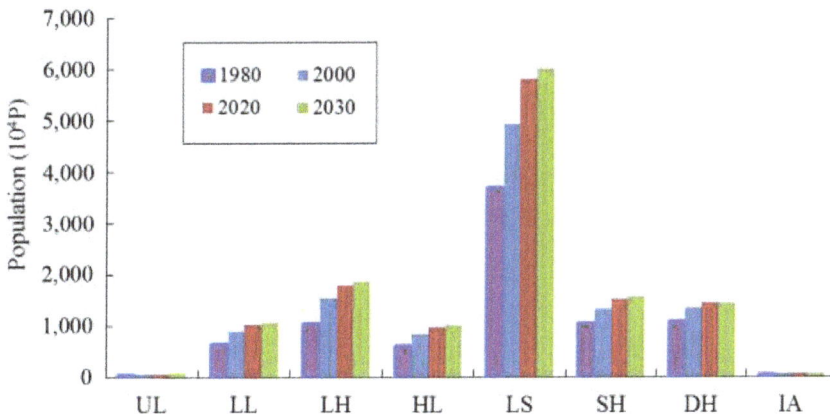

Figure 1. Historical and projected change in population in different sub-basins of the Yellow River

that the average per capita water consumption in the Basin has increased from only 73 l/p.d in 1980 to 103 l/p.d in 2000. It has been projected that per capita water demand will continue to increase in the Basin and will reach 115 l/p.d and 123.7 l/p.d in 2020 and 2030, respectively.

3.2 Projection of temperature

The climate models projected increase in temperature, but no significant change in rainfall in the Yellow River basin. The projections of temperature by GCMs BCC-CSM1-1, BNU-ESM, CNRM-CM5, GISS-E2-R, MIROC-ESM, PI-ESM-LR, MRI-CGCM3 in the Yellow River basin are shown in Figure 3. The figure shows a increase in temperature before 2030. As there is no change in rainfall, only temperature was considered to assess the impact of climate change on water demand in the Basin.

3.3 Domestic water demand due to climate change

The ensemble mean of the projected temperatures by different GCMs was used to estimate future changes in water demand in the Yellow River basin due to climate change. The changes in residential water demand in different sub-basins due to climate change for the years 2020 and 2030 are shown in Figures 4 and 5, respectively. The figures show increase in water demand in all

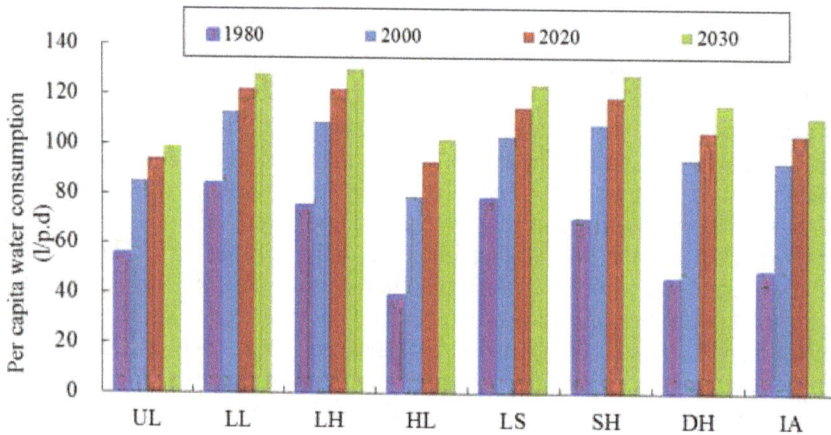

Figure 2. Changes in per capita water consumption in the Yellow River basin

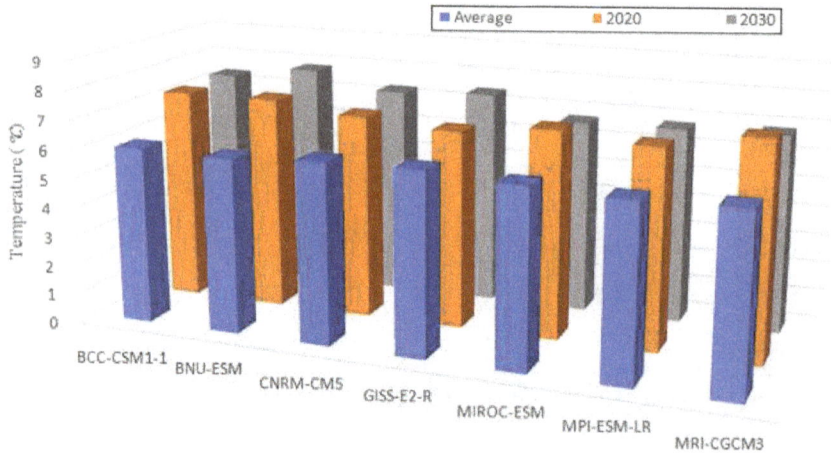

Figure 3. Temperature projections by different GCMs in the Yellow River basin under RCP4.5 in 2020 and 2030

the regions of the Yellow River basin. The highest increase was observed in LS, followed by DH. Domestic water demand in LS is already very high due to large population. Faster increase in water demand due to climate change may aggregate the condition of water scarcity and conflict among water users in the region.

3.4 Domestic water demand under changing environment

The projected growth in population and the advancement in water-saving technologies were also considered along with climate change to forecast domestic water demand in the Yellow River basin. Equation (1) was used to estimate the impact of population growth and urbanization on domestic water demand. Reduction of water demand through the advancement of water-saving technologies as planned by the Yellow River basin authority was then incorporated to estimate the net changes in water demand. Figures 6 and 7 show the projected change in domestic water demand in 2020 and 2030, respectively, due to climate change, population growth and technological advancement together.

Figure 6 shows that domestic water demand will increase in the range of 67.85×10^8 to 62.20×10^8 m^3 in 2020. The GCM BNU-ESM projected the highest increase, while CNRM-CM5 the lowest. The water demand was projected to increase in the range of 89.27×10^8 to 73.32×10^8 m^3 in 2030. The highest increase was projected by BNU-ESM and the lowest by MIROC-ESM. Among the Sub-Basins, the highest increase in water demand was projected in LS and the lowest in UL.

4. Conclusion

A statistical model was developed in this paper to forecast domestic water demand due to changes in climate, population and water-saving technologies. Results showed that

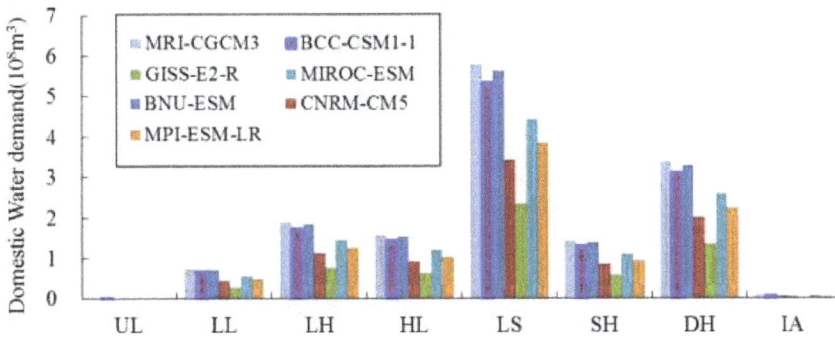

Figure 4. Projected domestic water demand in year 2020 under RCP4.5

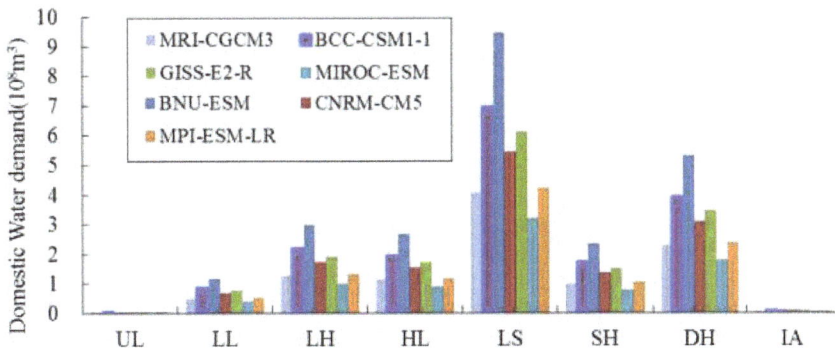

Figure 5. Projected domestic water demand in year 2030 under RCP4.5

domestic water demand in the Basin will increase in the range of 67.85×10^8 to $62.20 \times 10^8 \, m^3$ in 2020 and 89.27×10^8 to $73.32 \times 10^8 \, m^3$ in 2030. The highest increase will be in the LS and the lowest in UL. The LS is the most populated region of the Basin. High water demand due to large population has already made water in the Sub-Basin scarce. It can be anticipated that the rapid increase in water demand due to climate change and population growth will aggravate the situation in the future. It will also increase the competition among water users and, therefore, more conflicts.

The method proposed in the study can be used for rapid evaluation of possible changes in domestic water demand in the context of environmental changes. However, more attention is required for improvement of the method. Following points should be considered for future improvement of the model:

- domestic water demand for different in-house and outdoor activities are different and therefore, impacts of environmental changes on each activity should be evaluated for better projection of water demand;

- only the average temperature was considered to estimate the impacts of climate change on domestic water demand in this study and other climatic variables such as maximum and minimum temperature and humidity can be considered for the improvement of model accuracy; and

- furthermore, future study is required to assess uncertainties in water demand projection for better planning and management of water resources.

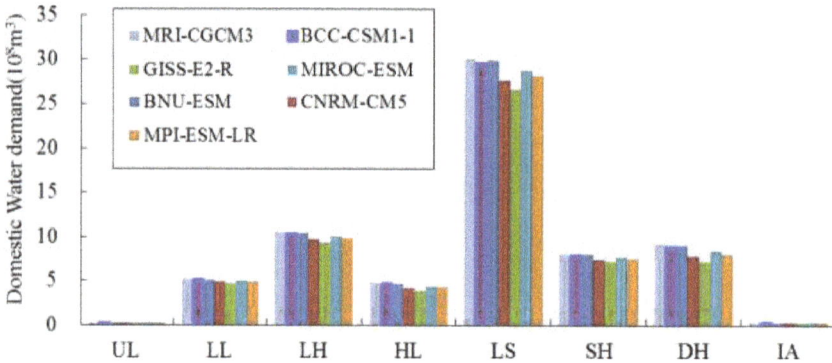

Figure 6. Changes in domestic water demand in year 2020 due to environmental changes

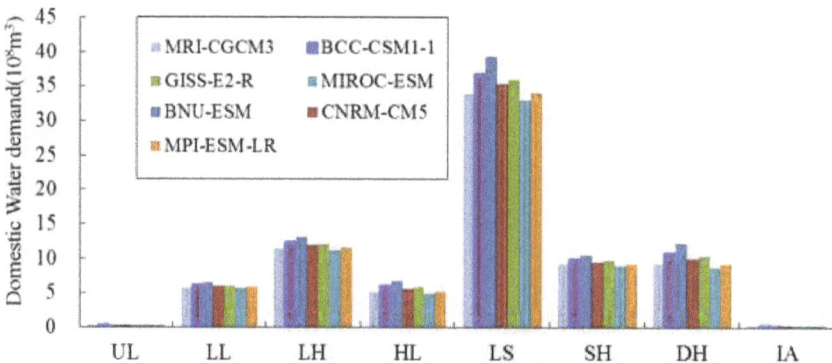

Figure 7. Changes in domestic water demand in year 2030 due to environmental changes

Acknowledgements

The authors are grateful to the National Key R&D Program of China (No. 2017YFC0403506), National Natural Science Foundation of China (No. 51309155), Strategic Consulting Projects of Chinese Academy of Engineering (No. 2016-ZD-08-05-02), China Water Resource Conservation and Protection Project (No. 126302001000150001) and Open Research Fund of State Key Laboratory of Simulation and Regulation of Water Cycle in River Basin China Institute of Water Resources and Hydropower Research (No. IWHR-SKL-201515) for providing financial support for this research. The authors are also thankful to anonymous reviewers and editors for their helpful comments and suggestions.

References

Adamowski, J.F. (2008), "Peak daily water demand forecast modeling using artificial neural networks", *Journal of Water Resources Planning and Management*, Vol. 134 No. 2, pp. 119-128.

Altukaynak, A.S., Özger, M. and Çakmakci, M. (2005), "Water consumption prediction of Istanbul city by using fuzzy logic approach", *Water Resources Management*, Vol. 19 No. 5, pp. 641-654.

Alvisi, S., Franchini, M. and Marinelli, A. (2003), "A stochastic model for representing drinking water demand at residential level", *Water Resources Management*, Vol. 17, pp. 197-222.

Babel, M.S., Das Gupta, A. and Pradhan, P. (2007), "A multivariate econometric approach for domestic water demand modeling: an application to Kathmandu, Nepal", *Water Resources Management*, Vol. 21 No. 3, pp. 573-589.

Barnett, J., Webber, M. and Wang, M. (2006), "Ten key questions about the management of water in the yellow river basin", *Environmental Management*, Vol. 38 No. 2, pp. 179-188.

Blokker, E.J.M., Vreeburg, J.H.G. and van Dijk, J.C. (2010), "Simulating residential water demand with a stochastic end-use model", *Journal of Water Resources Planning and Management*, Vol. 136 No. 2, pp. 19-26.

Bougadis, J., Adamowski, K. and Diduch, R. (2005), "Short-term municipal water demand forecasting", *Hydrological Processes*, Vol. 19 No. 1, pp. 137-148.

Butler, D. and Memon, F.A. (2006), *Water Demand Management*, IWA Publishing, London.

Caiado, J. (2010), "Performance of combined double seasonal univariate time series models for forecasting water demand", *Journal of Hydrologic Engineering*, Vol. 15 No. 3, pp. 215-222.

Combalicer, E.A., Cruz, R.V.O., Lee, S. and Im, S. (2010), "Assessing climate change impacts on water balance in the mount Makiling Forest, Philippines", *Journal of Earth System Science*, Vol. 119 No. 3, pp. 265-283.

Downing, T.E., Butterfield, R.E. and Edmonds, B. (2003), "Climate change and the demand for water", *Research Report*, Stockholm Environment Institute Oxford Office, Oxford.

Dursun, S. (2010), "Effect of global climate change on water balance of Beyşehir lake (Konya – turkey)", *Balwois 2010*, Ohrid, Republic of Macedonia, 25-29 May 2010.

Garcia, V.J., Garcia-Bartual, R., Cabrera, E. and Arregui, F. (2004), "Stochastic model to evaluate residential water demands", *Journal of Water Resources Planning and Management*, Vol. 130 No. 5, pp. 386-394.

Ghiassi, M., Zimbra, D.K. and Saidane, H. (2008), "Urban water demand forecasting with a dynamic artificial neural network model", *Journal of Water Resources Planning and Management*, Vol. 134 No. 4, pp. 138-146.

Giordano, M., Zhu, Z., Cai, X., Hong, S., Zhang, X. and Xue, Y. (2004), "Water management in the yellow river basin: background, current critical issues", *Comprehensive Assessment Research Report 3*, International Water Management Institute, Colombo, Sri Lanka.

Gonçalves, J.F.J., Graça, M.A.S. and Callisto, M. (2007), "Litter decomposition in a Cerrado savannah stream is retarded by leaf toughness, low dissolved nutrients and a low density of shredders", *Freshwater Biology*, Vol. 52 No. 8, pp. 1440-1451.

Howe, C., Jones, R.N. and Maheepala, S. (2005), "Implications of potential climate change for melbourne's water resources", *CSIRO Urban Water*, CSIRO Atmospheric Research, Melbourne.

Jakimavičius, D. and Kriaučiūnienė, J. (2013), "The climate change impact on the water balance of the Curonian Lagoon", *Water Resources*, Vol. 40 No. 2, pp. 120-132.

Karamouz, M., Yazdi, M.S.S., Ahmadi, B. and Zahraie, B. (2011), "A system dynamics approach to economic assessment of water supply and demand strategies", *Proceedings of the 2011 World Environmental and Water Resources Congress*, pp. 1194-1203.

Khatri, K.B. and Vairavamoorthy, K. (2009), "Water Demand Forecasting for the City of the Future against the Uncertainties and the Global Change Pressures: case of Birmingham", World Environmental and Water Resources Congress 2009: Great Rivers.

Kim, S., Kim, B.S. and Jun, H. (2014), "Assessment of future water resources and water scarcity considering the factors of climate change and social–environmental change in Han River basin, Korea", *Stochastic Environmental Research and Risk Assessment*, Vol. 28 No. 8, pp. 1999-2014, doi: 10.1007/s00477-014-0924-1.

Liu, C. and Zhang, S. (2002), "Drying up of the Yellow river: its impacts and counter-measures", *Mitigation and Adaptation Strategies for Global Change*, Vol. 7 No. 3, pp. 203-214.

MWR (2000-2012), *China Water Resources Bulletin. Ministry of Water Resources (MWR)*, MWR, Beijing, China.

Nigel, W.A. and Ben, L.-H. (2014), "The global-scale impacts of climate change on water resources and flooding under new climate and socio-economic scenarios", *Climatic Change*, Vol. 122 Nos 1/2, pp. 127-140.

Protopapas, A.L., Katchamart, S. and Platonova, A. (2000), "Weather effects on daily water use in New York City", *Journal of Hydrologic Engineering*, Vol. 5 No. 3, pp. 332-338.

Qi, C. and Chang, N. (2011), "System dynamics modeling for municipal water demand estimation in an urban region under uncertain economic impacts", *Journal of Environmental Management*, Vol. 92 No. 6, pp. 1628-1641.

Shahid, S., Minhans, A. and Puan, O.C. (2014), "Assessment of greenhouse gas emission reduction measures in transportation sector of Malaysia", *Jurnal Teknologi*, Vol. 70 No. 4, pp. 1-8.

Shahid, S., Pour, S.H., Wang, X.-J., Shourav, S.A., Minhans, A. and Ismail, T. (2017), "Impacts and adaptation to climate change in Malaysian real estate", *International Journal of Climate Change Strategies and Management*, Vol. 9 No. 1, pp. 87-103.

Shahid, S., Wang, X.-J., Harun, S.B., Shamsudin, S.B., Ismail, T. and Minhans, A. (2016), "Climate variability and changes in the major cities of Bangladesh: observations, possible impacts and adaptation", *Regional Environmental Change*, Vol. 16 No. 20, pp. 459-471.

Simonovic, P.S. (2002), "World water dynamics: global modeling of water resources", *Journal of Environmental Management*, Vol. 66 No. 3, pp. 249-267.

Vörösmarty, C.J., Green, P., Salisbury, J. and Lammers, R.B. (2000), "Global water resources: vulnerability from climate change and population growth", *Science*, Vol. 289 No. 5477, pp. 284-288.

Wang, X.-J., Zhang, J.-Y., Shahid, S., ElMahdi, A., He, R.-M., Wang, X.-G. and Ali, M. (2011), "Gini coefficient to assess equity in domestic water supply in the Yellow River", *Mitigation and Adaptation Strategies for Global Change*, Vol. 17 No. 1, pp. 65-75.

Wang, X.-J., Zhang, J.-Y., Shahid, S., ElMahdi, A., He, R.-M., Bao, Z.-X. and Ali, M. (2012a), "Water resources management strategy for adaptation to droughts in China", *Mitigation and Adaptation Strategies for Global Change*, Vol. 18 No. 8, pp. 923-937.

Wang, X.J., Zhang, J.Y., Ali, M., Shahid, S., He, R.-M., Xia, X.-H. and Jiang, Z. (2016b), "Impact of climate change on regional irrigation water demand in Baojixia irrigation district of China", *Mitigation and Adaptation Strategies for Global Change*, Vol. 21 No. 2, pp, pp. 233-247, doi: 10.1007/s11027-014-9594-z.

Wang, X.-J., Zhang, J.-Y., Shahid, S., Bi, S.-H., Elmahdi, A., Liao, C.-H. and Li, Y.-D. (2017), "Forecasting industrial water demand in Huaihe River Basin due to environmental changes", *Mitigation and Adaptation Strategies for Global Change*, pp. 1-15, doi: 10.1007/s11027-017-9744-1.

Wang, X.-J., Zhang, J.-Y., Shahid, S., ElMahdi, A., He, R.-M., Wang, X.-G. and Ali, M. (2012b), "Gini coefficient to assess equity in domestic water supply in the Yellow River", *Mitigation and Adaptation Strategies for Global Change*, Vol. 17 No. 1, pp. 65-75.

Wang, X.J., Zhang, J.Y., Shahid, S., Guan, E.-H., Wu, Y.-X., Gao, J. and He, R.-M. (2016a), "Adaptation to climate change impacts on water demand", *Mitigation and Adaptation Strategies for Global Change*, Vol. 21 No. 1, pp. 81-99, doi: 10.1007/s11027-014-9571-6.

Wang, X.-J., Zhang, J.-Y., Shahid, S., He, R.-M., Xia, X.-H. and Mou, X.-L. (2015), "Potential impact of climate change on future water demand in Yulin city, Northwest China", *Mitigation and Adaptation Strategies for Global Change*, Vol. 20 No. 1, pp. 1-19, doi: 10.1007/s11027-013-9476-9.

Wang, X.-J., Zhang, J.-Y., Shahid, S., Xia, X.-H., He, R.-M. and Shang, M.-T. (2014a), "Catastrophe theory to assess water security and adaptation strategy in the context of environmental change", *Mitigation and Adaptation Strategies for Global Change*, Vol. 19 No. No. 4, pp. 463-477, doi: 10.1007/s11027-012-9443-x.

Wang, X.-J., Zhang, J.-Y., Wang, J.-H., He, R.-M., ElMahdi, A., Liu, J.-H., Wang, X.-G., King, D. and Shahid, S. (2014b), "Climate change and water resources management in Tuwei river basin of Northwest China", *Mitigation and Adaptation Strategies for Global Change*, Vol. 19 No. No. 1, pp. 107-120.

YRCC (1998-2012), *Yellow River Water Resources Bulletin. Yellow River Conservancy Commission*, YRCC, Zhengzhou.

Zhou, S.L., McMahona, T.A., Waltonb, A. and Lewisb, J. (2000), "Forecasting daily urban water demand: a case study of Melbourne", *Journal of Hydrology*, Vol. 236 Nos 3/4, pp. 153-164.

Author affiliations

Xiao-jun Wang, State Key Laboratory of Hydrology-Water Resources and Hydraulic Engineering, Nanjing Hydraulic Research Institute, Nanjing, China and Research Center for Climate Change, Ministry of Water Resources, Nanjing, China

Jian-yun Zhang, State Key Laboratory of Hydrology-Water Resources and Hydraulic Engineering, Nanjing Hydraulic Research Institute, Nanjing, China and Research Center for Climate Change, Ministry of Water Resources, Nanjing, China
Shamsuddin Shahid, Faculty of Civil Engineering, Universiti Teknologi Malaysia (UTM), Johor Bahru, Malaysia
Lang Yu, China Institute of Water Resources and Hydropower Research, Beijing, China
Chen Xie, Yellow River Conservancy Commission, Zhengzhou, China
Bing-xuan Wang, Hohai University, Nanjing, China, and
Xu Zhang, State Key Laboratory of Hydrology-Water Resources and Hydraulic Engineering, Nanjing Hydraulic Research Institute, Nanjing, China and Research Center for Climate Change, Ministry of Water Resources, Nanjing, China

Corresponding author

Xiao-jun Wang can be contacted at: xjwang@nhri.cn

4

Effect of climate variability on crop income and indigenous adaptation strategies of households

Arega Shumetie
*Haramaya University, Dire Dawa, Ethiopia,
and Makerere University, Kampala, Uganda, and*

Molla Alemayehu Yismaw
Radboud University, Nijmegen, The Netherlands

Abstract

Purpose – This study aims to examine the effect of climate variability on smallholders' crop income and the determinants of indigenous adaptation strategies in three districts (Mieso, Goba-koricha and Doba) of West Hararghe Zone of Ethiopia. These three districts are located in high-moisture-stress areas because of crop season rainfall variability.

Design/methodology/approach – Primary data collected from 400 sample households were used for identifying factors that affect households' crop income. The study used ordinary least square (OLS) regression to examine the effect of climate variability. Given this, binary logit model was used to assess smallholders' adaptation behavior. Finally, the study used multinomial logistic regression to identify determinants of smallholders' indigenous adaptation strategies.

Findings – The OLS regression result shows that variability in rainfall during the cropping season has a significant and negative effect, and cropland and livestock level have a positive effect on farmers' crop income. The multinomial logistic regression result reveals that households adopt hybrid crops (maize and sorghum) and dry-sowing adaptation strategies if there is shortage during the cropping season. Variability in rainfall at the time of sowing and the growing are main factors in the area's crop production. Cropland increment has positive and significant effect on employing each adaptation strategy. The probability of adopting techniques such as water harvesting, hybrid seeds and dry sowing significantly reduces if a household has a large livestock.

Originality/value – The three districts are remote and accessibility is difficult without due support from institutions. Thus, this study was conducted on the basis of the primary data collected by the researchers after securing grant from Swedish International Development Agency (SIDA).

Keywords Household, Adaptation, Determinants, Climate variability, Crop income

1. Introduction

Climate change and variability are becoming a strong threat for food security in the twenty-first century, particularly for the agriculture-dependent sub-Saharan African countries (Eva, 2009). Global warming is projected to have a significant impact on factors affecting agriculture, including temperature, carbon dioxide emission and precipitation. Identifying the agricultural effect of climate change might help to properly anticipate and adapt to the problem and maximize smallholders' productivity (Fraser, 2008). Because most African countries lack the capacity of adapting to this problem, minor changes can spark a significant effect on the agriculture capability of the nations. While climate change has global impacts on agriculture, regional variations are significant depending on the geographic location. For most developing countries, these differences underscore the difficulty of proposing the general strategies required for adopting new agricultural technologies so that the problem can be dealt with (David, 1993). Changing the consistency and intensity of rainfall is one of the most widely spread and potentially devastating impacts of climate change in East Africa. Precipitation variation ultimately affects moisture availability that may result in a reduction in agricultural production and shortages in potentially widespread food supply (Michael, 2006).

Both insufficient rainfall and its erratic distribution are the main reasons that reduce crop production and thus income (Mertz *et al.*, 2010). Variability in crop production and animal husbandry owing to climate variability will directly result in smallholders' income variation. For Ethiopian smallholders, crop production is the area that is mostly susceptible to climatic variation.

Few countries in Africa are trying to conserve resources and integrate climate change adaptation strategies into their management plans (Hansen *et al.*, 2003). Some smallholders' plant drought-resistant crops to adapt moisture stress (Patt *et al.*, 2005). Almost all households in Ethiopia, Mali and Yemen use adaptation strategies including improved seeds and changed planting dates, to make yields less susceptible to climate variability. There is a remarkable difference among countries and between households in the type of adaptation strategies implemented. In Ethiopia, wealthier households participate mainly in adopting communal strategies, such as soil erosion, communal irrigation or reforestation, communal water harvesting and rangeland management, for which external help is necessary (Arjan *et al.*, 2011). Few households chose non-farm income sources, but the problem here is that rural areas lack auspicious environment to sustain those activities. This indicates that smallholders' climate variability adaptation strategies in Ethiopia are affected by different factors. Thus, the general objective of this study was to examine the effect of variability in climate on crop income. Given this, the study assessed factors that determine indigenous adaptation strategies of smallholder farmers in West Hararghe Zone, Ethiopia.

2. Methodology

2.1 Description of the study area

Western Hararghe is one of the 17 zones of Oromiya National Regional State. The Zone is bordered in the South by the Shebelle River, which separates it from Bale Zone, in the Southwest by Arsi, in the east by East Hararghe, in the Northwest by Afar and in the North by Somali Regional State. This Zone has good potential for coffee production of about 8,364 tons of coffee, accounting for about 7.3 per cent of the region-level production and 3.7 per cent of the national-level production in 2012 (CSA, 2013). This Zone had a total population of 1,871,706 with 15,065 km^2 area coverage, resulting in a population density of 124.23 persons/km^2 (CSA, 2007). There are about 395,127 households, with an average size of 4.74 persons per household and 380,019 housing units. More than 92 per cent of the zone has faced frequent bouts of malaria, as it was frequently drought-ridden (World Bank, 2004). Empirical studies showed that the eastern zones of the Oromiya Regional State (West and East Hararghe) are prone to chronic food insecurity problems (François, 2003). Non-remunerative international prices and outbreak of diseases together with high moisture stress made coffee production a less attractive business for households of the area.

This study was conducted in three districts of the West Hararghe Zone (Mieso, Goba-koricha and Doba) on the crop production sub-sector only. The three districts were selected purposively because they are highly dependent on crop production on the one hand and are suffering from climate variability, on the other hand.

2.2 Method of sampling and data collection
The study used both primary and secondary data to achieve the predefined objectives. Secondary data were collected from published and unpublished works of different governmental and non-governmental institutions. A multi-stage sampling technique was used to determine sample households and collect the data: first, sample districts that had severe moisture stress were selected purposively. Second, sample villages from these districts were selected randomly and the sampled households were finally selected using systematic sampling. A considerable sample size was used to make the sample more representative. The primary data were collected from farm households using questionnaire, which were distributed and collected by enumerators who know the culture and language of the research area. In addition, the study also conducted focus group discussions with the farming society to extract common problems shared by the society of the study area at large (Figure 1).

2.3 Method of data analysis
2.3.1 Ordinary least squares regression. Ordinary least squares (OLS) regression is a generalized linear modeling technique that is used to model a single response variable that has been recorded at an interval scale. The technique may be applied to either single or multiple explanatory- and categorical-type variables that have been properly coded (Hutcheson, 2011). Here in this research, the dependent variable was crop income collected

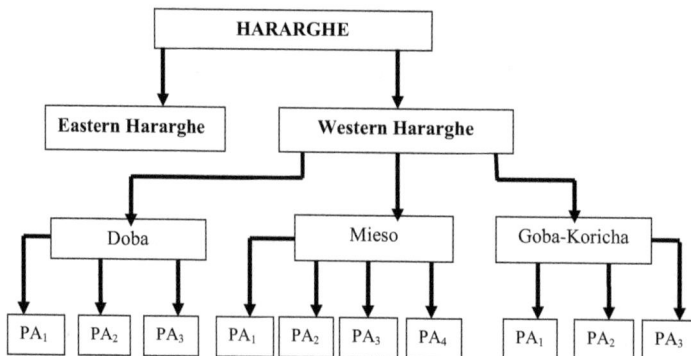

Source: Own formulation, 2015

Figure 1. The sample frame

by each sample household. Crop production inputs, crop price, natural factors and others may affect crop income of households.

$$Y = F(K, H, P, S, \ldots D) \tag{1}$$

where: Y is the crop income of households expressed in terms of birr, the local currency; K is the working capital budgeted for the crop production; H is human labor employed in the crop production sub-sector; P is the price of each crop depending on the nearest local market; S represents the climate-related shocks including rainfall inconsistencies D represents the demography-related variables like education, sex and age of the household head. The aforementioned mathematical expression would be transferred to a logarithmic functional form as follows:

$$\ln Y = \alpha + \beta_1 \ln K + \beta_2 \ln H + \beta_3 \ln P + \beta_4 \ln S + \mu_i \tag{2}$$

The unknown parameters in equation (2) will show the interaction of each explanatory variable with the dependent variable.

2.3.2 The logit regression. The model assumes that the data are case-specific; that is, each independent variable has a single value for each case and the dependent variable cannot be perfectly predicted from the independent variables for any case. As with other types of regression, there is no need for independent variables to be statistically independent from each other. The binary logit model for identifying determinants of adaptation behavior was specified as follows:

$$P(X) = \frac{\exp(\alpha + \beta X)}{1 + \exp(\alpha + \beta X)} \tag{3}$$

Where β is the coefficient of the covariates considered in the regression, and α refers to the value of the constant term.

The log of odds ratio, which is a linear function of parameter estimates, is:

$$\log\left(\frac{P_i}{1 - P_i}\right) = \alpha + \beta_j X_i \tag{4}$$

2.3.3 Multinomial logistic regression. This study addressed the responsiveness of households for climate variability by implementing the multinomial logistic regression model. This model can predict the probabilities of the different possible outcomes of a categorically distributed dependent variable, given a set of independent variables. The model can be implemented when the dependent variable in question is nominal (a set of categories which cannot be ordered in any meaningful way) and consists of more than two categories. The multinomial logit model assumes that the data are case-specific; that is, each independent variable has a fixed value for each case, and the dependent variable cannot be perfectly predicted from the independent variables for any case.

$$\Pr(y_i = j) = \frac{\exp\left(X_i \beta_j\right)}{1 + \sum_{j=1}^{J} \exp\left(X_i \beta_j\right)} \tag{5}$$

and

$$\Pr(y_i = 0) = \frac{1}{1 + \sum_{j=1}^{J} \exp(X_i \beta_j)}, \tag{6}$$

where: y_i is the observed outcome for the ith individual and X_i is a vector of explanatory variables.

3. Data analysis and discussion

3.1 Demographic characteristics of sampled households

Table I showed that the majority of the sampled households had a large family size, ranging from four to eight persons per household, and with an average family size of 6.03. In the study area, 75 per cent of the households had a family size of five to eight persons, which indicated that the majority had a large family size, and the Doba district had the largest family size compared with other districts. This large family size attributed to a relatively high dependency ratio, seen in the households of Goba-koricha.

Most household heads in the study area were under 50 years of age, which had a higher economic effect, especially for the crop production sub-sector that demands more labor. Most (86 per cent) of the sample household heads were at their economically active age (Table I), which may be very useful for efficient production and management of the crop production. Because the sub-sector is in need of active labor, having a younger household head would be a crucial input in performing better. More than half of the household heads were illiterate and did not have even basic education. These circumstances may create difficulty in expanding new technologies, as uneducated people are reluctant to adopt new technologies.

3.2 Landholding and farming system of households

Agriculture in the study area is purely practiced by an ox plough using traditional wooden materials because the area is one of the most densely populated regions of the country and here each household owns a very small plot of land, although they have a large family size to manage.

Table II showed that the majority of the sample households had no surplus draught power, given the other natural and uncontrolled factors that encumber crop production;

Table I. Family size and dependency ratio of sample households

Family size	Percent	District	Family size	Dependency ratio
≤2	1.5	Doba	6.5	1.72
2-4	20	Mieso	5.9	1.65
4-6	34.4	Goba-koricha	5.9	1.72
6-8	40.4	Average	6.03	1.69
8-10	3.4			

Source: Own computation, 2015

Table II. Cropland and livestock ownership of households

Land (ha)	Household (%)	Pair of oxen	Households (%)
Below 1	47.81	Below 1	40.0
1.1-2	36.57	1-2	51.6
2.1-3	12.82	2-3	7.5
Above 3	2.81	Above 3	0.9
Total	100	Total	100

Source: Own computation, 2015

draught power shortage was also the other bottleneck. Some households tried to generate income from renting oxen power; hence, those who did not have draught power could not take advantage of this benefit.

Average landholding in the study area was about 1.34 ha, ranging from a minimum landholding of 0.13 ha to a maximum of 5.50 ha. A significant proportion of sample households (about 48 per cent) had landholding of less than 1 ha, which was very small for the application of modern technologies and extensive farming system to collect potential gains from such applications. This low landholding forced farmers of the study area to practice intensification and inter-cropping as their main strategies for maintaining welfare. They had very low and fragmented plots of land as compared with the national-level landholding. In the West Hararghe Zone, the average landholding was 0.9 ha, in which around 66 per cent of the population owned less than 1 ha (CSA, 2007). Because the Mieso District is agro-pastoral, it had relatively larger livestock and a pair of oxen as compared to the other sampled districts.

Cereal crops took the predominant area of land (78.17 per cent) in the crop production of the country in 2013 (CSA, 2013). Similarly, these crops had the largest share in the quantity of grain crops produced in the study area. Farmers in the study area allocated most of their land for sorghum production, which is the main source of food and is also better drought-resistant. This crop takes the fourth largest share (15.58 per cent) of the national grain crop production. Maize takes the second largest share in land allocation of the sample districts because it is the main source of food and is somewhat adaptive to the climatic conditions of the area. The two cereals are major food crops both in terms of the area cultivated and volume of production. Farm households in these districts produced sorghum and maize in larger volume compared to other crops.

Most households allocated a small fraction of land to produce other cereals and the stimulant crop *K'ht*. It is a cash crop; households did not produce in plenty owing to land shortage and climate condition of the area. Farmers were highly dependent on few cereal crops because of their climate adaptability, in which there was a new variety of those crops that was drought-resistant and short-seasoned. Farmers of the Doba district also produce haricot bean and other vegetables because of relatively better moisture availability. They mainly produce haricot bean and some other cereals through intercropping with sorghum and maize. Households tried to cope with the land shortage problem through intercropping, which is a common practice adopted by most smallholders of the zone. Besides intercropping, farmers have a habit of producing more than once within one production season. This means farmers first cultivate short-seasoned crops and then they cover the land with other type of crops to use the available moisture and land (Table III).

Table III. Households' (%) cropland allocation for each crop

Land allocated	Sorghum	Maize	Other cereals	'khat'
0-0.5	55.63	86.25	95.94	98.75
0.56-1	28.75	10.63	4.06	1.25
1.1-1.5	8.75	0.94	0.00	0.00
1.56-2	5.31	1.88	0.00	0.00
2.1-2.5	1.25	0.31	0.00	0.00
2.56-3	0.31	0.00	0.00	0.00
Total	100	100	100	100

Source: Own computation, 2015

Farmers are highly dependent on the oxen plough, of which the majority of them possessed less than a pair. There are households that do not possess even a single ox, but based on their culture households who do not have ox would borrow for few days to plough their cropland and provide crop residue in exchange. Even though the area has this culture, more than half of the households responded that they have draught power shortage.

3.3 Climate variability and farmers' crop income

3.3.1 Crop season temperature and rainfall in the sample districts. Mieso District had the highest average temperature, in all the years considered, than the other two districts. The overall temperature condition of the study areas showed that there was a continuous increment over time, especially after early 1990s (Figure 2).

Although it became consistent recently, the rainfall variability in the Doba District was very high in previous years compared with other districts, especially in the 1980s. Recently, there was observed a drastic reduction in rainfall during the crop season of Goba-koricha and Doba districts. In the past three decades, greater variation has been observed regarding rainfall in the sampled three districts. In general, rainfall during the crop season of the Mieso district was lower than that of the other districts, and the circumstances were relatively similar in the other two districts (Figure 3).

Rainfall during the crop season of the Mieso district was consistent in 2013 and 2014 production years, which resulted in better productivity and higher food crop availability.

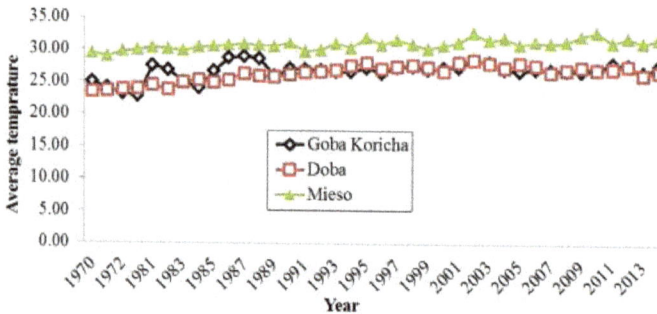

Source: Ethiopian National Meteorology Agency, 2014

Figure 2. Average crop season temperature in the three districts

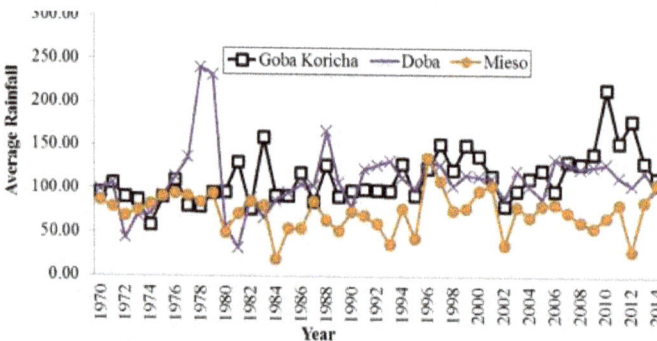

Source: Ethiopian National Meteorology Agency, 2014

Figure 3. Average crop season rainfall in the three districts

This contributed to reduce food insecurity problem. Doba district had a higher food insecurity problem because it received the lowest rainfall compared with the other two districts.

3.3.2 Weather condition and crop yield. Most studies revealed that climate change is likely to reduce agricultural productivity, production stability and household income in some areas that already have high levels of food insecurity (Greg *et al.*, 2011). Households of the study area, on average, lost about 8.06 quintals of crop because of insufficient rainfall during the cropping season in the previous production year. An increase in summer temperature had a negative effect on productivity, and it caused an average loss of 8 quintals for 86.88 per cent of the sample respondents. This result was consistent with the findings of Mariara and Karanja (2006) in Kenya and that of Eid *et al.* (2006) in Egypt. In connection to this, maize was the most sensitive crop as compared with other crops cultivated in the study area. Sample households viewed that sorghum had better resistance to weather condition of the area than maize. Respondents also replied that they had a habit of sowing sorghum on dry soil even under conditions of late-onset rain. They did this because of the drought-resistance nature of sorghum and because the seeds are small enough to wait safely in the soil without being damaged until the rain arrives.

Farmers developed a tradition of adopting new crops that had never been grown before and had good drought resistance with better yield on the existing water stress as an adaptation strategy. Rainfall oscillation, mainly shortage, which happened every five years had a devastating effect on crop yield. Table IV shows that sorghum was more productive as compared to other crops produced in the area. The two dominant crops of the area were more productive if one compares the national productivity with the study area. The national productivity of maize in 2012 and 2013 was 1,220 kg/ha and 1,379 kg/ha, respectively whereas that of sorghum was 2,054 kg/ha and 2,106 kg/ha for the 2012 and 2013, respectively. The yield of the two crops in 2013 was 2,688 kg/ha and 2,309 kg/ha for maize and sorghum, respectively, in the West Hararghe Zone (CSA, 2014). Sorghum yield in Doba district (2,550 kg/ha) was better than that in the other districts and even better than that at the zonal level (2,309 kg/ha). The average sorghum yield in Goba-koricha was also higher than that of Mieso. Besides growing the dominant crops of sorghum and maize, few farmers in the study area also produced wheat, wherein the yield of this cereal was better in Goba-koricha than in the other sampled districts.

Table IV showed that households of Doba faced extreme production losses (1,480 kg/ha) because of the previous year's rainfall shortage, which was the main reason for the reduction in yield as compared with the other two districts. Regarding the recorded production loss and information gained from group discussions, the previous year's weather condition was not favorable for crop production in Doba. All of the sampled households in the district faced production loss, although there was a difference in the amount of loss per hectare per household. Given this, only 78.35 per cent of the households in Mieso faced production loss,

Table IV. Crop yield (kg/ha) and loss in the study area

Districts	Sorghum	Maize	Other Cereals	Total production loss*
Doba	2552	1650	600	1480
Mieso	2430	1770	1180	580
Goba-koricha	2470	1460	1280	1003
Overall	2460	1638	1130	806

Source: Own computation, 2015; * The loss was due to previous year's rainfall shortage

which was not higher than the households of Doba. Farmers in the study area used both organic and inorganic fertilizers to increase crop productivity given the ongoing moisture stress. Organic fertilizer applications like compost and animal dung were common practice in each district. Farmers in Mieso district had a habit of practicing shifting cultivation to maintain soil fertility, as their landholding was relatively better.

3.3.3 Determinants of crop income. Household-level agricultural panel data would have been ideal for addressing inter-temporal effects, but such data sets are rare in developing countries. On the other hand, if sufficient variation is observed in the cross-sectional data, especially variation within what climate variability is likely to bring about, then it is reasonable to use such data to understand effects of the problem. However, instead of using farmland value, which is not available in most developing countries, it would be better to rely on net revenue per hectare as the dependent variable (Pradeep and Mohamed, 2006). Our specification was entirely driven from the Ricardian technique of determining farm net revenues. These empirical research studies used OLS estimation technique to identify determinants of net revenues per hectare using time-series data that include temperature and rainfall. The research included previous year's crop season rainfall condition as a dummy variable, which is one critical input to determine households' current year crop income.

Majority of the Ethiopian population is dependent on agriculture for its consumption and source of income. Thus, if any problem that harms farmers' agriculture occurs, then households may lose a lot of income and their food availability will considerably reduce. The crop production sub-sector of the country is the most susceptible for climate variability owing to its direct interaction and nature dependency. Based on the regression result, the model was strong enough to encompass the variations in the dependent variable. Cropland and livestock units had a positive and significant effect on households' crop income in the study area. A larger livestock unit implies possession of enough draught power, which is a critical input for plowing, harvesting and threshing activities, which significantly affects the crop income. Likewise, farmer's educational attainment had a positive and significant effect on household's crop income in the study area (Table V).

Previous year rainfall amount had a significant effect on households' crop income based the regression result. This implies that if rainfall reduces successively because of climate variability, smallholder households will not be able to produce even for their own subsistence. This result was consistent with the findings of Mano and Nhemachena (2006). Those authors concluded that if the temperature increases by 2.5°C and 5°C, the net farm revenue will decrease by approximately US$0.3bn and US$0.8bn, respectively based on the sensitivity analysis of alternative climatic scenarios.

Table V. Determinants of households' crop income

Variables	Coefficient	SE	Z
Ln Dependency ratio	−0.0727	0.3789	−0.19
Education level	0.1303	0.0705	1.85**
Ln Farm experience	−0.3055	0.3664	−0.83
Ln TLU	0.5568	0.2658	2.09**
Ln cropland	0.9253	0.3155	2.93***
Rainfall inconsistency	−1.3099	0.4976	−2.63***
Constant	5.9133	1.3760	4.30***

Sources: Model result, 2015; *** and ** represent 1 and 5% significance levels, respectively

3.4 Determinants of indigenous adaptation strategies

Although many of the farmers in the study area were not qualified enough to use modern technologies for sensing and understanding the extent and degree of climate change, they had better perception about the ongoing climate variability based on their traditional knowledge. They have argued that there is successive rainfall reduction and temperature increment in their surroundings. With time, climate variability has become more severe, and they are unable to produce some crops that were dominantly produced previously. Some farmers remembered that there was bimodal rainfall within the area and they had produced twice a year, but now they faced subsequent rainfall shortage even in the main rainy season to produce hybrid sorghum and maize that have better drought resistance.

About four-fifth (85.3 per cent) of sample households replied that the problem of climate variability, which can be explained by moisture stress, is easy to perceive. Deressa (2006) also concludes that decreasing precipitation appeared to be more damaging than increasing temperature. Table VI shows that water shortage is a critical problem in crop production of households in the study area. Shortage of cropland was the second cause affecting crop production of the study area. Given their culture sensing climate variability via moisture stress households of the area also perceived the direction of rainfall variation.

Table VI shows that early cessation of rainfall was the most frequent type of rainfall variability in the study area. Nearly 81 per cent of the households faced the problem of early rainfall cessation, even before the crops' grain-filling stage. Regarding the view of respondents in the study area, rain mostly ceased during the flowering stage of crops, resulting in significant production losses. Erratic distribution and insufficient amount of cropping season rainfall were also core bottlenecks of the crop production in the area. Although the main agricultural problem of the area is related to natural factors such as successive moisture stress and other unexpected windfall-type factors, there are also other bottlenecks such as shortages in the cropland area.

More than half of the households responded that factors related to climate variability are the main problems that triggered poor performance of agriculture. Interviews with household heads indicated that they frequently faced complete crop loss owing to moisture stress, especially if the problem happened at either the flowering or fruit formation stage of the crops. In such situations, households tried to supplement with irrigation from harvested water or the crop would be used as livestock feed.

Climate change adaptation is a two-step process. First, the household must perceive that the climate is changing. Next, they should respond to such changes through appropriate adaptation strategies. It is assumed that economic agents, including smallholder farmers, use adaptation methods only when they perceived that utility or net benefit from using such a method is significantly greater than without it (Maddison, 2006; Nhemachena and Hassan, 2008 and Deressa et al., 2008). Smallholders adopt different mechanisms to cope with climate

Table VI. Common direction of rainfall variation and crop production problems

Direction of rainfall variation	Frequency (%)	Common problems	Frequency (%)
Late start	7.6	Land shortage	26.7
Early cease	19.7	Water shortage	63.5
Insufficient amount	11.4	Over flood	5.1
High temperature	0.6	Low land productivity	4.8
Combination of all	60.7	Total	100.0
Total	100.0		

Source: Own computation, 2015

variability depending on their economic capacity. In connection with the ongoing climate variability that results in a late start and early cessation of rain, farmers adopted hybrid sorghum and maize, especially newly innovated sorghum varieties that have better drought resistance. About three-fourth (75 per cent) of the sampled farmers replied that they adopt mixed cropping strategies to mitigate the problem of crop failure. They hope that if one of the crops failed, the other one will substitute the failure and farmers will not totally be in loss.

The main reason behind the adaptation of different crop mixes was rainfall variability. They had responded that they frequently change their crop mix to cope with the rainfall variability, although the new crop mixes (66.3 per cent of sample households) do not fully mitigate the problem. If rains are delayed, then farmers of the area sow crops, especially sorghum in the dry land and wait until rain comes. At the same time, they construct terraces and collect rain water to supplement their crop when there is early cessation in the rainfall during the cropping season. Households of the area plough their land in a horizontal counter line way, and they build horizontally lined and parallel blocks from stone and mud to reduce water runoff. Thus, the land would remain moisturized for long periods and the plants use the water efficiently.

Households' ability to take advantage of climate change mitigation and adaptation technologies is also linked to their education level, cultural practices, skills acquired and access to financial assets (Greg et al., 2011). There are different constraints that hinder farmers to apply different adaptation mechanisms to reduce the negative effect of climate variability. Some of the constraints were cropland area shortage, low income, poor agricultural practices, low awareness, limited source of income, absence of off-farm and non-farm activities. The study area had no access to non-farm and off-farm employment, which may be sourced either from missing markets or entry barriers. Moreover, many of the household heads and their spouses lack a visionary outlook to identify the possible off-farm opportunities that may be managed with low education level. Being poor and not having buffered stock of money was another bottleneck for the absence of diversified sources of income and easy adaptation to the problem of climate variability (Table VII).

Households adopted different methods of reducing the negative effect of climate variability on their crop production. Adoption behavior and their capacity were hampered by different socioeconomic and natural factors. Deressa et al. (2008) indicated that there is a negative correlation between age and adoption of improved soil conservation practices. The same logic holds true here in this research. Households that receive better crop income and possessed relatively larger cropland areas would have better probability of adopting different mechanisms to reduce the effect of climate variability. Because the adaptation of new techniques may have risks, it may need a reserve to insure from the occurrence of risks

Table VII. Determinants of households' adoption behavior

Variables	Coefficient	SE	t-value
Dependency ratio	−0.1463	0.1608	−0.91
Education	0.0052	0.0588	0.09
TLU	−0.0820	0.0333	−2.46**
Total cropland	0.6730	0.2761	2.44**
agehhh^2	−0.0023	0.0014	−1.66*
Food security	0.4111	0.3286	1.25
Crop income	0.0012	0.0006	1.97*
Constant	1.2675	0.5475	2.32

Sources: Model result, 2015; ** and * indicates 5 and 10% levels of significance, respectively

connected to adoption. Cropland had a significant and positive relationship with farmers' adaptation behavior. Possessing relatively larger livestock reduced households' adoption behavior because of lower reliance on crop production that is more sensitive to climate variability (Table VIII).

Driving forces for the farmers to adopt one strategy in land-use choice are differences between the available crops for cultivation and the soil, climate and observable and unobservable characteristics of the farm (Leopoldo and Soto, 2005). Adaptation measures that farmers report may be profit-driven, rather than climate variability. Despite this missing link, Maddison (2006) and Nhemachena and Hassan (2008) assume that adaptation actions are driven by climatic factors, wherein Deressa et al. (2008) did the same. In the same logic, this study identified that smallholders adopt few strategies like water harvesting, using hybrid and drought-resistant crops, sowing short-seasoned crops and sowing on the dry soil to reduce the effect of climate variability. Ground water harvest technologies are well adopted as a means of managing moisture stress. Smallholders of the area collect rain water during the rainy season to supplement crops if rain ceased earlier.

Most scholars assumed that any change in the profitability of land will be immediately capitalized into land value. This implies that farmers can adapt to any changes in climate immediately and effortlessly. The belief is that farmers are keenly aware of changes in climate and they immediately select crops that are adaptable to the new climate conditions. This assumption effectively considers all constraints on the way farmers make their land-use decisions. Estimates produced under this model should be considered as a lower bound to the actual cost from climate variability, at least in the short run; there may be constraints that prevent the farmer from responding to climate variability with promptness.

This study considered five alternative levels such as no adaptation (0-level), using water harvesting techniques (1-level), adopting hybrid input crops (short-seasoned and drought-resistant) (2-level), applying afforestation programs (3-level) and dry sowing of the crop (4-level). The 0-level served as a base to simplify comparison among others. Because there was consideration of one alternative as the base, the model result represented for the rest four alternatives. Based on the value of χ^2, the model was strong enough and the covariates' explanatory power was very strong, which indicated that the variation in the dependent variable was strongly explained by the variations of independent variables included in the model.

Table IX showed that household's probability of adopting water harvesting technology increases on average with cropland area and vice versa, which means the probability of adopting water harvest technology increases when farmers have a larger landholding. Possession of large livestock would divert attention of households toward their livestock rather than the crop sub-sector and result in lower probability of adopting water harvesting technologies. Likewise, farming experience also decreased households' probability of adopting this technology in the study area. Household's tropical livestock unit (TLU) level

Table VIII. Marginal effect of variables on adoption behavior of household

Variables	dy/dx	SE	Z	Average value X
Dependency	−0.0218	0.0238	−0.91	1.8719
TLU	−0.0122	0.0049	−2.52***	6.2313
Total cropland	0.1000	0.0398	2.51***	1.3413
Agehhh^2	−0.0004	0.0002	−1.66	1615.76
Crop income	0.0002	0.0001	2.08**	2263.77

Sources: Model result, 2015; *** and ** indicates 1 and 5% levels of significance, respectively

Table IX. MNL Model result (0 = No adaptation is base outcome)

Strategies	Variables	Coefficient	SE	Z
1. Water harvest	Dependency ratio	−0.0106	0.2182	−0.05
	Education	−0.0616	0.0766	−0.80
	TLU	−0.0849	0.0427	−1.99**
	Cropland	1.7214	0.4225	4.07***
	Farming experience	−0.0536	0.0223	−2.41**
	Crop income	0.0005	0.0009	0.61
	Water shortage	0.5490	1.0079	0.54
	Constant	−0.0301	1.3075	−0.02
2. Hybrid seeds	Dependency ratio	−0.1010	0.2020	−0.50
	Education	−0.2015	0.0732	−2.75***
	TLU	−0.0678	0.0327	−2.07**
	Cropland	1.6032	0.4052	3.96***
	Farming experience	−0.0746	0.0206	−3.62***
	Crop income	0.0010	0.0010	0.82
	Water shortage	2.0783	0.8929	2.33**
	Constant	0.0656	1.1589	0.06
3. Afforestation	Dependency ratio	0.1774	0.3549	0.50
	Education	−0.2303	0.1720	−1.34
	TLU	−0.1076	0.1027	−1.05
	Cropland	1.6541	0.6343	2.61***
	Farming experience	−0.0116	0.0364	−0.32
	Crop income	0.0002	0.0013	0.12
	Water shortage	−11.461	499.165	−0.02
	Constant	9.196	499.167	0.02
4. Dry sowing	Dependency ratio	0.2044	0.3384	0.60
	Education	−0.2006	0.1379	−1.45
	TLU	−0.1673	0.0814	−2.06**
	Cropland	2.0926	0.5041	4.15***
	Farming experience	−0.0543	0.0327	−1.66*
	Crop income	0.0011	0.0010	1.14
	Water shortage	2.8747	1.0483	2.74***
	Constant	−4.7352	1.7552	−2.70***

Sources: Model result, 2015; ***, ** and * indicate 1, 5 and 10% statistically significance levels, respectively

also had a negative effect on probability of adopting hybrid seeds of sorghum and maize. Because hybrid crops are part of new technologies, experienced households with the crop production may hesitate to adopt them even though the inputs are better in their productivity as compared to the local breeds. Besides being less experienced, older household heads may have buffer stocks to cope with the risk that may source from adopting new technologies.

Previous year's rainfall shortage and expectation of the current problem initiated households to adopt hybrid seeds of the two cereals to reduce the effect of the problem. In the same fashion, households may sow seeds on the dry soil and wait until the rain comes if rains are delayed. They have practiced this method especially for sorghum, as the seeds are small and are not easily exposed to insects in the soil. Households that had relatively larger TLU would not adopt hybrid sorghum and maize, which may be because of the lower biomass of those crops to collect better livestock feed. It is well known that hybrid crops can withstand moisture stress and short seasoned crops have a lower biomass, which will result in having lower aftermath as a source of feed for the livestock. Because of lower crop residue supplied from hybrid sorghum and maize, households could not adopt those crops, especially if they have larger livestock.

If a household has a relatively larger cropland area, the probability of adopting hybrid sorghum and maize as means of mitigating the negative effect of rainfall shortage, which was considered as one dimension of climate variability, would increase. Possessing a large cropland also increased households' adaptation of this method to reduce production losses triggered by delayed rainfall. Having a larger cropland area also motivated farmers to use afforestation programs that could be a long-lasting means of mitigating the effect of climate variability. Possession of a large cropland may be one precondition to apply the method by mainly those households that have enough experience in rainfall forecasting.

4. Conclusions and recommendations

This paper addressed how agricultural production and farmers' incomes are affected by on-going climate variability, and the adaptation strategies practiced by smallholder farmers in three drought-prone districts of West Hararghe, Oromia National Regional State of Ethiopia. Most sampled farmers perceived and understood the effect of climate variability, which is a critical bottleneck for crop production of the area. Rainfall inconsistencies are common problems of the study area, creating a serious threat to smallholder households. The problem is becoming extreme dangerous unless protective measures like water harvesting and other long-lasting measures are taken by different stakeholders.

Most of the sample households had better understanding about the ongoing climate variability and they were trying to adopt different mechanisms for mitigating the likelihood effect of the problem on crop production. They had indigenous knowledge of water harvesting technologies, which are important technological advancements to make farmers more efficient in using scarce and the critical agricultural input, water. Thus, there should be rigorous efforts to enhance the local skill of smallholders and to make those strategies more efficient.

Adaptation of off-farm and non-farm activities is one method of reducing the effect of climate variability through having a diversified source of income. Continuous training and awareness session should be conducted for smallholders of the area to increase their participation in alternative sources of income.

Because smallholders in the study area had a relatively large livestock, each adaptation method should take into account the sub-sector that is highly dependent on the crop production as a source of food. For instance, hybrid sorghum and maize that are currently adopted by households have better moisture stress resistance with a lower biomass, which would result in having lower aftermath as a source of the livestock thereby smallholders were reluctant to adopt those crops especially if they have larger livestock units. Thus, more attention should be given to livestock feed when new hybrid crops are introduced in the study area to be easily accepted by farmers, as livestock feed shortage is the most appreciable problem equivalent to households' food shortage.

5. Limitations of the study

This study was based on primary data collected from three districts considering sample smallholder households, which has the limitation of considering some parameters of climate variability like long run rainfall and temperature. In addition, the study has the limitation of generalizability, as it was conducted in few districts that could not represent the national-level smallholder households.

References

Arjan, R., Mark, D.B., Minna, K.V.L. and Nico, P. (2011), *"Adaptation to climate variability: the role of past experience and institutions"*, Royal Haskoning, Nijmegen.

Central Statistics Agency (CSA) (2007), *"Population and housing census of Ethiopia"*, Addis Ababa.

Central Statistics Agency (CSA) (2013), *"Agricultural Sample Survey Report on Farm Management Practices (Private Peasant Holding Summer Season) 2013"*, Addis Ababa.

Central Statistics Agency (CSA) (2014), *"Agricultural Sample Survey Report on Farm Management Practices (Private peasant Holding Summer Season) 2014_2015"*, Addis Ababa.

David, P. (1993), "Climate changes and food supply", *Forum for Applied Research and Public Policy*, Vol. 8 No. 4, pp. 54-60.

Deressa, T.T. (2006), "Measuring the economic impact of climate change on Ethiopian agriculture, Ricardian approach", Centre for Environmental Economics and Policy in Africa, University of Pretoria, CEEPA Discussion Paper No. 25.

Deressa, T., Hassan, R.M., Tekie, A., Mahmud, Y. and Claudia, R. (2008), "Analyzing the determinants of farmers' choice of adaptation methods and perceptions of climate change in the Nile Basin of Ethiopia", Environment and Production Technology Division, IFPRI Discussion Paper 00798.

Eid, M.H., El-Marsafawy, S.M. and Ouda, S.A. (2006), "Assessing the economic impacts of climate change on agriculture in Egypt, A Ricardian Approach", Centre for Environmental Economics and Policy in Africa, University of Pretoria, CEEPA Discussion Paper No. 16.

Eva, L. (2009), *Climate Change, Water and Food Security*, Background Note, Overseas development institute, UK.

François, P. (2003), "Ethiopia: Hararghe food security hampered by long-term drought conditions and economic constraints", UNDP Emergencies Unit for Ethiopia.

Fraser, E. (2008), "Crop yield and climate change", Retrieved on 14 September 2009.

Greg, E., Anam, B.E., William, M.F. and Duru, E.J.C. (2011), "Climate change, food security and agricultural productivity in Africa: issues and policy directions", *International Journal of Humanities and Social Science*, Vol. 1 No. 21, p. 218.

Hansen, L.J., Biringer, J.L. and Hoffman, J.R. (2003), "buying time: a user's manual for building resistance and resilience to climate change in natural systems", World Wildlife Fund, Washington D.C.

Hutcheson, G.D. (2011), "Ordinary least-squares regression", The SAGE Dictionary of Quantitative Management Research, pp. 224-228.

Leopoldo, E. and Soto, A. (2005), "Optimal crop choice: farmer adaptation to climate change", University of California, Santa Barbara.

Maddison, D. (2006), "The Perception of and adaptation to climate change in Africa", Special Series on Climate Change and Agriculture in Africa: CEEPA Discussion Paper: No. 10 ISBN 1 920160-01-09.

Mano, R. and Nhemachena, C. (2006), "Assessment of the economic impacts of climate change on agriculture in Zimbabwe, A Ricardian approach", Centre for Environmental Economics and Policy in Africa, University of Pretoria, CEEPA Discussion Paper No. 11.

Mariara, K.J. and Karanja, F.K. (2006), "The economic impact of climate change on Kenyan crop agriculture, A Ricardian approach", Centre for Environmental Economics and Policy in Africa, University of Pretoria, CEEPA Discussion Paper No.12.

Mertz, O., Mbow, C., Nielsen, J.Ø., Maiga, A., Diallo, D., Reenberg, A., Diouf, A., Barbier, B., Moussa, I.B., Zorom, M., Ouattara, I. and Dabi, D. (2010), "Climate factors play a limited role for past adaptation strategies in West Africa", *Ecology and Society*, Vol. 15 No. 4, pp. 25.

Michael, C. (2006), *"Climate Change Impacts on East Africa a Review of the Scientific Literature"*, WWF-World Wide Fund for Nature, Gland.

Nhemachena, C. and Hassan, R. (2008), "Determinants of climate adaptation strategies of African farmers: multinomial choice analysis", *African Journal of Agricultural and Resource Economics*, Vol. 2 No. 1, pp. 83-104.

Patt, A., Suarez, P. and Gwata, C. (2005), "Effects of seasonal climate forecasts and participatory workshops among subsistence farmers in Zimbabwe", *Proceedings of the National Academy of Sciences of the United States of America*, Vol. 102 No. 35, pp. 12623-12628.

Pradeep, K. and Mohamed, I.A. (2006), "Application of the Ricardian technique to estimate: the impact of climate change on smallholder farming in Sri Lanka", *Climatic Change*, Vol. 81, pp. 39-59.

World Bank (2004), "*World Bank report on malaria expansion*".

Corresponding author

Arega Shumetie can be contacted at: ashueconomist@gmail.com

5

Climate change and variability: a review of what is known and ought to be known for Uganda

Francis Wasswa Nsubuga
Geography, Geoinformatics and Meteorology,
University of Pretoria, Pretoria, South Africa, and

Hannes Rautenbach
South African Weather Service, Pretoria, South Africa and
School of Health Systems and Public Health, University of Pretoria, South Africa

Abstract

Purpose – In view of the consensus that climate change is happening, scientists have documented several findings about Uganda's recent climate, as well as its variability and change. The purpose of this study is to review what has been documented, thus it gives an overview of what is known and seeks to explain the implications of a changing climate, hence what ought to be known to create a climate resilient environment.

Design/methodology/approach – Terms such as "climate", "climate change" and "climate variability" were identified in recent peer-reviewed published literature to find recent climate-related literature on Uganda. Findings from independent researchers and consultants are incorporated. Data obtained from rainfall and temperature observations and from COSMO-CLM Regional Climate Model-Coordinated Regional Climate Downscaling Experiment (CCLM CORDEX) data, European Centre for Medium-Range Weather Forecasts (ECMWF) Interim Re-Analysis (ERA-Interim) data and Global Precipitation Climatology Centre (GPCC) have been used to generate spatial maps, seasonal outputs and projections using GrADS 2.02 and Geographic Information System (GIS) software for visualization.

Findings – The climate of Uganda is tropical in nature and influenced by the Inter-Tropical Convergence Zone (ITCZ), varied relief, geo-location and inland lakes, among other factors. The impacts of severe weather and climate trends and variability have been documented substantially in the past 20-30 years. Most studies indicated a rainfall decline. Daily maximum and minimum temperatures are on the rise, while projections indicate a decrease in rainfall and increase in temperature both in the near and far future. The implication of these changes on society and the economy are discussed herein. Cost of inaction is expected to become huge, given factors like, the growing rate of the population and the slow expanding economy experienced in Uganda. Varied forms of adaptation to the impacts of climate change are being implemented, especially in the agricultural sector and at house hold level, though not systematically.

Originality/value – This review of scientific research findings aims to create a better understanding of the recent climate change and variability in Uganda and provides a baseline of summarized information for use in future research and actions.

Keywords Uganda, Climate projections, Recent climate, Representative concentration pathways

Introduction

Uganda is a landlocked country situated in the eastern part of Africa (Nsubuga *et al.*, 2014c). It covers an area of nearly 241,548 km^2, comprising 0.8 per cent of the total geographical area of the continent. Geographically the country is located between 1°30′S-4°N latitude and 29°30′E-34°E longitude, hence existing astride the Equator, making it more Equatorial than neighbouring countries. Studies for Uganda to detect changes in climate, especially with regard to the two parameters of rainfall and temperature; have been going on within the country, as well as in other parts of the world. The characteristics of climate and the impact of climate change in Uganda show similarities to other regional, national and global studies documented. Regional studies for Africa (such as those by Aguilar *et al.*, 2005; New *et al.*, 2006; Hulme *et al.*, 2001; Domroes and El-Tantawi, 2005) contextualize what has been found for the Uganda.

Considerable work has been done at a national level (Anyah and Semazzi, 2004; Osbahr *et al.*, 2011; Nsubuga *et al.*, 2011, 2014a, 2014b, 2014c, 2015; Bomuhangi *et al.*, 2016), which has revealed various climatic changes in Uganda. The present work attempts to review various works in peer-reviewed articles and other forms of publication which have documented climate change in Uganda. Hence forms a basic framework for further investigation of climate change in Uganda. It highlights what we might expect the climate to be like in future under different climate change scenarios and the associated impacts that ought to be known are also discussed. Hopefully, the information presented in this study will provide a basis for future action.

Methodology

This study reviews recent publications that document climate change in Uganda. It concentrates on recent literature published in times when the climate signal has intensified. Recent literature is also based on new scientific evidence and better climate change projections than have been available in the past. The review used word search for terms like *climate, climate variability* and *climate change* in Uganda. Such terms were found not only in peer-reviewed literature but also in published work of independent researchers and in consultants' reports. Secondly, the study linked references in classic articles and book chapters on the topic of climate change. However, our search was not expected to provide a comprehensive picture of climate change because of the limited research being undertaken in Uganda and lack of continuous climatic data. Thus, authors integrated iterative approaches of data analysis to visualize results. Observational data reflecting hydro-climatic conditions collected from 36 stations have been used to construct an overall picture of the mean monthly rainfall distribution. In addition, data obtained from the Global Precipitation Climatology Centre (GPCC) and from the European Centre for Medium Range Weather Forecasting Re-analysis Data (ECMWF) respectively were used for rainfall and temperature observations, in order to produce seasonal outputs. For climate change projections, average of data from an ensemble of four models of the Consortium for Small-scale Modelling Climate Mode (COSMO) and data from the Climate Limited-area Model (CLM), as well as simulations from the Coordinated Regional Climate Downscaling Experiment (CORDEX) initiative were used. Spatial maps were generated using GrADS and GIS software.

Mechanisms that affect Uganda's climate variability

A sound observational basis with a detailed description of the mechanisms that govern the space and time characteristics of rainfall and temperatures over Eastern Africa, where Uganda is located, can be constructed from information provided in a number of articles

(Ogallo, 1984, 1988, 1993; King'uyu *et al.*, 2000; Hastenrath, 2001; Schreck and Semazzi, 2004). Because of its geo-location (astride the Equator), temperatures Uganda are predominantly determined by heat emission from the Earth's surface.

On the other hand, the most important systems responsible for Uganda's rainfall include: the Inter-Tropical Convergence Zone (ITCZ), subtropical anticyclones, monsoonal winds and the moist westerly winds from the Congo basin (Nsubuga *et al.*, 2011). According to Beltrando (1990), Basalirwa (1995), Nicholson (1996), Schreck and Semazzi (2004), Anyah and Semazzi (2004) and Mubiru *et al.* (2012), local influences, such as large water masses, human activities, topography and other surface features also play a role in the climate experienced in the country. A combination of these factors, especially the bi-annual propagation of the ITCZ across the country, results in four broad rainfall seasonal patterns over most parts of Uganda (Figure 1). This pattern was widely reported in the report of the East African Meteorological Department in 1963 (Basalirwa, 1995) and supported by subsequent work by Indeje *et al.* (2000), Phillips and Mcintyre (2000), Hastenrath (2001), Kizza *et al.* (2009), USAID (2013) and the recent assessment of the economic impact of climate change in Uganda (GOU, 2015). For example, December, January and February (DJF) are generally dry (Figure 1-DJF), when the ITCZ is located to the south. Any rain during this time is locally driven (e.g. by Lake Victoria's lake/land breeze effect). March to May (MAM) is commonly known as the "long rains" season (Figure 1-MAM), which is

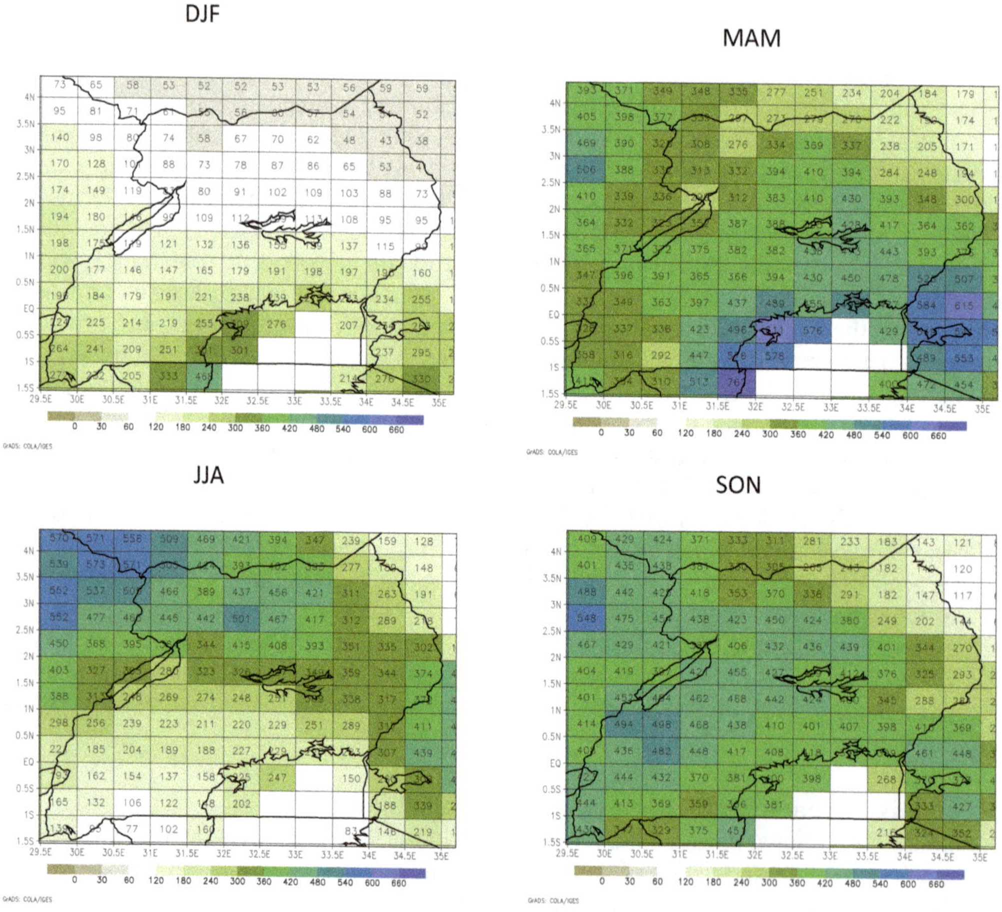

Figure 1. Estimated average (1981-2010) seasonal total rainfall (mm) for Uganda as derived by the Global Precipitation Climatology Center (GPCC)

modulated by synoptic scale circulation associated with the ITCZ and the Mascarene anticyclone (Basalirwa, 1995; Nicholson, 1996; Nsubuga *et al.*, 2011).

The Mascarene anticyclone is a high-pressure cell over the Indian Ocean, which is also the driving force behind the southeast monsoon, whose full force is felt during July and August. June to August (JJA) is generally a dry season, except for parts of northern Uganda where rains are associated with the influx of a moist westerly air mass from the Congo, modulated by the Atlantic anticyclone (see Figure 1-JJA). The September to November season (Figure 1-SON), also known as the "short rains", is the second rainy season and associated with meridional ITCZ propagation.

Because of these systems, most of Uganda has a mean annual rainfall of approximately 1,200 mm (Nicholson, 1996; Nsubuga *et al.*, 2014b). Previous studies provide some evidence that a bimodal rainfall regime dominates the south of Uganda, while a unimodal distribution is more apparent above 3° North (Komutunga and Musiitwa, 2001).

Rainfall

Uganda experiences varied rainfall, with some areas receiving heavy rains that in some instances have resulted in property destruction, while other areas have experienced drought. In some seasons, the rains start late and end late, while in other cases the rain comes early and stops when it is still expected to continue (Mubiru *et al.*, 2012). This pattern of variance has prompted studies aimed at detecting whether rainfall patterns have been changing. But few of these studies involve long-term series using daily observations and often use data which are aggregated to monthly, seasonal or annual time scales.

Nevertheless, a report produced by NEMA (2008), estimates the varying annual rainfall to be between 500 and 2,800 mm, with an average of more than 1,180 mm. The variability in distribution of rainfall according to Nsubuga *et al.* (2014c), arises from a series of interactions as indicated above. These mechanisms cause two rainfall seasons, also known as short and long rains, which often follow the movement of the ITCZ. This movement brings rainfall a month after the sun's migration north or south (Kizza *et al.*, 2009).

The seasonal pattern tends to be bimodal near the Equator and phases into a unimodal system as one moves away from the Equator (Conway, 2005; Asadullah *et al.*, 2008). Inter-annual variability of rainfall correlates with sea surface temperatures (SSTs) in the Pacific through atmospheric teleconnections and the ENSO phenomenon (Schreck and Semazzi, 2004; Mubiru *et al.*, 2012; Nsubuga *et al.*, 2014c). (El Niño–Southern Oscillation or ENSO is an irregularly periodic variation in winds and sea surface temperatures over the tropical eastern Pacific Ocean, affecting much of the tropics and subtropics.) Nicholson (1996) identifies a seasonal preference for the ENSO-related rainfall anomalies in East Africa where Uganda is located. She notes that positive anomalies are prevalent during the short rains of an ENSO year and drought during the long rains of the year that follows a finding supported by subsequent studies in Uganda. For example, periods of severe drought have since been identified by Phillips and Mcintyre (2000) and Nsubuga *et al.* (2014c) as 1945/1946, 1952-1924 and 1980-1984. This identification aligns with the ENSO- related anomalies.

A precipitation concentration index applied over Uganda by Nsubuga *et al.* (2014c), using the river basin approach, revealed that the north and north -west of the country (above 3° north of the Equator) experience a uniform to moderate seasonal distribution of rainfall. Other parts of the country experience a uniform seasonal rainfall, and this has not changed with time. There is no single month in Uganda where rainfall is not received (Figure 2). The highest percentage of rain comes in August, during the MAM and SON seasons. The highest rainfall totals are generally observed in mountainous regions of the central to north-west and central east of Uganda and Lake Victoria vicinity (Figure 2). The lowest rainfall totals

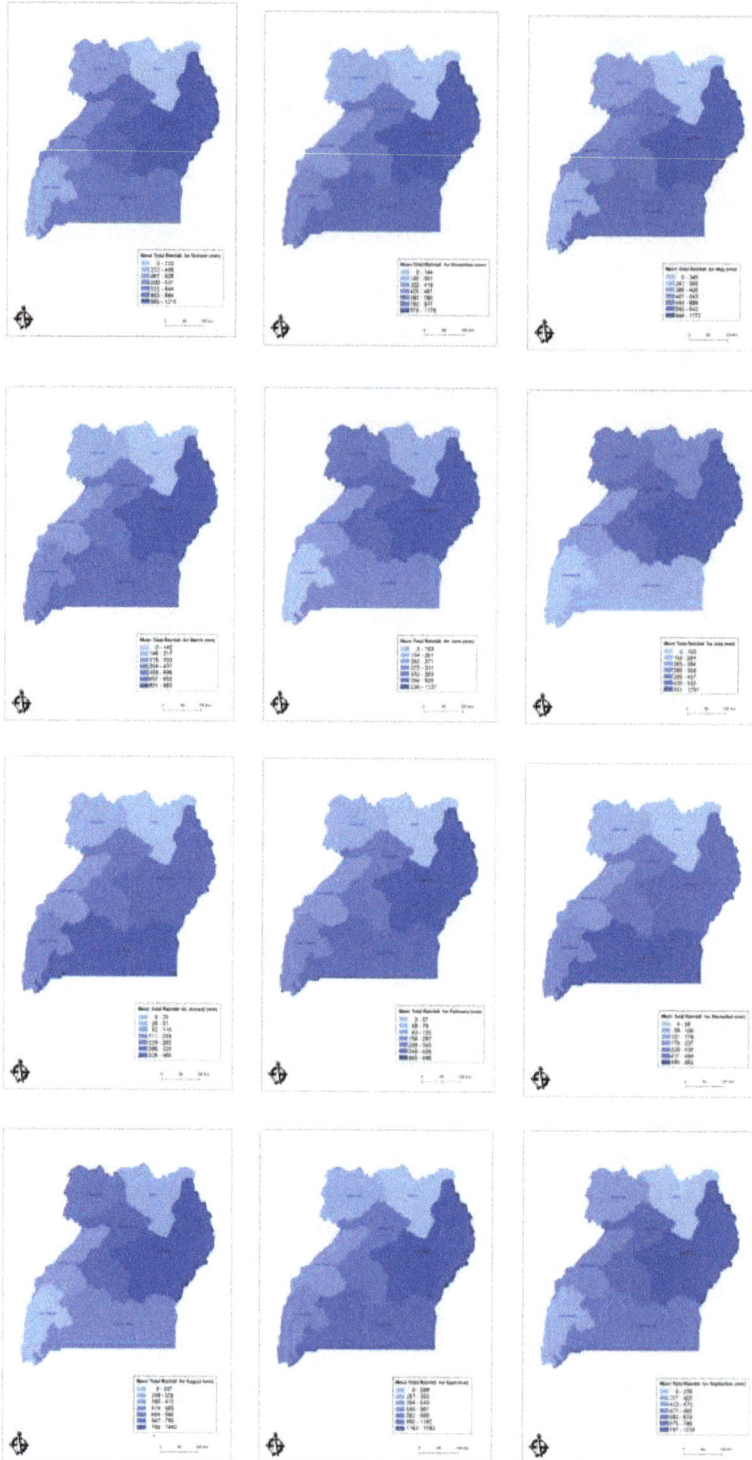

Figure 2. Spatial distribution of mean monthly total rainfall for Uganda, January to December

are found in the north-east of the country, along the border with Kenya and Sudan (Karamoja region) where drought is a common occurrence (NARO, 2001). It is also drier in the south-west where Uganda borders Rwanda.

Studies using daily records for Uganda are still scarce due to inhomogeneity found in station records. This problem has affected research using daily records on aspects such as the cessation and onset of rainfall. There is a need to establish the cessation and onset of rainfall changes so that we can determine the effects that may arise. Onset and cessation dates have been associated with the large-scale systems that influence regional weather, such as the El Niño Southern Oscillation (ENSO), cyclones and monsoons (Mubiru *et al.*, 2012).The onset and cessation of rainfall has an impact on agricultural practices in a country like Uganda. Rainfall studies also do not tell us much about the magnitude of change with significant confidence. However, it has been noted that rainfall may decrease by 5 mm (mostly in the north) to 10 mm (southern part) per month below the median (1985-2005). This is based on the assumption that there is a moderate level of mitigation taking place (Representative Concentration Pathways- RCPs4.5) in the next fifty years and up to 70 mm over lake Victoria in 80 years to come (GOU, 2015). Apuuli *et al.* (2000) also report about rainfall reduction in the northern districts and the cattle corridor, which extends to the southern part of Uganda. The GOU (2015) report further reveals that projected rainfall totals will differ from what the country is receiving at present. Seasonal rainfall will increase significantly in the DJF and could lead to a longer wet season. Detailed explanation of the modelling and science can be found in GOU (2015), and the subsequent projections for business as usual scenario are presented in the report.

Temperatures
Uganda is pleasantly cool with a long-term mean near-surface temperature of around 21°C. However, monthly temperatures range from a minimum of 15°C in July, to a maximum of 30°C in February (Nsubuga *et al.*, 2014a, 2014b, 2014c).The highest temperatures are observed especially in the north-east, while lower temperatures occur in the south. The JJA season is the coolest, while the DJF and MAM seasons are the warmest (Figure 3). Nyenje and Batelaan (2009) found an interesting relation of near-surface temperatures and their impact on ground water systems in the upper Ssezibwa catchment of Uganda, especially from 1990 onwards. This was confirmed by Nsubuga *et al.* (2014b) who demonstrated that Uganda experienced positive trends in minimum and maximum temperatures over the period 1960 to 2008. It was also found that the nights are warming faster than the day-time temperatures. The GOU (2015) report projects temperatures to increase by 2°C to 2.5°C in fifty to eighty years under the RCP4.5 scenario. The same report projects temperatures to increase more during the MAM and JJA seasons, compared to the DJF and SON seasons. The situation will be worse if business stays as usual (RCP8.5 scenario) for both daily and seasonal temperatures. However, smaller changes over Lake Victoria are expected (GOU, 2015). The effect of an increase in temperature, however small, will have a disastrous impact, for example on coffee growing (Jassogne *et al.*, 2013), fish stocks and fish-based livelihoods (Badjeck *et al.*, 2010) among others. Some studies (Lindsay and Martens, 1998; Hay *et al.*, 2002; Tanser *et al.*, 2003; Alonso *et al.*, 2010) have shown the role of temperature increase in the spread of mosquitoes and malaria. They also discuss the strengths and weaknesses of the past approaches to studying malaria transmission. Heal and Park (2014) underpin the unequal effects of higher temperatures on per capita output between warmer countries, of which Uganda is one and colder countries. Their study estimates lower output per capita for warmer countries compared to higher output per capita for nations with colder

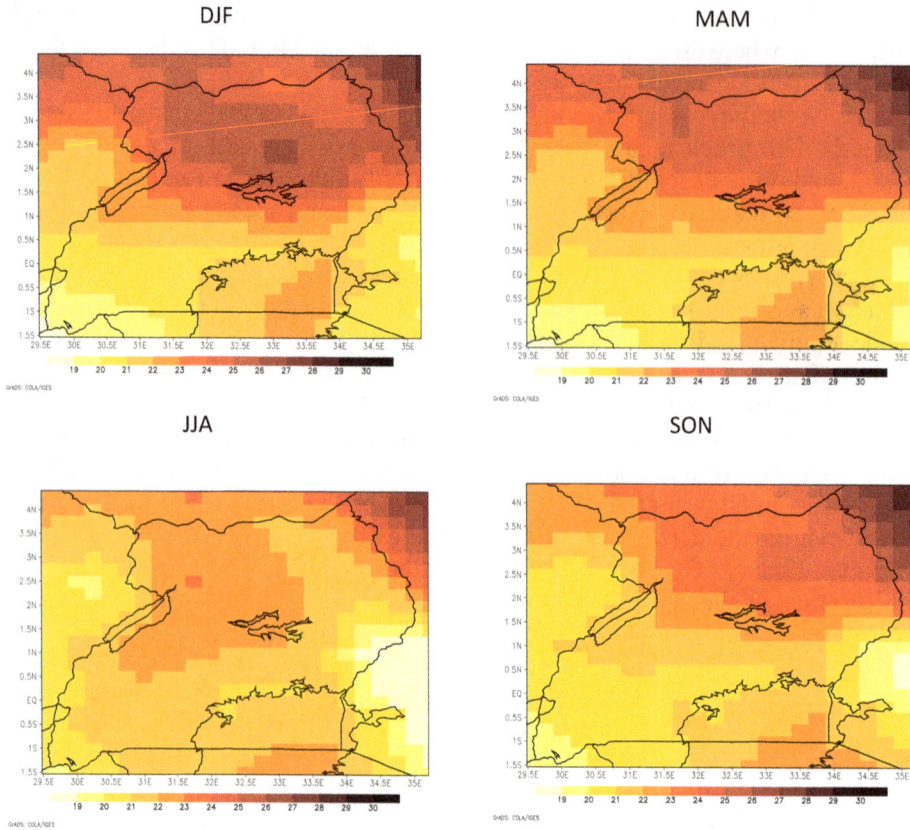

Sourcce: GOU (2015)

Figure 3. Average (1986-2005) seasonal near-surface (2 m above surface) temperature (°C) as captured using ERA Interim Reanalysis data

climates. This aspect requires further investigation to establish its relevancy on Uganda's economy.

Greenhouse gas emissions

The Ministry of Water and Environment's climate change department launched a national greenhouse gas (GHG) inventory system in October 2016. The system helps Uganda track, report and to prioritise emission reduction actions in key sectors to address climate change. Previously Apuuli *et al.* (2000), using 1990 as a base year, reported that carbon dioxide (CO_2) and methane emissions had amounted to 740 and 1,160 Gg, respectively, while noting that fossil fuels combustion contributed 0.78 million tonnes of carbon dioxide in 1990. Using the Global Protocol for Community-Scale Greenhouse Gas Emission Inventories (GPC), Lwasa (2017) reports that, using 2012 as the base year, a study by USAID put total GHG emissions for the country at $49MtCO_2e$ (0.10 per cent of the world total), with a per capita contribution of $1.36tCO_2e$. Much of these emissions were identified by world resources institute climate analysis indicator tools as coming from agriculture, forestry and other land use changes in 2012. In formulating policies to enforce Uganda's determination to control emissions, the country has to investigate and develop a combination of appropriate policies and measures.

Observed recent climate variability

A number of studies that we identified recently have informed us of the following about climate change and climate variability in Uganda, hence it forms part of what is known.

There is now evidence that the country as a whole is experiencing observable shifts in rainfall and temperatures (Kizza *et al.*, 2009; DEWPoint, 2012; FEWSNET, 2012; Mubiru *et al.*, 2012; Kilimani, 2013; Nsubuga *et al.*, 2014a; GOU, 2015). Less information is available on other variables such as wind, solar hours and humidity among others. As noted by King'uyu *et al.* (2000), most studies in the past have investigated temperatures and precipitation to explain present climate change patterns, thus giving a global picture that points to an increasing trend in temperature and reduced rainfall, which is not far from what has been observed in Uganda. Changes in rainfall and temperatures can influence changes in other factors of climate or the other way round.

In Uganda, impacts of extremes in weather are reported by the national media without a quantitative assignment, an aspect Apuuli *et al.* (2000) also recognised. For example, Hisali *et al.* (2011) refer to the unusual rains recorded in 1961/62, 1997/98 and 2007 and severe drought that hit the country in 1993/94. Work by Apuuli *et al.* (2000), however, gave some estimated values on how much rain fell during that drought period and also pointed out how people were displaced, that food was excessively priced and mentioned other negative effects of the drought. Similar findings are reported by Schreck and Semazzi (2004) but with no quantitative values of how severely the country had been affected. There are also indications that droughts have become more frequent, that the onset and cessation of rainfall is becoming more variable; and some models predict wetter conditions in the far future. Drought and heavy erratic rains are perceived by respondents to be extreme climate variability events. Events of high magnitudes are dated and reported by Bomuhangi *et al.* (2016) in the eastern districts of Uganda. Similar results have been reported by Osbahr *et al.* (2011) for south-west Uganda of farmers' perception of climate trends and variability according to their local knowledge. As far as extreme events are concerned, drought events represent an average annual damage in the past decade of US$237m (GOU, 2015).

Reports produced by the Ugandan Ministry of Water and Environment (2007) and LTS International (2008), anticipate that Uganda may experience changes in rainfall patterns with the second rains becoming more intense. A study by Mubiru *et al.* (2012), indicates that farmers in Uganda characterize the rains during the MAM season to be short, localized and with occasional hot and dry spells. Dry spells are reported by Nsubuga *et al.* (2014a) to be increasing in Uganda. This climate variability has been detected in fluctuations in the water resources, e.g. during the 2004/5 drought period, which correlated with Lake Victoria water levels dropping by a metre below the 10-year average (Kull, 2006). We have to remember that water resources are a proxy for climate variability (Nsubuga *et al.*, 2015).

Published peer-reviewed studies have found a decreasing trend in total rainfall during the long rains in MAM, which was also associated with a decrease in the number of wet days in and around Namulonge agricultural research station (Nsubuga *et al.*, 2011). Kizza *et al.* (2009) again identified positive rainfall trends at most stations located in the northern part of the Lake Victoria basin, especially during the short rains season of SON. However, no significant trends exist in annual total rainfall records. A study on community perceptions of variability in precipitation (Bomuhangi *et al.*, 2016) revealed that groups of farmers in eastern Uganda realised that rains came late in the season, and the seasons of rainfall were short but intensive and erratic. Anyah and Semazzi (2004) indicate that surface temperatures on Lake Victoria were warmer by more than 0.5°C during the 1990s compared with the 1960s.

While investigating the nature of rainfall in Uganda using historical data, Nsubuga *et al.* (2014c) identified decades of below-normal, normal and above-normal annual rainfall anomalies over the period 1940 to 2009. Three long epochs of below-normal rainfall occurred between 1940 and 1960, around the 1970s and again around the 1980s and 1990s. Above-normal rainfall periods occurred during the early 1960s and late 1970 and late 1990s. It is interesting to note that episodes of exceptionally high rainfall totals during the 1960s and 1970s were preceded by relatively long low rainfall periods (Nsubuga *et al.*, 2014c).

According to Hulme (1992), inter-annual variability or, on the other hand, consistency, of rainfall are important indicators of the risk of change (associated with higher variability) or reassurance of rainfall consistency or sustainability (associated with lower variability) in Africa. Percentages of coefficient of variation at 36 selected stations across Uganda are in the range of 13 to 29 per cent (Nsubuga *et al.*, 2014c).

High variations occurred at the Kakooge, Kotido and Kangole stations (south and north-east of Uganda), while the Aduku, Kirima Forest, Masindi Meteorological and Ngetta Farm stations experienced the lowest variations (south-west and north-west of Uganda) in inter-annual rainfall (Nsubuga *et al.*, 2014c). What is important from a similar study conducted in Turkey by Turkeş (1996) is that areas where percentages of the coefficient of variation is >20 per cent are more likely to experience frequent and severe droughts (or floods), while areas associated with lower coefficients of variation have more consistent or sustainable rainfall. Analysis indicated that rainfall, for example at the Kakooge station, has varied more over the past 30 years (1980-2009) than before (e.g. 1940-1969), consequently implying that the central region is at risk of experiencing more droughts in the future (Nsubuga *et al.*, 2014a). The USAID (2013) report contains a comprehensive analysis of the general characteristics of the observed climate over Uganda. Using data from 16 weather stations across Uganda, it concluded that no significant change in annual total rainfall occurred over the past 60 years. Another study (DEWPoint, 2012) came to the same conclusion as far as rainfall is concerned, but also found that near-surface temperatures appeared to have increased by approximately 0.2°C per decade in past 60 years. In addition, FEWSNET (2012), using the 1975-2009 climate normal conclude that temperatures increased by more than 0.8, while rainfall was approximated to have decreased by 8 per cent between 1900 and 2009 (Kilimani, 2013).

Climate variability in Uganda is multifaceted because Uganda lies astride the Equator and mechanisms identified above, control its climate. There is little information from models about changes in future variability on all climatic aspects. Nevertheless, the studies have given us some insight into what the situation is like in Uganda. Below are climate projections also reported in GOU (2015) which ought to be known.

Climate projections for Uganda

Climate projections are based on representative concentration pathways (RCPs), specifically scenarios RCP4.5 and RCP8.5. RCPs are four greenhouse gas concentration (not emissions) trajectories adopted by the Intergovernmental Panel on Climate Change (IPCC) for its fifth Assessment Report (AR5) (IPCC, 2013). RCP4.5 shows a moderate level of mitigation of greenhouse gases, resulting in some shifts in climate patterns globally, while under RCP8.5 far less mitigation takes place, resulting in much stronger changes in climate globally (Riahi *et al.*, 2011). In other words, RCP4.5 is an optimistic scenario, while RCP8.5 is more of a "business as usual" scenario in terms of carbon dioxide (CO_2) emissions.

Maps of the annual mean near-surface temperature and total rainfall change from the median, projected over 50 years and 80 years from the present, under both the RCP4.5 and the RCP8.5 concentration scenarios (Figures 4-7).

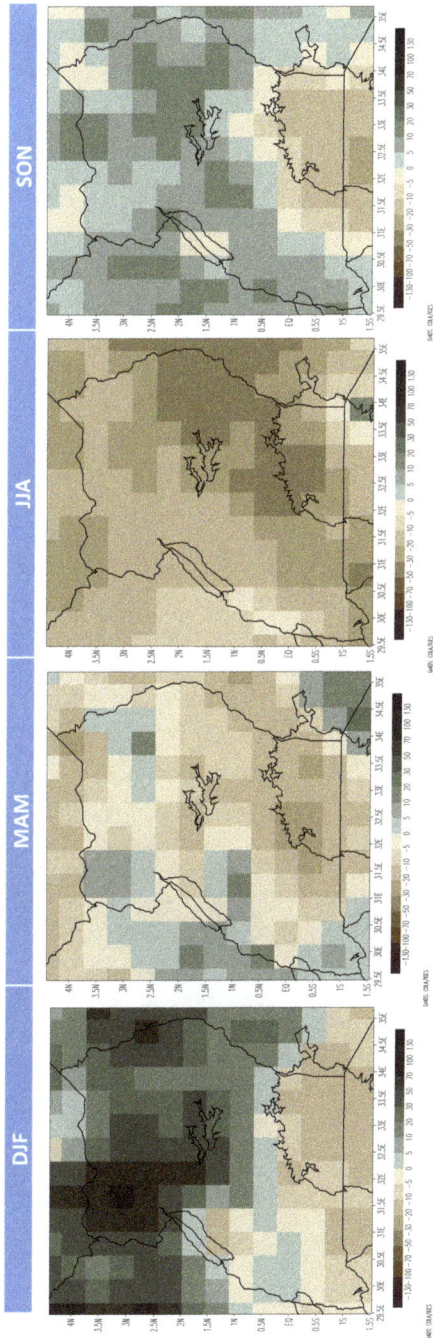

Figure 4. Projected percentage change of seasonal rainfall change for 2046 -2065 relative to 1985- 2005 of the RCP4.5 scenario

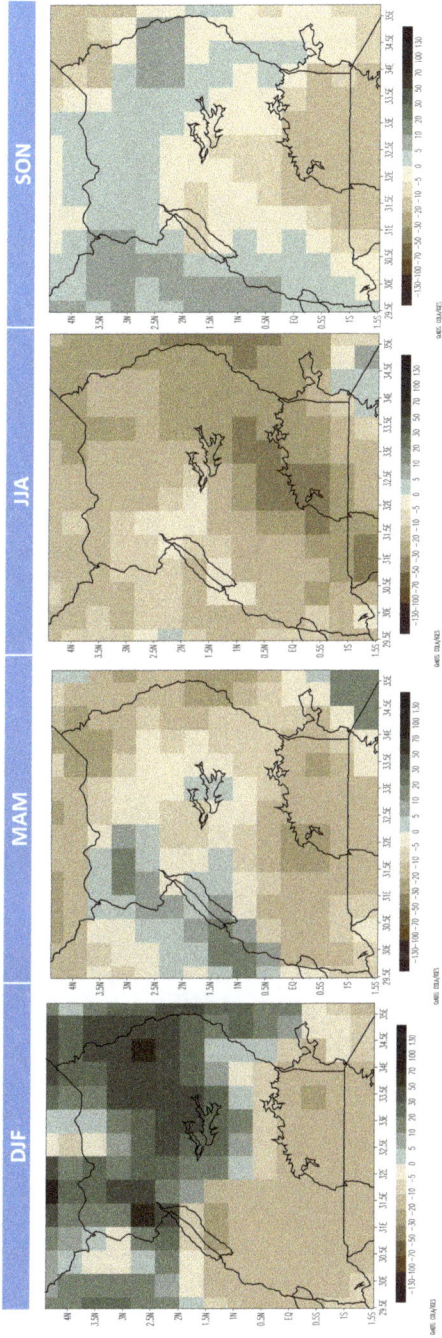

Figure 5. Projected percentage change of seasonal temperature change for 2046-2069 relative to 1985-2005 of the RCP4.5 scenario

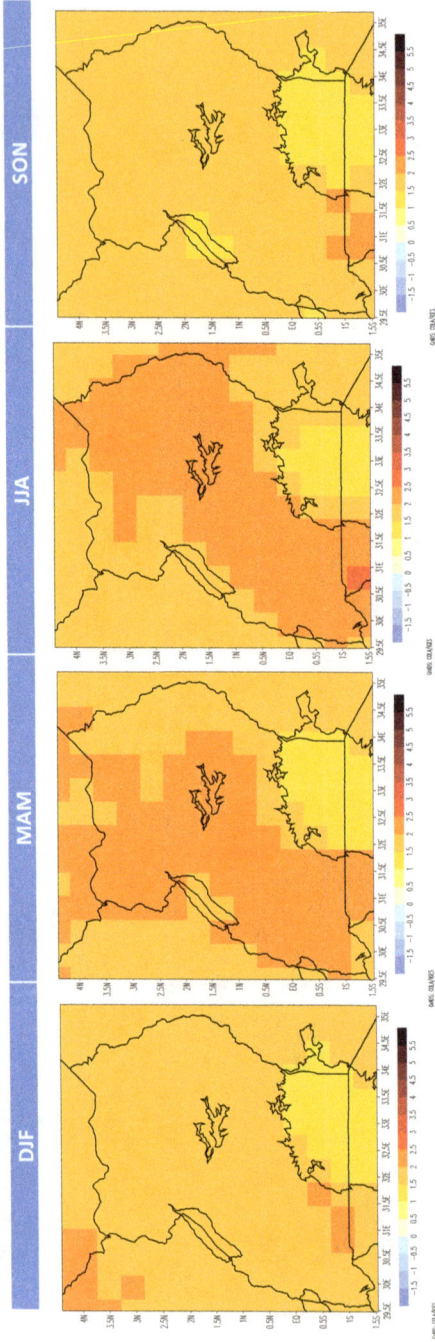

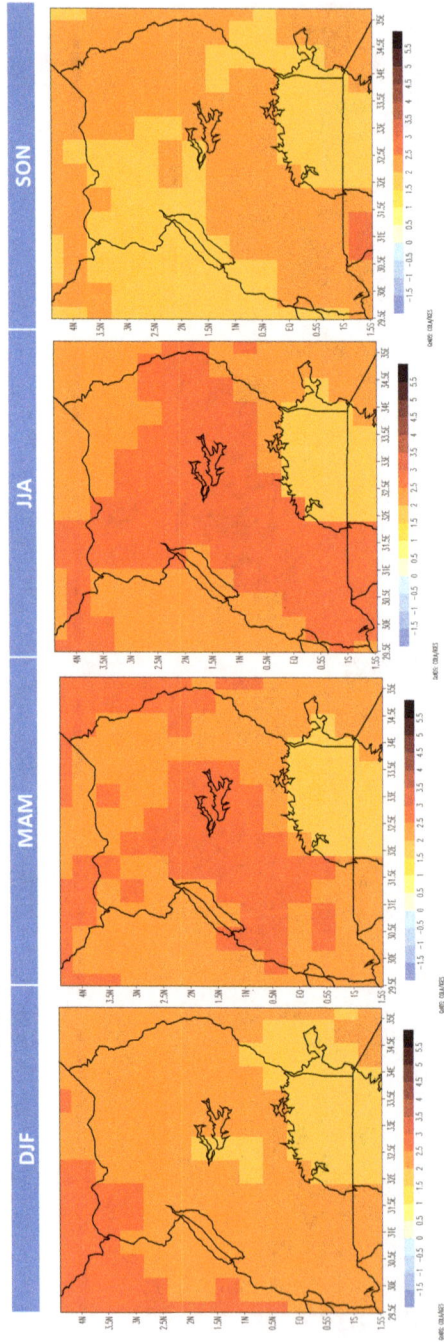

Projected change of the MEDIAN – RCP8.5

Seasonal rainfall change (mm/month) for 2046 – 2065 (+50 years) - relative to 1985-2005

Seasonal rainfall change (mm/month) for 2076 – 2095 (+80 years) - relative to 1985-2005

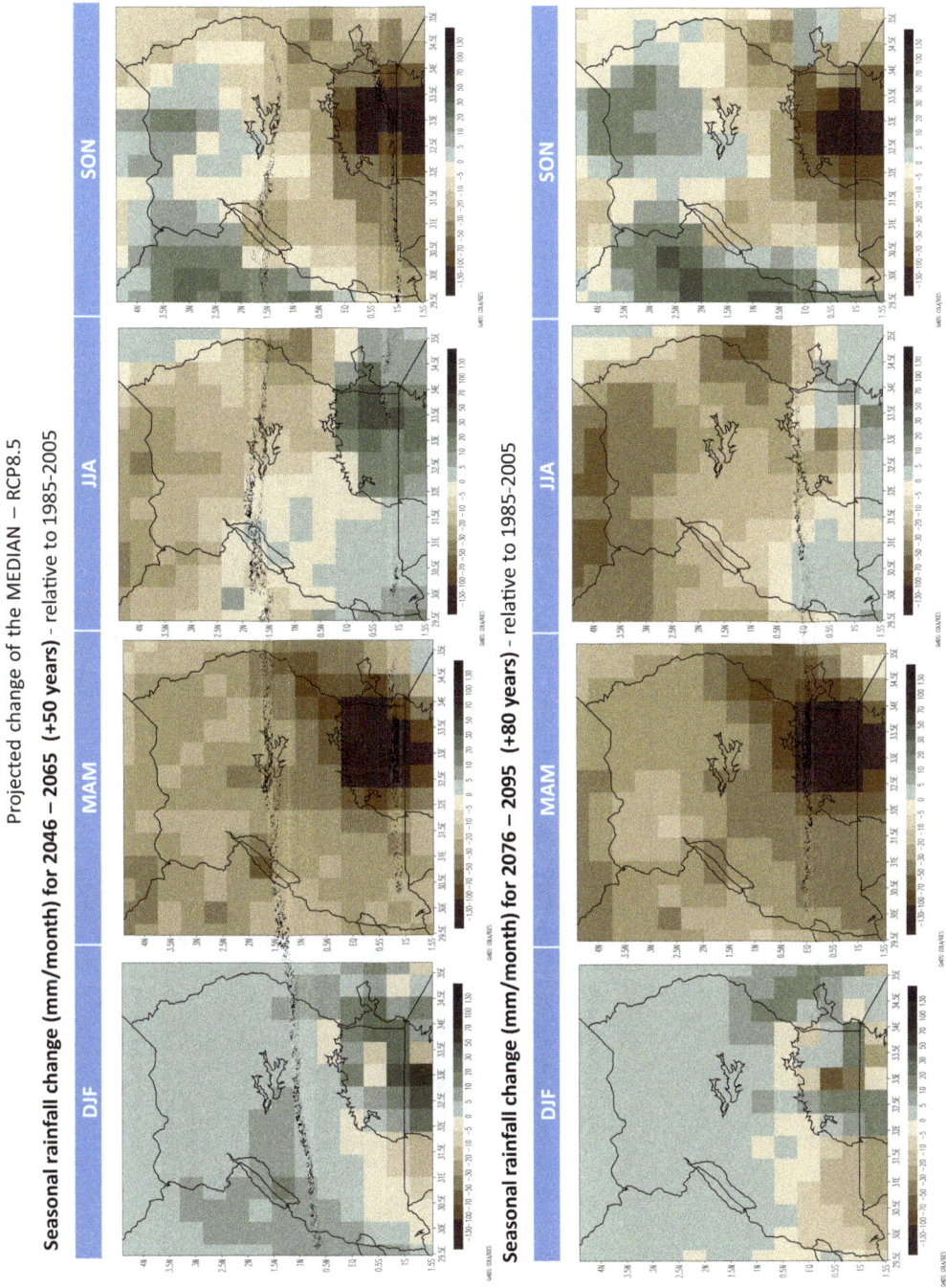

Figure 6. Projected percentage change of seasonal rainfall change for 2046-2065 relative to 1985-2005 and 2076-2095 relative to 1985-2005 of the RCP8.5 scenario

Projected change of the MEDIAN – RCP8.5

Seasonal temperature change (ºC) for 2046 – 2065 (+50 years) - relative to 1985-2005

| DJF | MAM | JJA | SON |

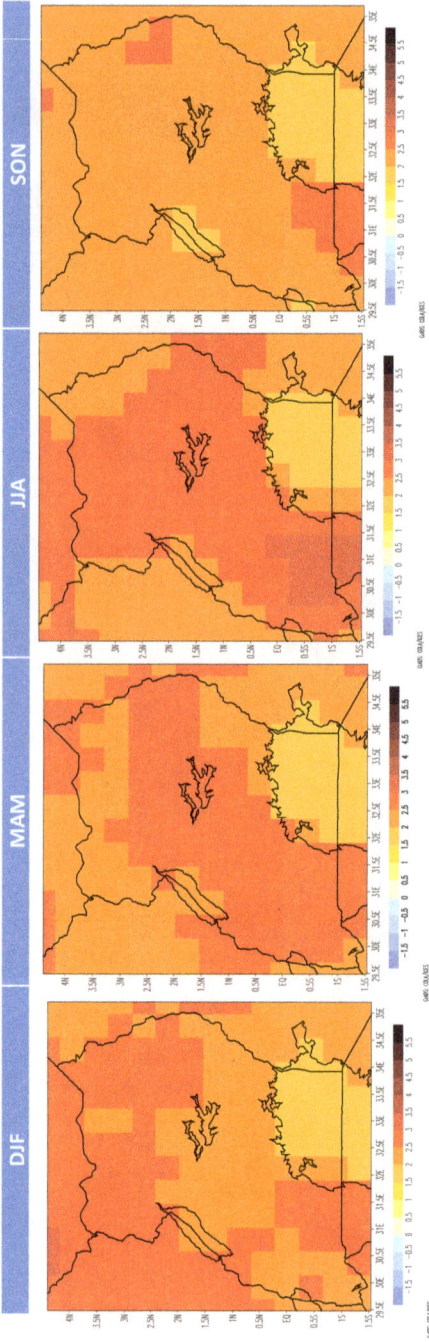

Seasonal temperature change (ºC) for 2076 – 2095 (+80 years) - relative to 1985-2005

| DJF | MAM | JJA | SON |

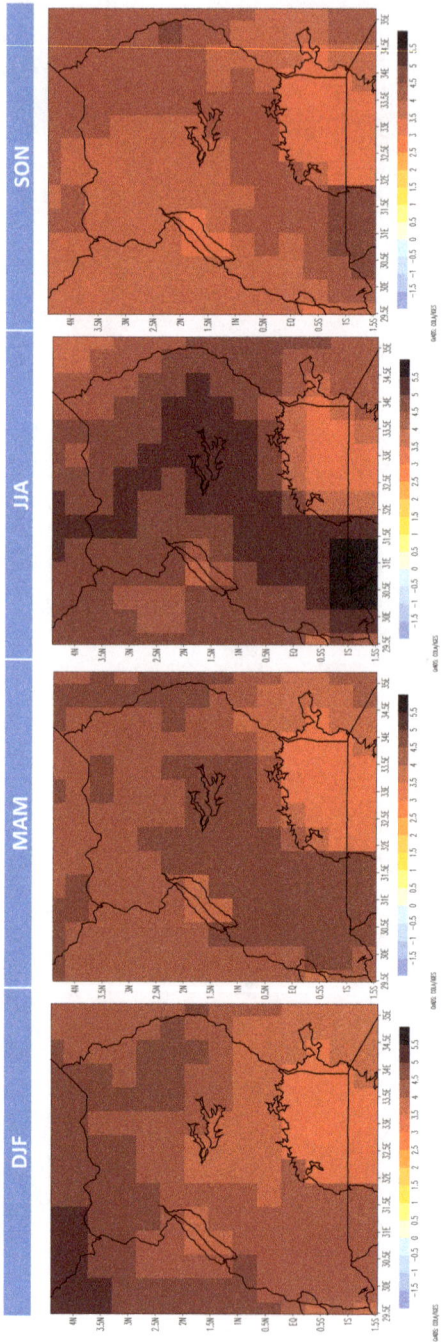

Figure 7. Projected percentage change of seasonal temperature change for 2046-2069 relative to 1985-2005 of the RCP8.5 scenario

Under the moderate RCP4.5, projected annual rainfall totals will differ a little from what is presently experienced. Projected changes within a range of less than plus or minus 10 per cent from present rainfall are expected. However, less rainfall is to occur over most of Uganda, with slightly wetter conditions over the west and north-west. Rainfall totals might drop by 20 per cent from the present levels over Lake Victoria. What is significant on a seasonal time scale, is the projected increase in seasonal rainfall for the DJF season (up to 100 per cent from the present), which is indicative of a longer wet season that extends from SON towards DJF (Figure 4).

Projected near-surface temperatures are of +2°C in 50 years from the present and of +2.5°C in 80 years from present averages. Temperatures will rise more during the MAM and JJA seasons in comparison to the DJF and MAM seasons (Figure 5) if moderate levels of mitigation are implemented. A lower temperature increases of about 1°C is expected for the Lake Victoria basin.

Under the extreme RCP8.5 scenario, projected annual rainfall total changes are very similar to that of the RCP4.5 projections, and therefore still close to what is currently observed. On a seasonal time scale, the MAM and JJA seasons might expect slightly less rainfall (Figure 6), while the percentage increase in DJF rainfall, as in the RCP4.5 projections, is again very significant. A similar drop (−20 per cent) over the Lake Victoria basin is projected.

In the "business as usual" scenario, projected near-surface temperatures are in the order of +3°C in 50 years from the present and in the order of +5°C in 80 years from the present. Seasonal temperatures will increase between +2°C and +3°C for DJF, MAM and JJA in 50 years from present values, with a slightly lower increase for SON (Figure 7). In 80 years from the present, temperatures might rise as much as +5.5°C during the JJA season (currently the coolest season), while increases of between +4°C and +5°C are expected for the seasons DJF, MAM and SON. Smaller changes over the Lake Victoria basin are expected.

What should be known

The above-summarized expected changes in near-surface temperatures and rainfall levels will feed into the economic performance of the country and affect the well-being of people, with serious implications for the present and the future. The main sectors that will be affected, and consequently have an impact on the economy, include water resources, agriculture, health, energy sources, infrastructure and tourism, with associated effects on livelihoods.

The decrease in rainfall in most of Uganda, combined with significant temperature increases, especially during the MAM and JJA seasons, will result in significantly drier conditions. A longer wet season that extends from SON towards DJF will have adverse implications for the country. For example, a significant drop of total rainfall over Lake Victoria (−20 per cent from present levels), combined with about 1°C temperature increase, will affect the Lake's water level and the associated livelihoods of nearby communities depending on activities such as fishing. Overall, the projected changes will require a number of adaptation strategies.

A study by Nsubuga et al. (2014b) points to the reoccurrence of extreme rainfall events similar to those of the 1960s. Given the current population growth, an overload of the sewer and storm-water drainage systems is inevitable, to the detriment of the environment. In East Africa, according to Nyenje and Batelaan (2009), runoff is to increase by 10 to 20 per cent by the 2050s, and Uganda is the wettest of the East-African countries. Floods are to increase in some parts of the country, which will have cost implications that require quantifying. Researchers such as Pillay and Van den Bergh (2016), quantified the cost in billions of

Euros, for treating depression due to flooding in 27 countries of the European Union. Uganda is increasingly adopting the use of fertilizers and pesticides in farming activities. Heavy rains will lead to surface overflows, fluid erosion and mobilization of agricultural fertilizers, pesticides, manure and animal waste into surface water resources, which affects stream and ecosystems health. Heavy rains can also cause earth mass movements, especially in the mountainous and hilly areas, which could even lead to dam and riverbank failures as observed on the Nyamwamba River in western Uganda and the Elgon mountains recently. Changes in weather patterns can have pronounced effects on food availability and on the income of agriculture-dependent households. There are situations currently in Uganda resulting from rainfall variability, which has reduced yields of major staple foods such as maize in the 2014/2015 growing season. Unfortunately, the consequences are not quantified in line with economic performance, but those are often reflected in inflationary tendencies. When it rains in Uganda, roads often become impassable and food does not reach the market easily. Similarly, when it does not rain, food is scarce on the market because of drought, which leads to price fluctuations reported in Uganda Bureau of Statistics (UBOS) quarterly bulletins. Given the declining trends in rainfall, new plant varieties that are able to withstand low rainfall were introduced but not evaluated. Climate variability has disrupted the growing conditions in the agricultural sector. Farmers often plant late or early, ending up with a failed crop. The disruption has serious implications for food security, water availability, agricultural productivity and exports.

The climatic differences in Uganda influence the geographic distribution of crops grown. For example, the coffee-banana system thrives in the wetter parts, while the sorghum- millet system grows well in the drier areas. A change in climate is going to create a shift in the cropping system so that new varieties, which are tolerant to less rain and warmer conditions, such as cassava, will replace the traditional systems. Crops like maize, grow in areas of high rainfall and require wet conditions upon being planted. When rainfall is unreliable, crops that require water at critical phases of development suffer greatly. As observed by Molua (2006), moisture stress during flowering, pollination and grain filling is harmful to staple crops such as maize and beans, thus making rain-dependent agriculture challenging. Where natural conditions required for the growth of food and cash crops are no longer suitable, irrigation options have to be explored. Locally designed drip irrigation technology is already in play especially among the coffee growers in the central parts of Uganda as an adaptation to changing climate. The demand for water in this case would also increase in a warming climate, resulting in competition among water uses. The water aspect reminds us of how, for example, 61 per cent of Uganda's safe water is from groundwater especially around Lake Victoria and in the southern and northern parts of the country (Nyenje and Batelaan, 2009). Total water demand is expected to increase from 408 million cubic metres a year (MCM/y) in 2010 to 3,963 million MCM/y in 2050 (GOU, 2015).

Like most traditional African societies, Ugandan farmers do not routinely use weather forecasts and early warning systems are inadequate, while climate information is usually not trusted or often not taken seriously. Worst of all, government involvement in planning what to grow and helping farmers to understand and adapt to climatic changes is very minimal.

The GOU (2015) study on the impact of climate change on the economy informs us of consequences that may result from a "business as usual" scenario. For example, the study has worked out that climate change could reduce biomass availability by 5 to 10 per cent between 2020 and 2050. This will be a difficult situation, especially if there are no efforts to electrify the rural areas, where most Ugandans use biomass for much of their energy requirements. In addition, there is a possibility that hydropower potential will decrease due to a reduction in precipitation, estimated to be around 26 per cent by 2050 according to the report.

There are a number of medical journals highlighted by Pillay and Van den Bergh (2016) which have published frequently on the relation between human health and climate change. The health risks associated with climate change vary according to age and gender (women, children and the elderly tend to be more at risk) and regions, especially in developing countries such as Uganda. Anticipated climate-related health risks can be either direct or indirect, such as those, which depend on climatic conditions (malaria, dengue fever, diarrhoea and cholera). Climate change can also affect allergic respiratory diseases, as argued by Ziello *et al.* (2012) and cardiovascular and respiratory diseases. Extreme heat and air pollution are important causes of health risks, especially in urban areas. Unfortunately, available economic studies of the costs of climate change have not emphasized this aspect, but Pillay and Van den Bergh (2016) expect the cost to be considerable.

The impact of climate change on infrastructure is felt in two dimensions:

(1)   lost resilience of buildings, roads and other artefacts owing to increased temperature and precipitation; and

(2)   damage caused by extreme events, including loss of life and injury, damage to assets, costs to persons due to displacement and inconvenience and expenses incurred for disaster relief.

There may be other problems related to climate change that we do not know about at present, but through research, we can inform ourselves about unforeseen problems and advise governments about necessary planning.

Concluding remarks

This review sought to place into context a diversity of evidence on climate change over the past years, and thereby to call attention to pending issues that need exploring in order to adapt to and mitigate possible threats. Pertinent are mechanisms that influence the climate of Uganda and long-term observations using gauging and modelled data. Of great interest is the seasonal and bimodal distribution of rains following the movement of the ITCZ. From reasonably well documented literature, an understanding of the annual, inter-annual and long-term variability of the climate has been reached. There is a clear signal in the climate data that temperature has been increasing and, to a lesser extent, evidence that the reliability of rains in the first season has decreased slightly. However, rainfall measurements do not show a downward trend in rainfall amount, a significant shift in the intensity of rainfall events or in the start and end of the rainy seasons (Osbahr *et al.*, 2011). The MAM rains are more abundant and will increase in the near future according to the RCP 4.5 scenario as well as the measured temperatures. Evidence from various studies is tenuous regarding the change in climate for the past three decades. Fieldwork by Osbahr *et al.* (2011), revealed that farmers felt that temperature had increased and seasonality and variability of rainfall had changed, with the first rainy season between March and May becoming more variable.

However, lack of adequate continuous ground-based observations hampers objective and comprehensive diagnostic and numerical characterization of climate change in Uganda. Having said that, scientists are circumnavigating the problem through iterative methods to produce credible outputs using the available data. Recent studies point to the significant costs of not taking action on climate variability and future climate change in Uganda. National studies e.g. GOU (2015), show annual costs that could be in the range of US$3.2 to 5.9bn within a decade. The biggest impacts will be on water, followed by energy, agriculture and infrastructure. In the case of energy and agriculture, the greater uncertainties relate to the physical effects of climate change, whereas in the case of infrastructure they relate more to the valuation of the consequences of climate change. Even if there were no further increases in temperature, precipitation or frequency of extreme events, the costs of inaction would rise over time, because of an increase in population (expected to grow by more than 2 per cent per annum over the next 40 years) and GDP (expected to grow at around 7 per cent to 8 per cent per annum over the same period).

Nevertheless, we foresee Ugandans taking on varied forms of adaptation to the effects of climate change in the near future-especially in the agricultural sector. These will include adapting to crop types that are drought tolerant and varieties that mature fast, using fertilizers, applying crop management techniques that limit soil moisture loss and finally, improving marketing and storage techniques.

Uganda being predominantly agricultural, the manufacturing sector will need to revisit the manner in which it acquires inputs. Financial institutions have to devise strategies to determine when to provide credit and crop insurance, while transporting companies have to improve efficiency in delivering inputs and outputs. Research institutions also need to step up and engage in similar pro-active efforts, especially in the problematic areas identified in this review.

References

Aguilar, E., Peterson, T.C., Ramirez Obando, P., Frutos, R., Retana, J.A., Solera, M., Soley, J., Gonzalez Garcia, I., Araujo, R.M., Rossa Santos, A., Valle, V.E., Brunet, M., Aguilar, L., Alvarez, L., Bautista, M., Castanon, C., Herrera, L., Ruano, E., Sinay, J.J., Sanchez, E., Harnandez Oviedo, G.I., Obed, F., Salgado, J.E., Vazquez, J.L., Baca, M., Gutierrez, M., Centella, C., Espinosa, J., Martinez, D., Olmendo, B., Ojeda Espinoza, C.E., Nunez, R., Haylock, M., Benavides, H. and Mayorga, R. (2005), "Changes in precipitation and temperature extremes in Central America and Northern South America, 1961-2003", *Journal of Geophysical Research*, Vol. 110 No. D23, p. D23107.

Alonso, D., Bouma, M.J. and Pascual, M. (2010), "Epidemic malaria and warmer temperatures in recent decades in an East African highland", *Proceedings of the Royal Society*, doi: 10.1098/rspb.2010.2020.

Anyah, R.O. and Semazzi, F.H.M. (2004), "Simulation of the sensitivity of Lake Victoria basin climate to lake surface temperatures", *Theoretical and Applied Climatology*, Vol. 79 Nos 1/2, pp. 55-69.

Apuuli, B., Wright, J., Elias, C. and Burton, I. (2000), "Reconciling national and global priorities in adaptation to climate change: with an illustration from Uganda", *Environmental Monitoring and Assessment*, Vol. 61 No. 1, pp. 145-159.

Asadullah, A., Mcintyre, N. and Kigobe, M. (2008), "Evaluation of five satellite products for estimation of rainfall over Uganda", *Hydrological Sciences Journal*, Vol. 53 No. 6, pp. 1137-1150.

Badjeck, M.-C., Allison, E.H., Halls, A.S. and Dulvy, N.K. (2010), "Impacts of climate variability and change on fishery-based livelihoods", *Marine Policy*, Vol. 34 No. 3, pp. 375-383.

Basalirwa, C.P.K. (1995), "Delineation of Uganda into climatological rainfall zones using the method of principal component analysis", *International Journal of Climatology*, Vol. 15 No. 10, pp. 1161-1177.

Beltrando, G. (1990), "Space–time variability of rainfall in April and October-November over East Africa during the period 1932-1983", *International Journal of Climatology*, Vol. 10 No. 7, pp. 691-702.

Bomuhangi, A., Nabanoga, G., Namaalwa, J.J., Jacobson, M.G. and Abwoli, B. (2016), "Local communities' perceptions of climate variability in the Mt Elgon region, Eastern Uganda", *Cogent Environmental Science*, Vol. 2 No. 1, p. 1168276.

Conway, D. (2005), "From headwater tributaries to international river: observing and adapting to climate variability and change in the Nile basin", *Global Environmental Change*, Vol. 15 No. 2, pp. 99-114.

DEWPoint (2012), "Support to the strategic programme review for climate change in Uganda: understanding the implications and appraising the response", Update to 2008 LTS Scoping Study [Steynor, A., Jack, C. and Smith, C.], DFID Resource Centre for Environment, Water and Sanitation, p. 1.

Domroes, M. and El-Tantawi, S. (2005), "Recent temporal and spatial temperature changes in Egypt", *International Journal of Climatology*, Vol. 25 No. 1, pp. 51-63.

FEWSNET (2012), *Famine Early Warning Systems Network. A Climate Trend Analysis of Uganda*, available at: http://pubs.usgs.gov/fs/2012/3062/FS2012-3062.pdf (accessed 9 January 2017 at 22: 45 CET).

Government of Uganda (GOU) (2015), *Economic Assessment of the Impacts of Climate Change in Uganda, Final Study Report. Ministry of Water and Environment*, Climate Change Department, Kampala.

Hastenrath, S. (2001), "Variations of east African climate during the past two centuries", *Climatic Change*, Vol. 50, pp. 209-217.

Hay, S.I., Cox, J., Rogers, D.J., Randolph, S.E., Stern, D.I., Shanks, G.D., Myers, M.F. and Snow, R.W. (2002), "Climate change and the resurgence of malaria in the East African highlands", *Nature*, Vol. 415 No. 6874, pp. 905-909.

Heal, G.M. and Park, J. (2014), *Feeling the Heat: Temperature, Physiology & the Wealth Nations*, Discussion Paper 2014-51, Harvard Environmental Economics program, Cambridge, MA, January 2014.

Hisali, E., Birungi, P. and Buyinza, F. (2011), "Adaptation to climate change in Uganda: evidence from micro level data", *Global Environmental Change*, Vol. 21 No. 4, pp. 1245-1261.

Hulme, M. (1992), "Rainfall changes in Africa: 1931-1960 to 1961-1990", *International Journal of Climatology*, Vol. 12 No. 7, pp. 685-699.

Hulme, M., Doherty, R., Ngara, T., New, M. and Lister, D. (2001), "African climate change: 1900-2100", *Climate Research*, Vol. 17, pp. 145-168.

Indeje, M., Semazzi, F.H.M. and Ogallo, L.J. (2000), "ENSO signals in East African rainfall seasons", *International Journal of Climatology*, Vol. 20 No. 1, pp. 19-46.

IPCC (2013), in Stocker, T.F., Qin, D., Plattner, G.-K., Tignor, M., Allen, S.K., Boschung, J., Nauels, A., Xia, Y., Bex, V. and Midgley, P.M. (Eds), *Climate Change 2013: The Physical Science Basis. Contribution of Working Group I to the Fifth Assessment Report of the Intergovernmental Panel on Climate Change*, Cambridge University Press, Cambridge, United Kingdom and New York, NY, p. 1535.

Jassogne, L., Laderach, P. and Van Asten,, P. (2013), "The impact of climate change on coffee in Uganda. Lessons from a case study in the Rwenzori Mountains", *Oxfam Research Reports*, IITA, CIAT, Oxfam, ISBN 978-1-78077-262-2, UK (accessed 10 January 2017 at 10:25 CET).

Kilimani, N. (2013), "Water resources accounts for Uganda: use and policy relevancy", ERSA working paper No. 365.

King'uyu, S.M., Ogallo, L.A. and Anyamba, E.K. (2000), "Recent trends of minimum and maximum surface temperatures over Eastern Africa", *Journal of Climate*, Vol. 13 No. 16, pp. 2876-2886.

Kizza, M., Rhode, A., Xu, Y.C., Ntale, H.K. and Halldin, S. (2009), "Temporal rainfall variability in the Lake Victoria Basin in East Africa during the twentieth century", *Theoretical and Applied Climatology*, Vol. 98 Nos 1/2, pp. 119-135.

Komutunga, E.T. and Musiitwa, F. (2001), "Agricultural systems", in Mukiibi, J.K. (Ed.), *Agriculture in Uganda*, Vol I General Information, Fountain Publ/CTA/NARO, pp. 220-230.

Kull, D. (2006), "Connections between recent water level drops in Lake Victoria, dam operations and drought", available at: www.irn.org/programs/nile/pdf/060208vic.pdf (accessed 24 March 2013).

Lindsay, S.W. and Martens, W.J.M. (1998), "Malaria in the African highlands: past, present and future", *Bulletin of the World Health Organisation*, Vol. 76 No. 1, pp. 33-45.

Lwasa, S. (2017), "Options for reduction of greenhouse gas emissions in the low-emitting city and metropolitan region of Kampala", *Carbon Management*, Vol. 8 No. 3, doi: 10.1080/17583004.2017.1330592.

Molua, E.L. (2006), "Climatic trends in Cameroon: implications for agricultural management", *Climate Research*, Vol. 30, pp. 255-262.

Mubiru, D.N., Komutunga, E., Agona, A., Apok, A. and Ngara, T. (2012), "Characterizing agro meteorological climate risks and uncertainties: crop production in Uganda", *South African Journal of Science*, Vol. 108 Nos 3/4, pp. 108-118.

National Agriculture Research Organisation (NARO) (2001), *Agriculture in Uganda:*, General information. Fountain publishers, ISBN 9789970022434, Vol. 1, p. 486.

National Environment Management Authority (NEMA) (2008), *State of Environment Report for Uganda, 2008*, National Environment Management Authority, Kampala, p. 282.

New, M., Hewitson, B., Stephenson, D.B., Tsiga, A., Kruger, A., Manhique, A., Gomez, B., Coelho, C.A.S., Masisi, D.N., Kululanga, E., Mbambalala, E., Adesina, F., Saleh, H., Kanyanga, J., Adosi, J., Bulane, L., Lubega, F., Mdoka, M.L. and Lajoie, R. (2006), "Evidence of trends in daily climate extremes over Southern and West Africa", *Journal of Geophysical Research*, Vol. 111 No. D14, p. D14102, doi: 10.1029/2005JD006289.

Nicholson, S.E. (1996), "A review of climate dynamics and climate variability in Eastern Africa", in Johnson, T.C. and Odada, E. (Eds), *The Limnology, Climatology and Paleo-Climatology of the East African Lakes*, Gordon and Breach, Amsterdam, pp. 25-56.

Nsubuga, F.W.N., Olwoch, J.M. and Rautenbach, C.J.deW. (2011), "Climatic trends at Namulonge in Uganda: 1947-2009", *Journal of Geography and Geology*, Vol. 3 No. 1, pp. 119-131.

Nsubuga, F.W.N., Olwoch, J.M. and Rautenbach, C.J.deW. (2014a), "Variability properties of daily and monthly observed near-surface temperatures in Uganda: 1960-2008", *International Journal of Climatology*, Vol. 34, pp. 303-314.

Nsubuga, F.W.N., Olwoch, J.M., Rautenbach, C.J.deW. and Botai, O.J. (2014b), "Analysis of mid-twentieth century rainfall trends and variability over southwestern Uganda", *Theoretical and Applied Climatology*, Vol. 115, pp. 53-71.

Nsubuga, F.W.N., Botai, O.J., Olwoch, J.M., Rautenbach, C.J.deW., Bevis, Y. and Adetunji, A.O. (2014c), "The nature of rainfall in the main drainage sub-basins of Uganda", *Hydrological Sciences Journal*, Vol. 59 No. 2, pp. 278-299.

Nsubuga, F.W.N., Botai, J.O., Olwoch, J.M., Rautenbach, C.J.deW., Kalumba, A.M., Tsela, P., Adeola, A.M., Sentongo, A.A. and Mearns, K.F. (2015), "Detecting changes in surface water area of lake Kyoga sub-basin using remotely sensed imagery in a changing climate", *Theoretical and Applied Climatology*, doi: 10.1007/s00704-015-1637-1.

Nyenje, P.M. and Batelaan, O. (2009), "Estimating the effects of climate change on groundwater recharge and base flow in the upper Ssezibwa catchment, Uganda", *Hydrological Sciences Journal*, Vol. 54 No. 4, pp. 713-726.

Ogallo, L.J. (1984), "Temporal fluctuations of seasonal rainfall patterns in East Africa", *Mausam*, Vol. 35, pp. 175-180.

Ogallo, L.J. (1988), "Relationships between seasonal rainfall in East Africa and the Southern oscillation", *Journal of Climatology*, Vol. 8 No. 1, pp. 31-43.

Ogallo, L.A. (1993), "Dynamics of East African climate. Proceedings Indian Academy of sciences", *Earth and Planetary Sciences*, Vol. 102 No. 1, pp. 203-217.

Osbahr, H., Dorward, P., Stern, R. and Cooper, S. (2011), "Supporting agricultural innovation in Uganda to respond to climate risk: linking climate change and variability with farmer perceptions", *Experimental Agriculture*, Vol. 47 No. 2, pp. 293-316.

Phillips, J. and Mcintyre, B. (2000), "ENSO and inter-annual rainfall variability in Uganda: implications for agricultural management", *International Journal of Climatology*, Vol. 20 No. 2, pp. 171-182.

Pillay, C. and Van den Bergh, J. (2016), "Human health impacts of climate change as a catalyst for public engagement: combining medical, economic and behavioral insights", *International Journal of Climate Change Strategies and Management*, Vol. 8 No. 5, pp. 578-596.

Riahi, K., Roa, S., Krey, V., Cho, C., Chirkov, V., Fisher, G., Kindermann, G., Nakicenovic, N. and Rafaj, P. (2011), "RCP 8.5–a scenario of comparatively high greenhouse gas emissions", *Climate Change*, Vol. 109, pp. 33-57.

Schreck, C.J. and Semazzi, F.H.M. (2004), "Variabilility of the recent climate of Eastern Africa", *International Journal of Climatology*, Vol. 24 No. 6, pp. 681-701.

Tanser, F.C., Sharp, B. and le Sueur, D. (2003), "Potential effect of climate change on malaria transmission in Africa", *The Lancet*, Vol. 362 No. 9398, pp. 1792-1798.

Turkeş, M. (1996), "Spatial and temporal analysis of annual rainfall variation in Turkey", *International Journal of Climatology*, Vol. 16 No. 9, pp. 1057-1076.

USAID (2013), in Caffrey, P., Finan, T., Trzaska, S., Miller, D., Laker-Ojok, R. and Huston, S. (Eds), *Uganda Climate Change Vulnerability Assessment Report*, Report to the United States Agency for International Development by Tetra Tech ARD who is responsible for the content, p. 76.

Ziello, C., Sparks, T.H., Estrella, N., Belmonte, J., Bergmann, K.C., Bucher, E. and Menzel, A. (2012), "Changes to airborne pollen counts across Europe", *PLoS ONE*, Vol. 7 No. 4, pp. e34076.

Further reading

Oxfam (2008), *Turning up the Heat: Climate Change and Poverty in Uganda*, Oxfam online imprint, Oxfam GB, Uganda and Oxfam GB, UK. Kampala, Uganda and Oxford, United Kingdom, available at: www.oxfam.org.uk/publication

Jarmillo, J., Muchungu, E., Vega, F.E., Davis, A., Borgemeister, C. and Chabi-Olaye, A. (2011), "Some like it hot: the influence of climate change on coffee berry borer (hypothenemus hampeil) and coffee production in East Africa", *PLoS ONE*, Vol. 6 No. 9, pp. e24528, doi: 10.1371/journal.pone.0024528

Nardone, A., Ronchi, B., Lacetera, N., Ranieri, M.S. and Bernabucci, A. (2010), "Effects of climate changes on animal production and sustainability of livestock systems", *Livestock Science*, Vol. 310, pp. 57-66.

Nkondze, M.S., Masuku, M.B. and Manyatsi, A.M. (2014), "The impact of climate change on livestock production in Swaziland: the case of Mpolonjeni area development programme", *Journal of Agricultural Studies*, Vol. 2 No. 1, available at: https://doi.org/10.5296/jas.v2i1.4416

Ogallo, L.A. (1981), "Trend of rainfall in East Africa", *Kenya Journal of Science and Technology*, Vol. 2, pp. 83-90.

Samson, H.W. (1952), *The Trend of Rainfall in East Africa*, East African Meteorological Department Series paper.

Wu, X., Lu, Y., Zhou, S., Chen, L. and Xu, B. (2016), "Impact of climate change on human infectious diseases: empirical evidence and human adaptation", *Environment International*, Vol. 86, pp. 14-23.

Corresponding author

Francis Wasswa Nsubuga can be contacted at: francis.nsubuga@up.ac.za

6

Perspectives of artist–practitioners on the communication of climate change in the Pacific

Stuart Capstick
School of Psychology and Tyndall Centre for Climate Change Research, Cardiff University, Cardiff, UK

Sarah Hemstock
Secretariat of the Pacific Community (SPC), Suva Regional Office, Suva, Fiji, and

Ruci Senikula
University of the South Pacific, Suva, Fiji

Abstract

Purpose – This study aims to investigate the role of the visual arts for communicating climate change in the context of the Pacific islands, through the perspectives of artists and climate change practitioners.

Design/methodology/approach – As part of an "Eco Arts" project carried out in Fiji, semi-structured research interviews were undertaken with artists and climate change practitioners.

Findings – Participants' motivations to produce art reflected their personal concerns about, and experiences of, climate change. There was an intention to use art-based approaches to raise awareness and promote action on climate change. The artwork produced drew on metaphors and storytelling to convey future climate impacts and aspects of climate change relevant to Fijian and Pacific communities.

Research limitations/implications – The study reports the perspectives of participants and discusses the potential uses of arts communication. Conclusions cannot be drawn from the findings regarding the effectiveness of specific artwork or of arts communication as a general approach.

Practical implications – The research offers suggestions for the inclusion of creative approaches to climate change communication within education and vocational training. A consideration of the perspectives of artist–practitioners has implications for the design and conduct of climate change communication.

Social implications – The involvement of artist–practitioners in the communication of climate change offers the potential for novel discussions and interpretations of climate change with individuals and within communities, which complement more formal or scientific communication.

Originality/value – The present study identifies the motivations and objectives of artist–practitioners involved in climate change communication. The authors highlight the role of personal experience and their use of artistic concepts and creative considerations pertinent to the geography and culture of the Pacific region.

Keywords Communication, Climate change, Visual arts, Environmental communication, Pacific islands

Introduction

Climate change and a changing environment are among the Pacific region's greatest contemporary challenges, with impacts upon its societies and cultures which are far reaching and already underway (Savo *et al.*, 2016). Geographic isolation, the fragility of ecosystems, limited land resources and depleted marine resources, all contribute to the vulnerability of climate change across the Pacific islands (Weir *et al.*, 2016; Taylor and Kumar, 2016). For those living in the islands of the Pacific, the consequences of climate change include risks to health and livelihoods from extreme weather events and changing weather patterns, as well as threats to food and water security (McIver *et al.*, 2016). Some evidence suggests that climate change impacts are already leading directly to migration within and beyond the region (Locke, 2009) although, for many people, it is not yet seen as a principal concern (Mortreux and Barnett, 2009). A range of other problems are faced across the Pacific in connection with climate change, including inundation of low-lying islands by the sea and beach erosion, flooding and saltwater intrusion to drinking supplies, and impacts on infrastructure (McLean and Kench, 2015; Mimura, 1999; Ellison and Fiu, 2010).

Because of the particular vulnerability of islands in the Pacific to the consequences of climate change, regional bodies such as the Pacific Island Development Forum (PIDF), together with international groupings such as the Association of Small Island States (AOSIS), were along the most vocal advocates for the inclusion of an "aspiration" to keep global temperature rise to within 1.5°C at the 2015 Paris climate talks (Hoad, 2016; although this still fell short of a desire among many for 1.5°C to be set as a harder limit). Indeed, a key driver of the pressure for this target came from the Pacific-wide political demands laid out in the Suva Declaration on climate change, signed in Suva, Fiji, just before the Paris talks (PIDF, 2015).

Although Pacific nations have been in the vanguard of pressing for ambitious action to mitigate climate change, as well as for climate finance to enable adaptation, work commissioned by the United Nations Development Program (UNDP) has concluded that most people in the Pacific islands have a limited understanding of climate change and its potential for harm, with an attendant lack of community participation and decision-making taking place on adaptation (Nunn, 2012).

Because of these shortcomings, Nunn (2012) argues that levels of awareness about climate change need to be raised among individuals and communities in the region. Furthermore, if communities are to access available funding for climate change adaptation and be active participants in national and international climate change debates, then it is essential for people to have both an awareness and informed leadership on these issues at local, national, regional and international levels (Moser, 2014).

Where communities are actively involved in projects that seek to adapt to the impacts of climate change on their livelihoods and resources, this has the potential to lead to positive outcomes such as enhanced food and water security (McNamara, 2013). By contrast, a lack of community involvement can lead to inappropriate or maladaptive changes (Barnett and O'Neill, 2010). Despite this, there has been little practical guidance available on how to effectively communicate climate change specifically in ways that increase community resilience and capacity to adapt (McNaught *et al.*, 2014; Moser, 2014). Indeed, some commonly used climate change communication strategies have now been shown to be counter-productive, for example where they lead to fear or disempowerment (O'Neill and Nicholson-Cole, 2009).

One way in which public audiences can be drawn to be interested in the subject of climate change is through artistic approaches. Art that addresses sustainability and climate change has been found variously to emphasise an intent to educate, to bring an experiential element to people's understanding, or to encourage mitigation and behaviour change (Giannachi, 2012). Arts projects that consider climate change have typically engaged with the scientific basis and physical realities of climate change, while finding opportunities to explore ways of responding to climate change as individuals and societies (Gabrys and Yusoff, 2012). The prominent, international project *Cape Farewell* has approached this by bringing together artists and scientists during expeditions, leading to numerous exhibitions and events that have set out to communicate the urgency of climate change and to promote societal change (Buckland, 2012; Knebusch, 2008). Other high-profile festivals and programmes of work have also brought together artists' responses to climate change. The Australian *Climarte* festival has included work which sets out to personalise and engender conversation about climate change (van Renssen, 2017). In parallel to the international climate change negotiations in Paris in 2015, a programme of work termed *ArtCOP21* likewise highlighted the role of culture in addressing climate change, including through installations that drew attention to the plight of climate change refugees and visual illustrations of the scale of individuals' carbon emissions (Sommer and Klöckner, 2017; van Renssen, 2017).

In addition to the exhibition of work by professional artists, artistic approaches have been used to raise awareness of climate change at the community level and using participatory methods. By encouraging members of the public to contribute waste plastic to a street sculpture, a local arts group in Austin, USA, has set out to draw attention to the interrelated issues of climate change and ocean pollution; a parallel project in Seattle, USA, entailed local volunteers raising markers at a shoreline to show potential for raised sea levels in the future, a work that Hall *et al.* (2009) describe as conveying a sense of change and foreboding.

Arts projects such as these often incorporate educational or activist aims, as well as an emphasis on finding effective responses to climate change. For example, Randerson (2007) discusses a project based in Rarotonga (Cook Islands), which encouraged schoolchildren to design posters about how to lessen the impact of climate change; their submissions included imagery that drew attention to the negative effects of extreme weather, industrialisation, and livestock introduction upon island life. Similarly, a community arts project carried out in Tasmania, Australia, enabled children to work with artists and educators to produce and showcase artwork on their understanding of climate change and to propose solutions and explore their own agency in addressing this issue (Stratford and Low, 2015).

As the examples highlighted above illustrate, climate change art has the capacity to ask people to think through issues for themselves and to engage with these in a creative, personal and immediate way (Duxbury, 2010; Gabrys and Yusoff, 2012). At its most powerful, art that addresses climate change has the potential to engage society in new and meaningful ways that promote positive changes in people's behaviours and outlooks (Duxbury, 2010; Curtis, 2009).

Of particular relevance for the present study is the nature of art across the Pacific region; this has been described by Watanabe (2013) as reflecting aspects of the collective self and shared identity. Much Pacific art recognises traditional approaches, including performance and use of materials such as barkcloth, while reflecting influences from elsewhere

(Cochrane, 2016). As a response to climate change in the region, Koya (2012) argues that the arts offer the opportunity for social learning as well as cultural expression, for example through discussions on how to reconcile traditional lifestyles with contemporary commercial and individualised pressures. The application of art practice to these ends includes the work of *Laje Rotuma*, a Fijian NGO whose activities incorporate the use of storytelling and visual arts to reflect people's experiences and to raise awareness of marine conservation, as well as projects by communities which have interpreted and addressed environmental issues through styles of music and dance characteristic to the region (Koya, 2012).

Background to the research and methods

There is now a substantial and growing literature on climate change communication and public engagement, including studies that have assessed the general properties that can render it effective or meaningful to particular audiences (Whitmarsh *et al.*, 2012). Largely absent from this literature has been research that explores the role and perspectives of artist-practitioners themselves as contributors to climate change communication. The personal and creative interpretation of climate change is, however, a critical part of the process of making meaning of this complex subject matter. The present study focuses upon the perspectives of a group of artists and communicators working to engage a wider audience with climate change. In doing so, we seek to provide insights into the motives and inspirations that influenced their work.

The research was undertaken during the course of a week-long "Eco Arts" practicum held in Fiji during November 2014. The practicum was funded and enabled by the University of the South Pacific (USP) in conjunction with the Oceania Centre for Arts, Culture and Pacific Studies.

The objectives of the practicum were to develop and evaluate the potential for the visual arts to communicate and enhance public engagement with climate change issues. The project sought to bring together individuals with an interest both in climate change and the visual arts. As such, those participating comprised both professional artists whose work considered environmental issues, as well as sustainability practitioners (including academics) interested in communicating climate change. The arts practicum permitted those present to exchange ideas and to experiment with different techniques for representing climate change. Through formal activities (e.g., activities at a local school and workshops) as well as unstructured creative practice, a series of artworks were developed. These included material produced using digital media, sculpture, painting and photography, some of which were subsequently displayed at an exhibition in April 2015 at the Oceania Centre gallery[1].

Participants at the practicum were all residing in Fiji when interviews took place, with a majority being Fijian citizens; Fiji-resident practitioners from elsewhere in the world (Tonga, UK and Spain) also participated. Seven of ten participants had a primarily arts-based or communications background; the remaining three individuals were employed in climate change research or policy-making but with an active interest in the use of art for communication of climate change. Anonymised summary descriptions of the backgrounds of research participants are given in Table I. Where we refer to participants subsequently, this is by number (e.g. Participant 3 or P3).

The aim of the present study is to identify and articulate the perspectives of those participating in the practicum. This was done to gain insights into their motivations, assumptions and expectations concerning the use of visual arts to communicate climate change, particularly in a Pacific context.

Table I. Participants in the arts practicum and research interviews

Participant code	Area of expertise and professional background
P1	Professional graphic designer and climate change communications officer from Fiji. Works using photography and digital media
P2	Professional artist originally from Spain, resident in Fiji. Works using painting and sculpture
P3	Professional climate change adviser and doctoral student from Tonga, studying impacts of climate change on health and well-being in Tonga
P4	Professional artist from Fiji. Works using paint, lino prints and carving
P5	Professional climate change and bioenergy researcher, with experience exhibiting artworks using installations, painting and sculpture
P6	Professional artist from Fiji. Works using wood carving and painting
P7	Professional artist and art curator from Fiji. Works using metal sculptures, print making and wood carving
P8	Professional climate change officer working in local government, based in the UK
P9	Professional artist and art curator based in Fiji. Works using print making, wood and recycling "found" materials
P10	Professional artist from Fiji. Works using paint and wood carving

The first author of the present study conducted a series of ten individual interviews *in situ* towards the end of the arts practicum. We sought to examine subject matter ranging from the general (e.g., the role of art in communicating climate change) to the personal and specific (e.g., the role of people's own experiences in informing their approaches) to explore the range of influences upon, and objectives of, art-based communication. As such, the interviews were designed to address the following research questions:

RQ1. What are participants' personal motivations for engaging in climate change communication via the arts?

RQ2. How have participants' own experiences informed their approach to communicating climate change?

RQ3. What aspects of climate change are participants seeking to convey through their art; what are they aiming to achieve through the approaches adopted?

RQ4. What are considered by participants to be the advantages and drawbacks of using art for the communication of climate change?

RQ5. How do participants understand the tensions and interactions between arts and science-based ways of representing environmental problems such as climate change?

RQ6. What do participants consider to be the practical and cultural considerations relevant for the effective and appropriate communication of climate change in the Pacific?

The advantage of using research interviews to address these questions is that it enables us to examine in detail participants' subjective positions on the use of art to communicate climate change – including, but not limited to, the role of their own direct experiences, inspirations, expectations, motives and intentions. Given the centrality of subjectivity and the creative process in the generation of art (Lubart, 2001), we argue that participants' positions on these matters represent an important component within the broader topic of the use of artistic approaches in the communication of environmental issues (Klöckner, 2015;

Dunkley, 2014; Curtis, 2009). Nevertheless, the use of research interviews in this manner is not directly able to assess whether – and in which ways – the visual arts may influence audiences who encounter them, for example, by changing their attitudes towards climate change. A further limitation of the present study to which we draw attention is the relatively small sample size of 10 interviewees; while this has enabled diverse insights to be presented, we cannot consider these to represent an exhaustive set of practitioners' perspectives. We also note that interviewees have differing backgrounds; in particular, some interviewees bring to bear their lifetime experiences as Pacific Islanders, whereas others consider the research questions from a position as formal "experts" based in the region. Our focus throughout the analysis has been to draw out a range of viewpoints concerning key research themes, rather than to separate or compare these according to particular participant characteristics.

A semi-structured interview technique was used, enabling key themes and questions to be addressed, but allowing flexibility for more extensive elaboration or articulation of ideas in line with participants' own reflections and interests (Wolgemuth et al., 2015). Interviews lasted between 60 and 90 min and followed a protocol which began with general, open-ended questions and moved on to asking about more specific topics germane to the research questions. Participants were asked about their own background and interests, their points of view as to the role of the arts in communicating climate change, and their own experiences and expectations.

Participant interviews were recorded and subsequently transcribed for analysis. This was undertaken using template analysis, a form of thematic analysis entailing a systematic and structured approach to assessment of textual data (i.e. the interview transcripts) (King, 2012). Template analysis permits flexibility as regards the particular purpose of a study, allowing themes to be identified that relate both to *a priori* researcher categories – in this case, pertaining to the research questions of the present study – as well as salient themes that emerge from the data (King, 2012).

The content and aesthetic properties of artwork produced by participants for the Eco Arts project is not the principal focus of the present study. However, we refer to the details of individual artworks where this is relevant to draw out the context of a participant's comments, and to provide further background to the project.

Themes arising from research interviews

Here, we outline five themes derived from the analysis of interviews; within each of these we identify sub-themes in relation to the perspectives provided by project participants.

The primary themes identified relate closely to the research questions of the study, as well as salient topics emerging from discussions. We consider each of these in turn, providing supporting evidence and material in the form of participant quotes. Across these themes, we consider participants' perspectives on: the motivations and purposes of using visual art to communicate climate change; the concepts and ideas behind specific pieces of art; the benefits and limitations of using arts-based communication; and additional cultural and regional considerations relevant to participants' practice.

Motivations for undertaking arts communication

The first theme of interest across the interviews concerned participants' motivations for pursuing artistic approaches to the communication of climate change. Within this theme, we refer to participants' rationales for using visual arts in this context; the purposes for which arts communication is used, and the concepts and ideas applied by participants, are considered in subsequent themes as outlined below.

First, Participant 1 articulated his motivations in terms of a desire to highlight areas of life that he valued and that he considered could be lost or affected by climate change. In particular, this participant made reference to his sense that experiences and circumstances taken for granted in his own life were less certain in the future:

> I've been thinking about what I love and what I'll miss about it. For example, if you watch [a child] play on the beach, you think that in future people might not have that same opportunity.

In a similar manner, Participant 2 stressed her concern for those people across the Pacific who are especially vulnerable to climate change. In particular, this participant referred to her anxiety at the potential trauma for those in the islands of Kiribati at risk of losing their land and needing to relocate:

> Something that I'm really concerned about here is that in 50 years this country [Kiribati] may disappear. It's incredible. So the people will have to go to other places; they will lose their identity, their country. What about this psychologically? How will they feel? Imagine you live in Kiribati and someone told you that your country will disappear.

In many instances, participants referred to their own direct experiences of a changing climate in Fiji and the wider Pacific. Although not always articulated as directly motivating art, these were often spoken of with concern and as a preface to participants' interest in communicating and portraying climate change.

One striking example of such an occurrence was given by Participant 3, who discussed at some length the impact of the tropical cyclone that had hit the Ha'apai island group in Tonga, the participant's home country, earlier in the year. As well as a large part of these islands being devastated by this event, Participant 3 referred in particular to the more personal impacts upon people living there:

> So that means there's no home; the plantations were destroyed. So it's a misery. It's so miserable. People are so stressed they don't know how to do – what to do [...] You have no idea how stressed the people are. They are really stressed about climate change.

Other smaller-scale but no less personal accounts of the impacts of climate change (as perceived by participants) were also provided. Participant 1 talked both of the flooding of his aunt's house in Fiji and of a nearby town where several graveyards had been flooded by rising sea levels. Participant 4 similarly spoke of how the walkway he used to take to school in his village was now beneath the sea.

A personal concern for the direct impacts of climate change was referred to elsewhere within the research interviews, including in the context of the links between climate change and vulnerable livelihoods in the Pacific. At other times, however, participants referred to more pragmatic and professional motivations for developing artistic approaches. Participant 5, for example, indicated an interest in enabling others to make connections between the causes and consequences of climate change:

> A lot of the issues and the problems that are happening here are not made here. We're having to deal with other people's [actions] basically. So it would be great if people made those connections.

Besides expressing a motivation to use art to communicate climate change, there was the view expressed that the reverse held true, as articulated by Participant 9: "You could look at it the other way around as well – that climate change or environmental problems could trigger or could bring about good art".

A related perspective for developing artistic approaches in this area was offered by two of the Fijian artists involved in the project, in terms of their need to obtain recognition and, ultimately, income from their art. Participant 6 discussed their life-long interest in promoting traditional Fijian designs in their carving and painting, but also explained the need for their art to sell. Likewise, Participant 4 observed that "it's really hard to be an artist in Fiji [...] it's a struggle to do this". Given the difficulties of being able to manage financially, Participant 4 suggested that artistic interpretations of climate change provided an opportunity to achieve an income:

> We Fijians are only a few artists that have come up with the idea to work on communicating climate change [...] There are only a few people doing artworks on climate change. So maybe the demand [for] climate change in art may increase my market.

Purpose of art communication

In addition to these motivations, participants also articulated a range of more specific purposes and objectives regarding the use of art to communicate climate change.

Principal among these was an intention to raise awareness about the issue of climate change. This was mentioned by participants both in the context of raising general awareness – for example that climate change is "real" – and with respect to more specific issues such as the need to adapt to particular problems. Participant 4, for example, asserts that he sees his role as an artist to encourage people to "understand" the impacts of climate change.

Although participants at times presented a role for arts communication that was educational or instructive, at other times the use of art to communicate about climate change was presented as an opportunity to catalyse a process of thinking or reflection on the topic. For example, Participant 2 suggested that her art was intended as a form of storytelling, including in reference to a sculpture produced during the practicum, which addressed the stress on freshwater systems in the Pacific (this is discussed further below).

Participant 7 spoke in terms of using arts communication to prompt people to ask critical questions both about the role of human action in causing climate change, and ways of responding to it:

> I'd like to ask the questions, put it out there for people to think about climate change, think about how they are contributing to climate change, think about how they could respond.

The use of an artistic approach to inspire action was mentioned by participants in several places. Participant 3, for example, refers indirectly to adaptation where asserting that "communicating [...] through arts is a powerful way of convincing people to respond to the effects of climate change". Participant 4, likewise, suggests that as well as climate change art reflecting a sense of loss, this can also "push people to act at the same time".

Other participants referred to art having the potential to inspire positive responses and emotions in order to better address climate change. Participant 8, for example, mentions that "I would prefer to have a positive response out of any artwork [...] I don't feel that negative reactions are helpful, because people will just turn away from them". Similarly, Participant 1 (whose work also encompasses projects to help communities adapt to climate change) expresses the view that "hope" is a desirable response to seek from the use of arts communication:

> At the end of the day climate change is happening, so you can't really feel [angry] [...] But you can have hope [...] I do feel bitter about it sometimes, but that's not going to work. So it's just giving community members hope that you're doing this not just for now but for your future generation. So what we're doing is trying to give people hope, basically.

Concepts and ideas used in participants' artwork

Having considered what participants see as the more general purposes of using art to communicate climate change, we now turn to the types of concepts and ideas used in their own work. We focus in this section on participants' perspectives on four pieces of art produced during the practicum, each of which presented a different set of conceptual underpinnings in their communication of issues relating to climate change.

One of the artworks we consider here was a sculpture made from found objects, produced primarily by Participant 7. This was an approximately life-size model of an elderly man, assembled from plastic bottles and other wood and metal items found close to the island shore. The figure was designed to appear stooped and decrepit, and clutched a walking stick in one hand.

The participant explained that the sculpture was intended to represent a person who has damaged himself through living an unhealthy life; this was, in turn, presented as a metaphor for society harming itself through climate change as a result of its own actions. The sculpture is further contextualised to the Pacific in terms of illnesses and unhealthy habits endemic to the region:

> There are people who, the doctor says, should stop drinking because it is contributing to your sugar level. The guy [presented in the sculpture] comes home, listens to the doctor because he's in pain [but] after two weeks or so he starts feeling better, he starts taking a drink or two, or three, and then he continues until he gets to feel those pains again. Then the wife takes him to the doctors [. . .] When he feels better he repeats this same thing again [. . .] until the day he gets his leg amputated.

The relationship between this recalcitrant patient and the wider actions of society are explained via the sculpture in terms of a refusal to act in our own interests:

> For this sculpture, although we know that climate change will hit us, that it's hitting us, we won't – some of us won't [change] [. . .] I wanted to show how vulnerable we are to climate change, which we contribute to. Although we're getting weaker, weaker, we won't change our ways for the better. I wanted people to see that if we don't change, if we don't ask ourselves those questions, we don't – we may die, not improving our wellbeing as well as those of others around us.

Further parallels are drawn between the unhealthy character and the damaging aspects of society, again in the context of the Pacific, where Participant 7 relates poor food choices both to individual ill-health and wasteful production practices linked to climate change.

A different approach was taken by three of the project participants in the creation of a sand sculpture of a traditional village close to the sea edge, which was filmed using time-lapse video as the tide covered it. As Participant 6 explained, this enabled the creation of a small-scale simulation of the effects of climate change in a Fijian context:

> [There is] the shape of the Fijian *bure* [dwelling], the playing ground, the pig pen – a set-up of a real Fijian village. Sand is really easy to explain the message of sea rise. What will happen to a village? [. . .] I feel sorry for this village. Let's put ourselves in this situation. If it was a real village, this is exactly what will happen.

It is notable in this instance that although this simple model is based on a potential consequence of physical climate change impacts, Participant 6's reflections on the process were both emotional and personal:

> It really touched me because I was looking at the villages start to wash away, my village, and I can imagine people running away, taking – looking for shelter [. . .] The place where they were

Dunkley, R.A. (2014), "Reimagining a sustainable future through artistic events: a case study from Wales", in Moufakkir, O. and Perneck, T. (Eds), *Ideological, Social and Cultural Aspects of Events*, CAB International, Wallingford, pp. 100-109.

Duxbury, L. (2010), "A change in the climate: New interpretations and perceptions of climate change through artistic interventions and representations", *Weather, Climate and Society*, Vol. 2 No. 4, pp. 294-299.

Ellison, J. and Fiu, M. (2010), *Vulnerability of Fiji's Mangroves and Associated Coral Reefs to Climate Change: A Review*, WWF, Suva, Fiji.

Fløttum, K. and Gjerstad, Ø. (2017), "Narratives in climate change discourse", *Wiley Interdisciplinary Reviews: Climate Change*, Vol. 8 No. 1,

Gabrys, J. and Yusoff, K. (2012), "Arts, sciences and climate change: practices and politics at the threshold", *Science as Culture*, Vol. 21 No. 1, pp. 1-24.

Giannachi, G. (2012), "Representing, performing and mitigating climate change in contemporary art practice", *Leonardo*, Vol. 45 No. 2, pp. 124-131.

Hall, D., Bernacchi, L.A., Milstein, T.O. and Rai, T. (2009), "Calling all artists: moving climate change from my space to my place", in Endres, D., Sprain, L. and Peterson, T. (Eds), *Social Movement to Address Climate Change: Local Steps for Global Action*, Cambria, Amherst, pp. 53-80.

Haluza-DeLay, R. (2014), "Religion and climate change: varieties in viewpoints and practices", *Wiley Interdisciplinary Reviews: Climate Change*, Vol. 5 No. 2, pp. 261-279.

Hemstock, S., Des Combes, H.J., Martin, T., Vaike, F.L., Maitava, K., Buliruarua, L.A., Satiki, V., Kua, N. and Marawa, T. (2017), "A case for formal education in the Technical, Vocational Education and Training (TVET) sector for climate change adaptation and disaster risk reduction in the Pacific Islands region", in Filho, W. (Ed.), *Climate Change Adaptation in Pacific Countries*, Springer, Berlin, pp. 309-324.

Hoad, D. (2016), "The 2015 Paris climate agreement: outcomes and their impacts on small island states", *Island Studies Journal*, Vol. 11 No. 1, pp. 315-321.

King, N. (2012), "Doing template analysis", in Symon, G. and Cassell, C. (Eds), *Qualitative Organizational Research*, Sage, London, pp. 426-450.

Klöckner, C.A. (2015), *The Psychology of Pro-environmental Communication: Beyond Standard Information Strategies*, Palgrave Macmillan, London.

Knebusch, J. (2008), "Art and climate (change) perception", in Kagan, S. and Kirchberg, V. (Eds), *Sustainability: A New Frontier for the Arts and Cultures*, Verlag für Akademische Schriften, Frankfurt, pp. 242-261.

Koya, C.F. (2012), "In the absence of Land all we have is each other: art, culture & climate change in the Pacific", unpublished manuscript, available at: http://repository.usp.ac.fj/5755/ (accessed 12 September 2017).

Locke, J.T. (2009), "Climate change-induced migration in the Pacific Region: sudden crisis and long-term developments", *The Geographical Journal*, Vol. 175 No. 3, pp. 171-180.

Lubart, T.I. (2001), "Models of the creative process: past, present and future", *Creativity Research Journal*, Vol. 13 Nos 3/4, pp. 295-308.

McKenzie, B. (2008), "Cultural education for a changed planet", *Engage: The International Journal of Visual Art and Gallery Education*, Vol. 21, pp. 19-23.

McIver, L., Kim, R., Woodward, A., Hales, S., Spickett, J., Katscherian, D., Hashizume, M., Honda, Y., Kim, H., Iddings, S., Naicker, J., Bambrick, H., McMichael, A.J. and Ebi, K.L. (2016), "Health impacts of climate change in Pacific island countries: a regional assessment of vulnerabilities and adaptation priorities", *Environmental Health Perspectives*, Vol. 124 No. 11, pp. 1707-1714.

McLean, R. and Kench, P. (2015), "Destruction or persistence of coral atoll islands in the face of 20th and 21st century sea-level rise?", *Wiley Interdisciplinary Reviews: Climate Change*, Vol. 6 No. 5, pp. 445-463.

born and brought up the whole of their life, and it is covered by water. This is your village. It's under water. It's really sad. Especially for me, because I'm an islander.

Another sculpture produced, in this instance by Participant 2, was used to articulate concerns about freshwater use and risks to supply. The artist in this case used dead coral, twine and a metal tap (faucet) to produce a scene in which a dehydrated face and gasping mouth is shown trying to obtain a last drop of water. In conceiving of this piece, the artist explained her thought process in terms of a story of what occurs as water supplies run dry: "I think it's a story [...] the man [face] and many animals are fighting for the same drop, the final drop [...] the face has become dry, converting it to stone".

An artwork developed in the practicum in which elements of sculpture and photography were used, was designed by two participants to illustrate the relationship between consumerism and climate change. In this case, plastic soda bottles were filled with concrete and photographed being set alight against the backdrop of a well-known seascape in the capital Suva. According to the participants who worked on this (Participants 5 and 10), several concepts were incorporated: the parallels between consumption (in economic terms) and the bottles consumed by flames; the remnants of dirty carbon and pollutants from these processes; and the setting of the sea with its significance for island life, as well as the risks from rising sea levels.

Benefits and limitations of arts communication
The advantages of using artistic approaches to communicate climate change were typically articulated in terms of the unique capacity of art to incorporate a role for emotion, to speak to audiences in a different way to technical or scientific material, and for artistic approaches to be able to be left open to interpretation.

Participant 1 discussed the advantages of using art in the context of audiences who may be unfamiliar with, or unreceptive to, scientific communication:

> If you can use art and [...] present it in a way where you're driven with power and emotion and expression, people will be captivated [...] But if a scientist were to come and [talk], it's just going to fall on deaf ears, unless you're a scientist [yourself].

A distinction between the capacities of art and science to communicate climate change was also made by Participant 8, who argued similarly that:

> You've got to have a variety of ways of getting information over to different people. Art is one way. There are certain people that respond to [academic] papers and figures and statistics, and there are people that respond to getting involved with making art.

Again, the capacity for artistic approaches to address topics in ways that may be more accessible to those unfamiliar with scientific approaches and terminology – particularly in a Fijian context – was outlined by Participant 4:

> Actually, a lot of people don't understand what terms in science mean. So we normally understand English, but a lot of climate change things are coming through in terms of science [...] So I think villages, from my point of view as a villager, they understand more in art.

While the emotional component and potential for interpretation and provocation was referred to across the interviews, for Participant 4 this was also considered to have both benefits and drawbacks. In the latter sense, it was argued that differing perspectives on a piece of work inhibited the ability to convey a straightforward message about a topic such as climate change:

As an artist if you've got an idea to put inside an artwork, you don't know if that is the idea of somebody else [viewing the work] [. . .] So you can't use artwork to send your message through, unless you talk to him or you just write a piece of paper you stick on the side of your artwork [. . .] Because we have different views of what is going on inside that canvas.

The limitations and drawbacks of using artistic approaches were also seen often in more practical and pragmatic terms. In particular, several participants referred to the limited opportunities and space for audience interaction with artwork in Fiji and across the Pacific region. In this vein, Participant 1 argued: "Especially if we're talking about the Pacific and Fiji, the art community is so small, so you only reach a fraction of [the people] you want to reach".

Cultural and regional considerations
Some important features of culture in Fiji and the Pacific we have considered already – for example, the impacts already experienced and expected in this part of the world and how these were integrated in participants' artwork. It is also important to note that "culture" is not a phenomenon that can somehow be separated either from the lines of argument used by participants or people's experiences. Nevertheless, it is useful to consider here some additional aspects referred to directly by participants, in order to place the project and the role of arts communication in an appropriate cultural and regional context.

First, although the practicum itself and participants' approaches were focussed upon climate change, for many participants this held meaning and associations beyond the scientific conception of this term. In particular, across the interviews participants referred to their concerns about local pollution of the environment and their emphasis upon this in their art. The problem of plastic was a common thread referred to as requiring awareness and action, and as illustrative more broadly of a lamentable disregard shown for the environment – both on land and for the sea. Participant 7, for example, remarked that "for us in the village setting, with plastics you either burn it or you bury them [. . .] without realising that it's our plastic bags that kill the turtles". Likewise, Participant 6 indicated that a message he wished to convey through his art was "to minimise pollution, whatever you dump in the sea, be careful of dumping rubbish because the marine life you can damage, you can kill fish".

The importance of Christianity in the Pacific islands was also stressed by several participants. Participant 3 referred to his own experience that a large proportion of people in Tonga understood climate change to be under the control of God. Participant 6 also placed climate change firmly in a religious context, extending to the assertion that climate change is in this sense "normal" and to the need to pray:

I'm a very religious person. You know what? If you want to get away from this [tackle climate change] you'd better learn. Climate change – learn it properly and pray to God. Because this is all God's creation [. . .] [The bible says that] things are going to change slowly and surely and it marks a coming of Christ. So if [the climate] changes, from my point of view, it's normal, because it's been there in the bible. For a Christian, if you read your bible you'll know what's happening.

In an extension of his metaphor used to depict the "old man" sculpture, as discussed above, Participant 7 made broader reference to the tensions between traditional island, village life, and modern, urban living. In this sense, the issues relating to climate change, environmental problems and unhealthy or unwise practices, become bound up in the way of life of many people in contemporary Fiji as Participant 7 sees them:

If you live on an island, the trees will be growing; you'll always have fruit all year round. But most of us, despite this, will leave the islands. We'll go to the urban centres and we become like that old man. We drink excessively or we smoke excessively, we eat noodles and canned fish instead of fresh fish.

Perspectives such as these illustrate that the notion of "climate change" and indeed what constitutes art communication cannot be easily delineated. It may, however, be the case that interpretation and portrayal through the arts have a unique opportunity to consider where climate change and other environmental problems intersect with cultural, contemporary and regional issues.

Discussion

Through a series of research interviews undertaken with artists and climate change practitioners, the present study has uncovered a range of perspectives concerning the motivations, hopes and creativity of this group of individuals, who have set out to use artistic approaches in their communication of this pressing issue. We have outlined some of the ways in which the different facets of climate change – whether in terms of physical impacts or in relation to personal experiences or wider social contexts – have informed their points of view. The aims and aspirations of participants were explored, both in terms of the purposes for which they have developed artwork, and the underlying concepts and ideas on which they have drawn. Finally, we have set out to identify participants' views on the benefits and drawbacks of the use of visual arts to communicate climate change, and the cultural contexts within which this has occurred.

An important unifying feature across participant interviews was the way in which climate change is recognised to be a part of the contemporary reality of the Pacific. Whether through reflections upon small-scale, personal changes or by reference to current and future impacts at country and regional level, participants' experience and understanding of climate change informed and motivated their artistic approaches. Such a connection between people's "experience" of climate change and their subsequent beliefs and actions has now been observed more generally across a range of research studies (Reser *et al.*, 2014). Although we do not make strong claims here about a direct causal effect of experience, the present study points to a potentially important but, to date, overlooked role for experience in informing creative approaches and communication of climate change.

Several participants also identified a different set of reasons for engaging in these approaches. This may be summed up as a reversal of a premise that "art" should act in the service of climate change: for some participants, there was recognition of an opportunity instead for this prominent issue to inform their art, or to make it marketable. In the case of climate change inspiring interesting or provocative art, this has been discussed previously by writers such as Nurmis (2016) who stresses the value of climate change art as "essentially artistic" rather than serving an "instrumental" or activist function, such as promoting action or change in attitudes. Here, the role of the arts may also be conceived as enabling new types of understanding (Gabrys and Yusoff, 2012) or offering different ways of imagining and encountering the environment (Duxbury, 2010).

In instances where participants did affirm they wished to pursue particular purposes or ends, a common thread was to raise awareness and/or to prompt people to act on climate change. This view of environmental art as seeking to effect positive change is reflected often in the wider literature, to the extent that it may be seen even as a typical feature of the work of artists operating in this area. Dunkley (2014) for example has argued that a growing trend in arts-based movements in the UK has been an intention to enable a more sustainable society and way of life. Reflecting the position of many of our research participants, McKenzie (2008) has argued that art is invaluable as a form of "cultural education" in the context of climate change, serving such functions as communicating its likely impacts and helping to support people reducing their own carbon emissions.

As for the concepts applied within the artworks themselves, interviews with participants revealed rich and complex considerations of the relationships between individuals, communities, contemporary ways of life, and the consequences of climate change in the Pacific. In the case of the "old man" sculpture, for example, the role of individual action in causing climate change was recognised, but in terms that saw human behaviour and habits as incorporating tendencies to self-destruction and folly, that were not exclusive to environmental problems.

Both in the case of this piece of art, and other approaches that were presented in terms of storytelling and the use of metaphorical devices, can be found a potentially fruitful approach to engaging an audience with climate change. A series of studies now affirms the advantages of narrative approaches to communicate scientific concepts (Dahlstrom, 2014), including climate change (Fløttum and Gjerstad, 2017). Indeed, the advantages of using an artistic approach were presented typically as enabling different aspects of climate change to be communicated; in particular, to express and use emotion, and to connect with people and ideas in ways that formal, scientific communication could not. In this way, there exists the potential for wider discussions and interpretations to be enabled. The types of concerns articulated across the research interviews – the contradictions inherent within us, the sense of loss from a changing world, and our reliance on consumer society – would seem to be particularly amenable to artistic consideration.

As Crimmin (2008) argues, artists have often seen it as their role to question received notions of progress as well as the relationships between society and the environment. Notions of progress – and the alternatives that may be left behind – were indeed raised directly during the course of the research interviews in the present study and used to highlight aspects of Fijian culture pertaining to the relationship between people and their environment.

It is important, finally, to note the centrality of Christianity to Fijian and wider Pacific culture, and the likely implications this holds for understanding and addressing climate change. Although none of the artwork produced during the practicum had an overtly religious basis, references were made by participants to their own and others' Christian perspectives. Given the implicit ethical and moral dimensions to climate change, including with respect to personal responsibilities and duty towards the natural environment (Posas, 2007), this connection may be a valuable area for exploration in future research and arts practice. The argument for a closer consideration of the role of religion (including but not limited to Christianity) in tackling climate change in the Pacific islands has been taken up by Haluza-DeLay (2014) who suggested that the role of faith is important for understanding how people make sense of climate change, and for developing effective communication in this area.

The present study has focussed upon the perspectives of artist-practitioners, rather than the reaction or interpretation of an audience to this manner of climate change communication. While this is itself a limitation to the research, we suggest that important questions are nevertheless raised by the arguments put forward by the artists, which warrant attention in future research with audiences. In particular, many of the interviewees assert a clear intention to raise awareness, promote understanding, or to convince people to respond to climate change. Despite these aspirations, it cannot be assumed, however, that the artwork produced had the potential to bring about such effects. Indeed, there may be a disjuncture between the desire to inspire understanding and positive responses, and the typical tenor of pieces produced, which for the most part emphasised and represented the

negative or upsetting consequences of climate change. Although the portrayal of harmful climate impacts is commonplace more generally in the communication of climate change, including in visual formats (O'Neill and Smith, 2014), it remains far from clear whether this inspires action or may even be counter-productive (Moser and Dilling, 2012; O'Neill and Nicholson-Cole, 2009). With respect to portrayals of climate change within the visual arts, in particular, Nurmis (2016: 501) expresses alarm about a tendency to utilise "apocalyptic" imagery; this author argues that this "results in art that may be poignant, but falls out of step with the professed motivations of artists". A clear need for future research, therefore, is an assessment of the consequences of artistic communication for those who encounter it, in order to ascertain its efficacy or lack thereof.

Even in a situation where arts communication approaches are able to engage both artists and audiences, a broader question remains as to the potential for such approaches to be utilised and expanded in a practical sense. While it has been argued that the arts thrive in Pacific communities, this has often been in a manner that is poorly integrated with development funds and objectives (Teaiwa and Huffer, 2017).

A recent approach that has sought to integrate the visual arts in a Pacific community context, with these trialled in Vanuatu, is the European Union-funded Pacific Vocational Education and Training in Sustainable Energy and Climate Change Adaptation project (EU PacTVET; Hemstock *et al.*, 2017; Bartlett, 2016). This project has developed teaching resources that aim to explore and apply the perceived personal experiences that learners have of climate change, in order to inform their behavioural responses (such as installing water tanks for rainwater collection) or beliefs (such as whether climate change is a human-induced or supernatural phenomenon). It is hoped that the inclusion of creative approaches to climate change communication in a formal educational environment may yield positive results in this current and ongoing programme of work. In this way, the types of perspectives and insights revealed in the present study may have the potential to offer broader implications for communities in the region.

Note

1. See www.usp.ac.fj/news/story.php?id=1782 for archive of this exhibition's opening.

References

Barnett, J. and O'Neill, S. (2010), "Maladaptation", *Global Environmental Change*, Vol. 20 No. 2, pp. 211-213.

Bartlett, C. (2016), "SPC-GIZ coping with climate change in the Pacific Islands region (CCCPIR): Vanuatu programme annual report", available at: www.nab.vu/document/2016-spc-giz-cccpir-annual-report (accessed 18 September 2017).

Buckland, D. (2012), "Climate is culture", *Nature Climate Change*, Vol. 2 No. 3, p. 137.

Cochrane, S. (2016), "Art in the contemporary Pacific", *Junctures: The Journal for Thematic Dialogue* Vol. 17, pp. 78-95.

Crimmin, M. (2008), "Treading lightly", *Engage: The International Journal of Visual Art and Gallery Education*, Vol. 21, pp. 24-28.

Curtis, D. (2009), "Creating inspiration: the role of the arts in creating empathy for ecological restoration", *Ecological Management & Restoration*, Vol. 10 No. 3, pp. 174-184.

Dahlstrom, M.F. (2014), "Using narratives and storytelling to communicate science with nonexpert audiences", *Proceedings of the National Academy of Sciences*, Vol. 111 No. S4, pp. 13614-13620.

Dunkley, R.A. (2014), "Reimagining a sustainable future through artistic events: a case study from Wales", in Moufakkir, O. and Perneck, T. (Eds), *Ideological, Social and Cultural Aspects of Events*, CAB International, Wallingford, pp. 100-109.

Duxbury, L. (2010), "A change in the climate: New interpretations and perceptions of climate change through artistic interventions and representations", *Weather, Climate and Society*, Vol. 2 No. 4, pp. 294-299.

Ellison, J. and Fiu, M. (2010), *Vulnerability of Fiji's Mangroves and Associated Coral Reefs to Climate Change: A Review*, WWF, Suva, Fiji.

Fløttum, K. and Gjerstad, Ø. (2017), "Narratives in climate change discourse", *Wiley Interdisciplinary Reviews: Climate Change*, Vol. 8 No. 1,

Gabrys, J. and Yusoff, K. (2012), "Arts, sciences and climate change: practices and politics at the threshold", *Science as Culture*, Vol. 21 No. 1, pp. 1-24.

Giannachi, G. (2012), "Representing, performing and mitigating climate change in contemporary art practice", *Leonardo*, Vol. 45 No. 2, pp. 124-131.

Hall, D., Bernacchi, L.A., Milstein, T.O. and Rai, T. (2009), "Calling all artists: moving climate change from my space to my place", in Endres, D., Sprain, L. and Peterson, T. (Eds), *Social Movement to Address Climate Change: Local Steps for Global Action*, Cambria, Amherst, pp. 53-80.

Haluza-DeLay, R. (2014), "Religion and climate change: varieties in viewpoints and practices", *Wiley Interdisciplinary Reviews: Climate Change*, Vol. 5 No. 2, pp. 261-279.

Hemstock, S., Des Combes, H.J., Martin, T., Vaike, F.L., Maitava, K., Buliruarua, L.A., Satiki, V., Kua, N. and Marawa, T. (2017), "A case for formal education in the Technical, Vocational Education and Training (TVET) sector for climate change adaptation and disaster risk reduction in the Pacific Islands region", in Filho, W. (Ed.), *Climate Change Adaptation in Pacific Countries*, Springer, Berlin, pp. 309-324.

Hoad, D. (2016), "The 2015 Paris climate agreement: outcomes and their impacts on small island states", *Island Studies Journal*, Vol. 11 No. 1, pp. 315-321.

King, N. (2012), "Doing template analysis", in Symon, G. and Cassell, C. (Eds), *Qualitative Organizational Research*, Sage, London, pp. 426-450.

Klöckner, C.A. (2015), *The Psychology of Pro-environmental Communication: Beyond Standard Information Strategies*, Palgrave Macmillan, London.

Knebusch, J. (2008), "Art and climate (change) perception", in Kagan, S. and Kirchberg, V. (Eds), *Sustainability: A New Frontier for the Arts and Cultures*, Verlag für Akademische Schriften, Frankfurt, pp. 242-261.

Koya, C.F. (2012), "In the absence of Land all we have is each other: art, culture & climate change in the Pacific", unpublished manuscript, available at: http://repository.usp.ac.fj/5755/ (accessed 12 September 2017).

Locke, J.T. (2009), "Climate change-induced migration in the Pacific Region: sudden crisis and long-term developments", *The Geographical Journal*, Vol. 175 No. 3, pp. 171-180.

Lubart, T.I. (2001), "Models of the creative process: past, present and future", *Creativity Research Journal*, Vol. 13 Nos 3/4, pp. 295-308.

McKenzie, B. (2008), "Cultural education for a changed planet", *Engage: The International Journal of Visual Art and Gallery Education*, Vol. 21, pp. 19-23.

McIver, L., Kim, R., Woodward, A., Hales, S., Spickett, J., Katscherian, D., Hashizume, M., Honda, Y., Kim, H., Iddings, S., Naicker, J., Bambrick, H., McMichael, A.J. and Ebi, K.L. (2016), "Health impacts of climate change in Pacific island countries: a regional assessment of vulnerabilities and adaptation priorities", *Environmental Health Perspectives*, Vol. 124 No. 11, pp. 1707-1714.

McLean, R. and Kench, P. (2015), "Destruction or persistence of coral atoll islands in the face of 20th and 21st century sea-level rise?", *Wiley Interdisciplinary Reviews: Climate Change*, Vol. 6 No. 5, pp. 445-463.

McNamara, K.E. (2013), "Taking stock of community-based climate-change adaptation projects in the Pacific", *Asia Pacific Viewpoint*, Vol. 54 No. 3, pp. 398-405.

McNaught, R., Warrick, O. and Cooper, A. (2014), "Communicating climate change for adaptation in rural communities: a Pacific study", *Regional Environmental Change*, Vol. 14 No. 4, pp. 1491-1503.

Mimura, N. (1999), "Vulnerability of island countries in the South Pacific to sea level rise and climate change", *Climate Research*, Vol. 12 Nos 2/3, pp. 137-143.

Moser, S.C. (2014), "Communicating adaptation to climate change: the art and science of public engagement when climate change comes home", *Wiley Interdisciplinary Reviews: Climate Change*, Vol. 5 No. 3, pp. 337-358.

Moser, S.C. and Dilling, L. (2012), "Communicating climate change: closing the science-action gap", in Norgaard, R., Schlosberg, D., Dryzek, J. (Eds), *Oxford Handbook of Climate Change and Society*, Oxford University Press, Oxford, pp. 161-174.

Mortreux, C. and Barnett, J. (2009), "Climate change, migration and adaptation in Funafuti, Tuvalu", *Global Environmental Change*, Vol. 19 No. 1, pp. 105-112.

Nunn, P.D. (2012), "Climate change and Pacific Island countries", Asia Pacific Human Development Report, Background Papers Series 2012/07, United Nations Development Programme.

Nurmis, J. (2016), "Visual climate change art 2005–2015: discourse and practice", *Wiley Interdisciplinary Reviews: Climate Change*, Vol. 7 No. 4, pp. 501-516.

O'Neill, S. and Nicholson-Cole, S. (2009), "Fear won't do it" promoting positive engagement with climate change through visual and iconic representations", *Science Communication*, Vol. 30 No. 3, pp. 355-379.

O'Neill, S.J. and Smith, N. (2014), "Climate change and visual imagery", *Wiley Interdisciplinary Reviews: Climate Change*, Vol. 5 No. 1, pp. 73-87.

PIDF (2015), "The Suva declaration", available at: http://pacificidf.org/wp-content/uploads/2016/02/ecopy-Declaration.pdf (accessed 10 March 2017).

Posas, P.J. (2007), "Roles of religion and ethics in addressing climate change", *Ethics in Science and Environmental Politics*, Vol. 9, pp. 31-49.

Randerson, J. (2007), "Between reason and sensation: antipodean artists and climate change", *Leonardo*, Vol. 40 No. 5, pp. 442-448.

Reser, J.P., Bradley, G.L. and Ellul, M.C. (2014), "Encountering climate change: 'seeing' is more than 'believing'", *Wiley Interdisciplinary Reviews: Climate Change*, Vol. 5 No. 4, pp. 521-537.

Savo, V., Lepofsky, D., Benner, J.P., Kohfeld, K.E., Bailey, J. and Lertzman, K. (2016), "Observations of climate change among subsistence-oriented communities around the world", *Nature Climate Change*, Vol. 6 No. 5, pp. 462-473.

Sommer, L. and Klöckner, C. (2017), "What characterises climate change related art that activates the audience? A study on the ArtCOP21 event in Paris", paper presented at the International Conference on Environmental Psychology, A Coruña, Spain, 30 August-1 September.

Stratford, E. and Low, N. (2015), "Young islanders, the meteorological imagination and the art of geopolitical engagement", *Children's Geographies*, Vol. 13 No. 2, pp. 164-180.

Taylor, S. and Kumar, L. (2016), "Global climate change impacts on pacific islands terrestrial biodiversity: a review", *Tropical Conservation Science*, Vol. 9 No. 1, pp. 203-223.

Teaiwa, K. and Huffer, E. (2017), "Structuring the culture sector in the Pacific Islands", in Stupples, P. and Teaiwa, K. (Eds), *Contemporary Perspectives on Art and International Development*, Routledge, New York, NY, pp. 64-80.

van Renssen, S. (2017), "The visceral climate experience", *Nature Climate Change*, Vol. 7 No. 3, pp. 168-171.

Watanabe, F. (2013), "Red wave art in Oceania", *Global Ethnographic*, Vol. 1, pp. 1-11.

Weir, T., Dovey, L. and Orcherton, D. (2016), "Social and cultural issues raised by climate change in Pacific Island countries: an overview", *Regional Environmental Change*.

Whitmarsh, L., Lorenzoni, I. and O'Neill, S. (2012), *Engaging the Public with Climate Change: Behaviour Change and Communication*, Routledge, London.

Wolgemuth, J.R., Erdil-Moody, Z., Opsal, T., Cross, J.E., Kaanta, T Dickmann, E.M. and Colomer, S. (2015), "Participants' experiences of the qualitative interview: considering the importance of research paradigms", *Qualitative Research*, Vol. 15 No. 3, pp. 351-372.

Corresponding author

Stuart Capstick can be contacted at: capsticksb@cardiff.ac.uk

7

Characterization of European cities' climate shift – an exploratory study based on climate analogues

Guillaume Rohat

Institute for Environmental Sciences, University of Geneva, Geneva, Switzerland and Faculty of Geo-Information Science and Earth Observation, University of Twente, Enschede, The Netherlands

Stéphane Goyette

Institute for Environmental Sciences, University of Geneva, Geneva, Switzerland, and

Johannes Flacke

Faculty of Geo-Information Science and Earth Observation, University of Twente, Enschede, The Netherlands

Abstract

Purpose – Climate analogues have been extensively used in ecological studies to assess the shift of ecoregions due to climate change and the associated impacts on species survival and displacement, but they have hardly been applied to urban areas and their climate shift. This paper aims to use climate analogues to characterize the climate shift of cities and to explore its implications as well as potential applications of this approach.

Design/methodology/approach – The authors propose a methodology to match the current climate of cities with the future climate of other locations and to characterize cities' climate shift velocity. Employing a sample of 90 European cities, the authors demonstrate the applicability of this method and characterize their climate shift from 1951 to 2100.

Findings – Results show that cities' climate shift follows rather strictly north-to-south transects over the European continent and that the average southward velocity is expected to double throughout the twenty-first century. These rapid shifts will have direct implications for urban infrastructure, risk management and public health services.

Originality/value – These findings appear to be potentially useful for raising awareness of stakeholders and urban dwellers about the pace, magnitude and dynamics of climate change, supporting identification of the future climate impacts and vulnerabilities and implementation of readily available adaptation options, and strengthening cities' cooperation within climate-related networks.

Keywords Awareness-raising, Climate analogues, Climate shift, Climate velocity, Urban adaptation

1. Introduction

It is by now widely acknowledged that climate change will pose significant threats to both urban systems and city dwellers (Bulkeley, 2013). Because urban areas hold more than half of the world's population and most of people's assets, it is of utmost importance to define adequate adaptation strategies (Lee and Lee, 2016). Their strict implementation at the urban level is supposed to significantly reduce the inhabitants' vulnerability to climate change and ensure the quality of life for future generations. Nevertheless, despite an overwhelming scientific evidence of increasing climatic threats, urban adaptation strategies are more often absent than present, even in countries of the global North. Although a certain number of cities self-reported to be actively engaged in climate adaptation and mitigation at the local scale (Aylett, 2015), Reckien *et al.* (2013) found that 72 per cent of 200 European major cities have not yet implemented a climate adaptation plan. Such lack of political commitment is explained by numerous factors (Juhola, 2016), including insufficient funding, time-scale mismatches between political mandate and climate change (Bicknell *et al.*, 2009; Hallegatte, 2009), underlying uncertainties of climate projections (Schneider, 2006) and misunderstanding of the forthcoming climate impacts (Van der Linden *et al.*, 2014). Moreover, among the great number of factors identified as drivers of urban adaptation planning (Reckien *et al.*, 2015), efficient and easy-to-understand scientific information and knowledge (Archie *et al.*, 2014; Mycoo, 2015), involvement in climate-related cities' networks and strong community engagement (Bulkeley *et al.*, 2011) are identified to play an important role. Consequently, there is a growing need of new and innovative methods that: (i) raise urban residents and stakeholders' awareness about the potential impacts of climate change; (ii) provide easily understandable scientific information about the future impacts and adequate adaptation options; (iii) foster cities' collaboration within climate-related networks.

The climate analogues approach has the potential to address this need. This method – also known as the "climate twins approach" (Ungar *et al.*, 2011) – is designed to match the future (or past) climate of a given location with the current climate of another location. This way, a pair of climate analogues is made of two different geographical locations sharing a significantly similar climate for a different time period. Such approach has been initially developed in the field of ecological studies, with the purpose of investigating climate change impacts on the shift of ecological communities and species habitat and the appearance of novel climate and ecoregions (Saxon *et al.*, 2005; Peacock and Worner, 2006; Williams and Jackson, 2007, Veloz *et al.*, 2012a, 2012b), as well as the implications of such shift for species' survival and abundance (Anderson *et al.*, 2013; Leibing *et al.*, 2013). Climate analogues have also been used in agricultural studies to identify potential cultivars better suited to future climatic conditions (Webb *et al.*, 2013) and to investigate adaptation solutions existing today, based on the assumption that the future of one farmer is similar to the present of another one, located in a different region (Ramirez-Villegas *et al.*, 2011).

This approach has also shown a great potential for raising awareness about the magnitude and pace of climate change. For instance, Ungar *et al.* (2011), CSIRO-Bureau of Meteorology (2016) and Rohat *et al.* (2016) developed user-friendly climate analogue tools which provide an intuitive visualization of potential climate change impacts. In the same line, Kopf *et al.* (2008) and Climate Communication (2014) used climate analogues to communicate about the amplitude of climate change to a lay audience, whereas Beniston (2013) matched the past and current climates to provide easy-to-understand information about the celerity of climate change in the past decades.

However, the application of this approach in urban areas has largely been underused so far. The few climate analogues studies focusing on cities have shown that climate analogues can help assessing economic damages of climate change (Hallegatte *et al.*, 2007) and identifying both adequate adaptation policies (Kellett *et al.*, 2011) and best practices of climate adaptation (Rohat *et al.*, 2016). Nevertheless, none of these studies used climate analogues to characterize the velocity of cities' future climate shift – i.e. the speed and orientation of the geographical displacement over time – and to explore its potential implications on urban dwellers and on the design of adaptation strategies.

In this interdisciplinary effort, we propose a climate-matching method that reliably matches the current and future climates of any location worldwide, and we show how it can be used to assess the associated shift velocity. Employing a large sample – 90 different cities – we exemplify the applicability of this method and characterize the climate shift of European cities from 1951 to 2100. We then discuss the potential implications of such cities' climate shift and provide insights into the possible use of the proposed approach, e.g. for raising awareness of both city dwellers and decision-makers about the pace, magnitude, and dynamics of climate change, for supporting the identification and implementation of adequate adaptation strategies, and for enhancing cities' cooperation within transnational climate-related networks.

2. Methods and materials

2.1 Climate-matching approach

In the past few years, two main methods to match one climate with another have been described. One is based on the aggregation of different climate statistics within a similarity index – e.g. the CCAFS index (Climate Change, Agriculture and Food Security; Ramirez-Villegas *et al.*, 2011; Leibing *et al.*, 2013) or a simpler index using the standardized Euclidean distances (SEDs) (Williams and Jackson, 2007; Veloz *et al.*, 2012a) – whereas the other is based on a comparison between a set of univariate climatic criteria and a set of arbitrarily established thresholds (Hallegatte *et al.*, 2007; Ungar *et al.*, 2011; Rohat *et al.*, 2016). While the latter allows an easy control of the climate analogues' quality – in terms of climatic proximity – the use of a similarity index allows ranking them and hence identifying the climatically closest one. Nevertheless, Grenier *et al.* (2013) showed that the uncertainty associated with the choice of climate models and scenarios is largely superior to the variation resulting from the use of different climate-matching approaches.

In this study, we applied a combination of the two foregoing methods to:

(1) match the climate of any location of interest (LOI) with other locations sharing similar climate – but at a different time period – which we named the LOI's climate analogues; and

(2) determine the best climate analogue – i.e. the one sharing the most similar climate – for a given LOI and time period.

Although climate has been traditionally characterized by a specific combination of various variables (IPCC, 2001), matching one climate with another requires relaxing this definition. Because the climate-matching method developed in this study is used to investigate cities' climate shift, we took into account climate variables that both represent the overall climate and have a major influence on the functioning of urban areas. Trade-offs have to be made between including the numerous climatic variables that determine a city's climate and keeping low the number of climatic variables to identify a substantial number of climate analogues. This led us to select the five following quantities: monthly mean temperature and monthly mean precipitation, which are the two most essential climatic determinants (Holdridge, 1947); monthly minimal temperature for winter months (December, January and February) and monthly maximal temperature for summer months (June, July and August), which are, respectively, the indicators of cold and warm spells (Ungar *et al.*, 2011); and annual total precipitation, which is an important climatic factor for water management in cities (Hallegatte *et al.*, 2007). These variables were computed monthly (or annually in case of the annual total precipitation variable) and averaged over five 30-year periods, namely, P1 (1951-1980), P2 (1981-2010), P3 (2011-2040), P4 (2041-2070) and P5 (2071-2100).

To identify the climate analogues of a given LOI and time period, we first computed and averaged (as per grid points in the computational domain) the Euclidean distances between the LOI's current climate (P1) and the future climate (P2, P3, P4 or P5) of all the grid points, for the five climatic variables (methodology available as Appendix). Second, we compared the averaged Euclidean distances (five per grid points) with specific thresholds. We arbitrarily fixed these thresholds at 1°C for the three temperature variables and 25 per cent of the LOI's mean value (over the reference time period) for the two precipitation variables. If the averaged Euclidean distances for the five climate variables are under their respective thresholds, then the grid point's future climate is considered as similar to the current LOI's climate. Third, we applied an altitude filter to select the grid points that are located within a 200-meter range (above or below) of the LOI's altitude. Although applying such altitude filter is uncommon in climate analogues studies (Hallegatte *et al.*, 2007; Beniston, 2014), we argue here that it enables a more precise computation of the velocity of latitudinal climate shifts (Section 2.4). The remaining grid points – i.e. those which share significantly similar climate to the LOI and which have passed through the altitude filter – are considered as the LOI's climate analogues (for a given future time period). Finally, we computed their similarity index based on an unweighted SED metric (Appendix) and ranked those to identify the best one, in terms of climatic proximity. Such workflow (Figure 1) is repeated for every LOI and for each of the four 30-year time periods (i.e. P2, P3, P4 and P5).

2.2 Climatic data

Data sets for the case study presented in this paper were extracted from the European project ENSEMBLES (2009), which provides daily values at a horizontal grid-spacing of 25 km, from 1951 to 2100, under the A1B scenario of Special Report on Emission Scenarios of the Intergovernmental Panel on Climate Change. To reduce the uncertainties associated with the use of a single regional climate model (RCM), we computed multimodel means of the five climatic variables used in the climate-matching method. Climatic projections originated from seven different RCMs, namely, CNRM-RM4.5 (CNRM, 2008), KNMI-RACMO2 (van Meijgaard *et al.*, 2008), OURANOS-MRCC4.2.1 (Plummer *et al.*, 2006), SMHI-RCA3 (Kjellström *et al.*, 2005), DMI-HIRHAM5 (Christensen *et al.*, 2007), GKSS-CCLM4.8 (Böhm *et al.*, 2006) and METEO-HC-HadRMQ0 (Collins *et al.*, 2006). The five climatic variables were computed monthly (and annually for the variable of annual total precipitation) for the five 30-year periods and for all the grid points (32,300 in total) of the 25-km grid-spacing computational domain.

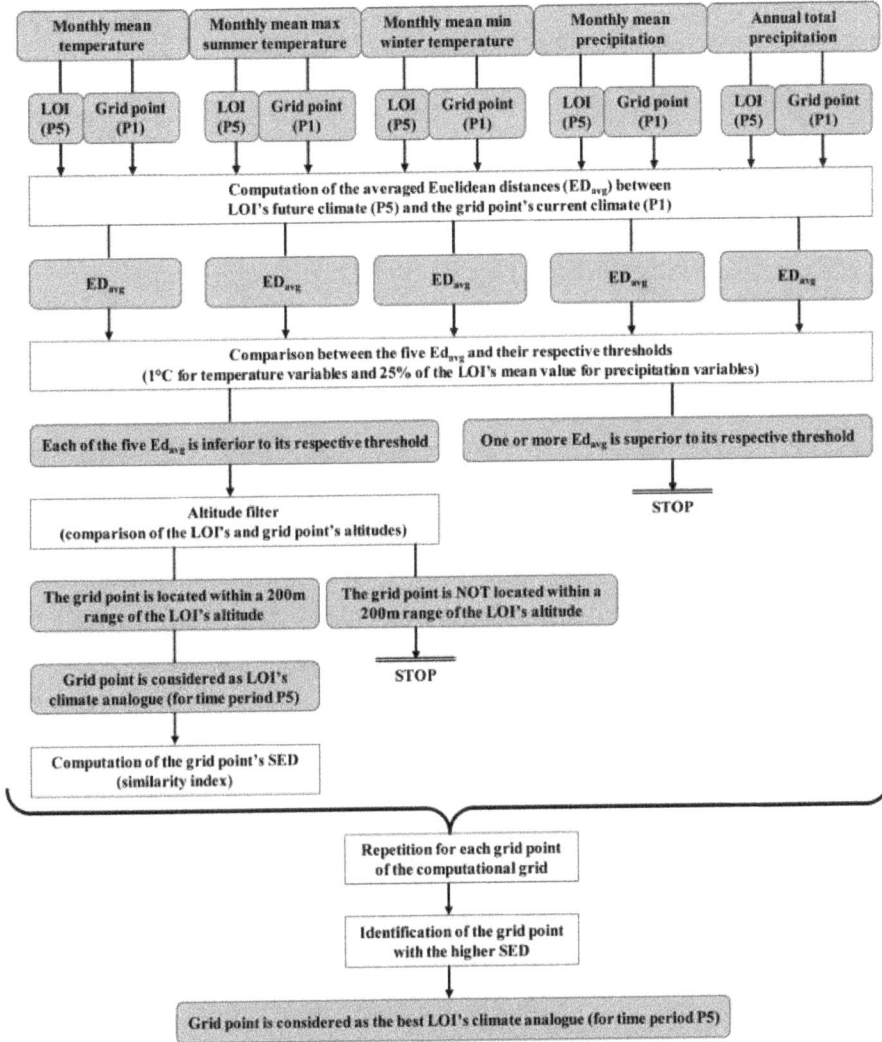

Note: Here, climate shift from P1 to P5 as an example

Figure 1. Schema of the workflow that has been applied to identify the best – in terms of climatic proximity – climate analogue of each LOI, for each shift time period

2.3 Transects

According to studies applying the Köppen climate classification in Europe (de Castro *et al.*, 2007; Gerstengarbe and Werner, 2008), the European historical climate is represented by a temperate climate in Western Europe, a continental climate in Eastern Europe and a subtropical climate in the Southern part. Jylhä *et al.* (2010) recently showed that European climates tend to move northeastwards. In this study, there is no attempt to assess the shift of European climatic zones, but rather the positional shift – mainly southwards – of European cities' climate. Beniston (2013) showed that European isotherms have been moving northwards in the past decades, along several north-to-south transects. In the same line and following the existing studies assessing the European climate shift (Jylhä *et al.*, 2010; Beniston, 2013, 2014, 2015), we developed three north-to-south transects, namely an Eastern Europe transect, a Continental transect, and a Maritime transect (Figure 2). These allow investigating the potential differences of climate shift over the European continent. Each of these is made of 30 different cities that have been chosen with regard to both their geographical location (proximity with a selected transect and distance with other cities) and regional importance (size of the population and administrative role). Overall, these 90 cities (Appendix Table AI) are located across 22 European countries, are distributed within several climatic zones, and host approximately 416 million inhabitants, i.e. more than half of the European population (Eurostat, 2012).

2.4 Southward velocity

To assess the southward climate shift velocity – i.e. the speed (in kilometres per year) of the expected southward positional change – of each city of the three transects, we first

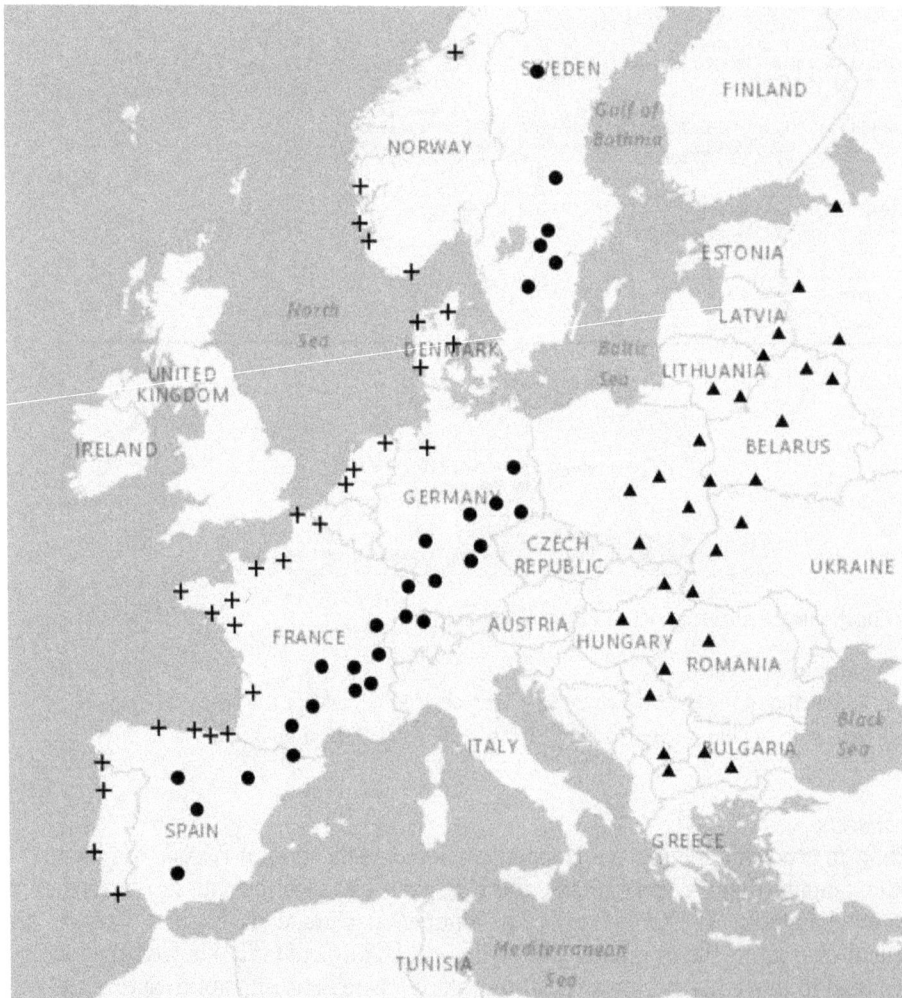

Figure 2. Map displaying the location of the 90 cities forming the three north-to-south transects, namely, Maritime transect (+), Continental transect (●) and Eastern Europe transect (▲)

computed the latitudinal distance between the city and its best climate analogue (for each 30-year time period), using the Haversine formula (Sinnott, 1984). We then divided the latitudinal distance by the number of years between the reference period and the projected period, which can vary from 30 years up to 120 years. Applying this method, we estimated the southward velocity of every city's climate shift for the seven following shifts:

(1)   P1-P2: From 1951-1980 to 1981-2010 (30-year shift).

(2)   P2-P3: From 1981-2010 to 2011-2040 (30-year shift).

(3)   P3-P4: From 2011-2040 to 2041-2070 (30-year shift).

(4)   P4-P5: From 2041-2070 to 2071-2100 (30-year shift).

(5)   P1-P3: From 1951-1980 to 2011-2040 (60-year shift).

(6)   P3-P5: From 2011-2040 to 2071-2100 (60-year shift).

(7)   P1-P5: From 1951-1980 to 2071-2100 (120-year shift).

3. Results

3.1 Applicability of the method

Out of the 360 different attempts (90 cities and four future 30-year time periods) to identify a climate analogue, 304 were successful (success rate of 84 per cent), highlighting the applicability of this method over the European continent. Among the 90 investigated cities, 70 cities were found to have reliable climate analogues for each of the four 30-year future time periods. For the other 20 cities, no climate analogue was found for at least one future time period. Among them, two cities, namely Geneva (Switzerland) and Sofia (Bulgaria), did not have any climate analogues for the four future time periods. Most of these 20 cities are located at the edge of the European domain; hence their respective climate analogues are presumably located outside Europe. For instance, climate analogues of the cities located in the Iberian Peninsula (extreme south of the computational grid), e.g. Vigo (Spain), Faro (Portugal), and Porto (Portugal), are presumably located in North Africa. However, for other cities such as Geneva (Switzerland), no climate analogue was found simply because its future climate does not currently exist in Europe. This may be because of the appearance of novel climates in a changing climate context (Williams and Jackson, 2007).

3.2 Direction of shifts

Results showed that climate analogues are always located southwards of their respective city of reference. This rather expected result highlights the well-known equator-ward displacement of the climatic zones (i.e. from north to south in case of Europe). One of the added values of the method as applied here lies in the identification of transects of climate shift. To assess whether or not European cities' climate shift follows the three predetermined transects, we computed the longitudinal distance between each climate analogue and their reference transect (Maritime, Continental, or Eastern Europe). Results showed that the longitudinal distance between climate analogues and their reference transect ranges from 0 to 218 km (68 km in average) for climate analogues of the Continental transect, from 0 to 715 km (88 km in average) for the ones of the Maritime transect, and from 0 to 428 km (average of 75 km) for the ones of the Eastern Europe transect. In addition to this great spatial proximity – in terms of longitudinal distance – between climate analogues and their reference transect, spatial analysis of the results showed that these three transects of climate shift very rarely overlap with each other (Figure 3). This emphasizes the future north-to-south transect-oriented shift of European cities' climate.

3.3 Speed of southward velocity

Table I summarizes the main findings resulting from the southward velocity computation carried out for European cities' climate shift, for the three north-to-south transects and for all the shift time periods (see Appendix Table AII for all detailed results). Overall, the southward velocity of European cities' climate shift greatly differs depending on both their geographical location and the shift time period.

Among the four 30-year shifts, the slowest southward velocity was found for the cities of Jönköping (Sweden) and Cordoba (Spain), with a speed of 0.9 km year^{-1} for the P2-P3 shift.

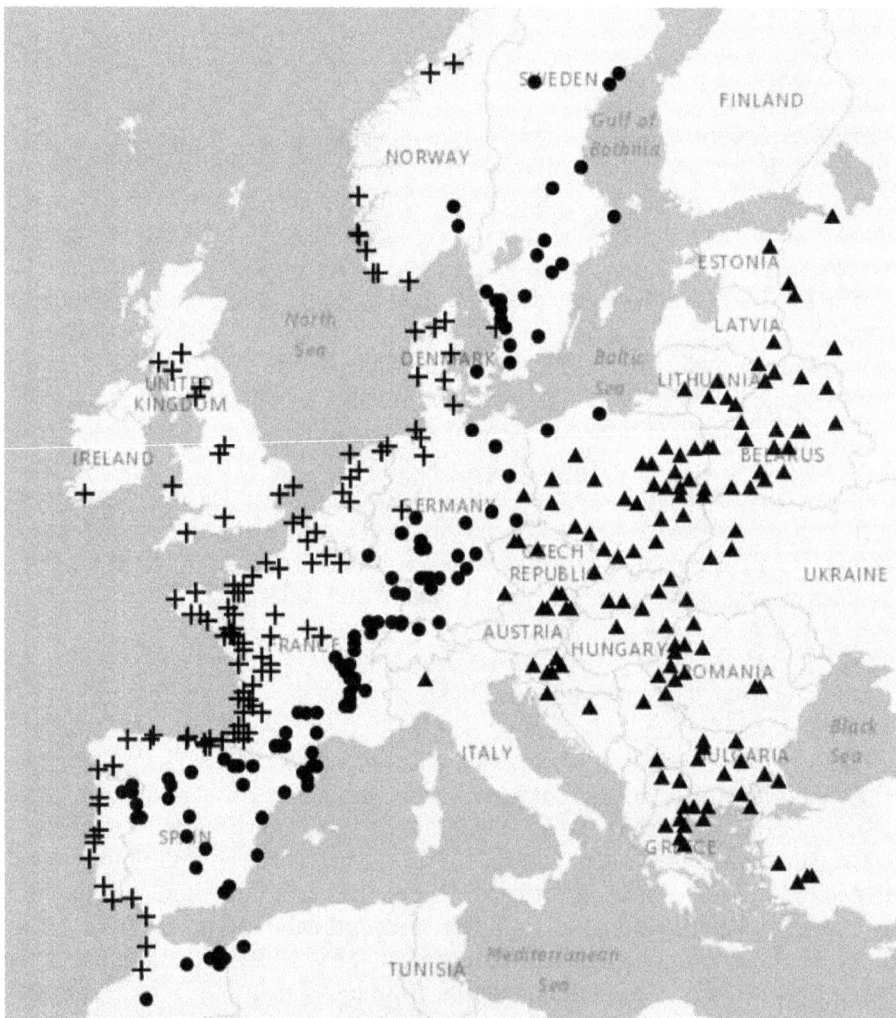

Figure 3. Map displaying the investigated cities and their respective climate analogues for the four future 30-year time periods, according to the Maritime (+), Continental (•) and Eastern Europe (▲) transects

Zurich (Switzerland) and Cracow (Poland) exhibited the fastest southward velocity, 34.0 and 38.1 km year^{-1}, respectively, for the P4-P5 shift.

Throughout the entire study period P1-P5 (i.e. from 1951-1980 to 2071-2100), Andorra-la-Vella (Andorra) and Berlin (Germany) showed the slowest (2.9 km year^{-1}) and quickest (13.2 km year^{-1}) climate shift, respectively. Such pace results in considerable displacement of climate in space over the European continent throughout the twenty-first century. As an example, Berlin's climate in 2071-2100 (P5) will be located not less than 1,584 km southwards (South Spain) than its climate in 1951-1980 (P1).

When averaging the southward velocity of cities' climate shift per transects, results showed that the cities of the Maritime transect tend to migrate southwards slower (7.3 km year^{-1} for the P1-P5 shift) than the cities of the two other transects (8.2 and 8.0 km year^{-1}). Such conclusion would need to be strengthened by integrating more cities, although it corroborates findings from an earlier study based on rather similar transects (Beniston, 2013).

Results also indicated that the southward velocity of European cities' climate shift is not constant from 1951 to 2100, and instead significantly accelerates throughout the twenty-first century (Table I and Figure 4). It starts from an average of 7.0 km year^{-1} for the P1-P2 shift and almost doubles to reach an average of 13.4 km year^{-1} for the P4-P5 shift. Computations of the averaged southward velocity for the two 60-year shifts also confirmed this finding. It starts from an average of 5.9 km year^{-1} for the P1-P3 shift and almost doubles to reach 11.3 km year^{-1} for the P3-P5 shift. Such doubling of speed is exemplified in Figure 5 for cities of the Continental transect.

A sensitivity analysis has also been performed – using the second best climate analogue of each LOIs rather than their best climate analogues – and showed similar results, in particular for the averaged result over transects and shift time periods (Appendix Table AIII).

4. Discussion

4.1 Implications

Up until now, most of the research on climate analogues and climate shift has aimed to assess the survival and abundance of species as well as the ecological changes of their habitat in response to the shift of climatic conditions. Recent findings (Ash et al., 2016)

Table I. Minimum, maximum and mean southward climate shift velocity for the three transects and the seven shift time periods

Shift time period	Velocity (km year^{-1})	Transects			Average (km year^{-1})
		Continental	E. Europe	Maritime	
P1-P2	Min-max	1.9-15.0	2.0-20.0	0.8-22.3	
	Mean	7.3	7.8	5.9	7.0
P2-P3	Min-max	0.9-15.3	3.9-14.2	2.8-24.5	
	Mean	5.3	7.1	7.5	6.6
P3-P4	Min-max	2.0-29.6	5.0-30.1	2.8-34.3	
	Mean	11.3	11.3	10.4	11
P4-P5	Min-max	5.6-34.0	2.0-38.1	3.9-25.6	
	Mean	14.1	15.2	10.7	13.4
P1-P3	Min-max	0.9-11.4	2.4-11.3	2.0-12.1	
	Mean	5.5	6.3	5.9	5.9
P3-P5	Min-max	5.1-19.7	5.1-20.2	3.9-21.1	
	Mean	11.9	11.9	9.9	11.3
P1-P5	Min-max	3.0-13.2	5.4-11.3	4.0-10.9	
	Mean	8.2	8.0	7.3	7.9

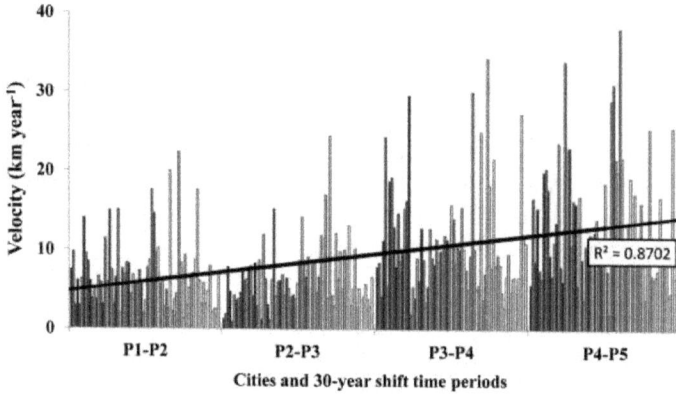

Figure 4. Southward velocity of the investigated cities for each 30-year shift time period

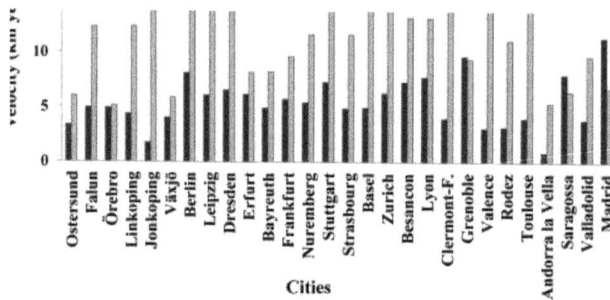

Figure 5. Southward velocity of the Continental transect's cities for the two 60-year shift
time periods, namely, P1-P3 and P3-P5

highlighted that species either shift their distribution to track climate change or adapt
to the changes in their local environmental conditions. Similarly, climate shift in cities
will threaten urban dwellers' quality of life and alter cities' functioning because of the
new climate-related issues that will appear along the climate shift. Nevertheless,
contrary to plants and other animal species, cities' residents are unlikely to shift their
distribution to track climate change, and hence will rather have to adapt to these
changing conditions.

We have shown that European cities' climate will shift southwards with an average
speed of 7.9 km year^{-1} from 1951-1980 to 2071-2100 (P1-P5), under the A1B IPCC SRES
scenario. This means that within one human generation (i.e. 25 years), European cities'
climate will shift 200 km southwards in average. Such rapid climate shift will
undoubtedly have negative implications on the 416 million inhabitants of the 90
investigated cities, and potentially also on many more in similar cases. Moreover, cities'
residents will have to not only face the changing climatic conditions but also cope with

the acceleration of the rate at which these changes occur, which is expected to double throughout the twenty-first century.

Such finding emphasizes on the strong dynamics of climate change, underlining that climatic conditions will change faster in the near future, without ever reaching an equilibrium state. Hence, an adaptation measure that is efficient at a certain period of time will not necessarily be efficient at another future time period. This brings attention to the fact that both dynamics and acceleration of the rate at which climatic conditions are changing must be taken into account when designing adaptation strategies in urban areas. Although we conducted the analysis over Europe only, findings are likely to be of similar magnitude in other continents.

4.2 Potential uses

In addition of being scientifically based, findings of climate analogues studies are thought to be easily understandable, hence can successfully raise awareness of a lay audience about climate change issues (Kopf *et al.*, 2008; Jylhä *et al.*, 2010; Beniston, 2014; Rohat *et al.*, 2016). Despite being based on a rather complex method, our study is no exception. Its findings, straightforward and readily comprehensible, can potentially raise awareness of urban dwellers and decision-makers about both the magnitude and the pace of climate change, particularly when graphically displayed at city scale (Figure 6). Indeed, when cities' residents and stakeholders visualize on a map that their city's climate is shifting at several

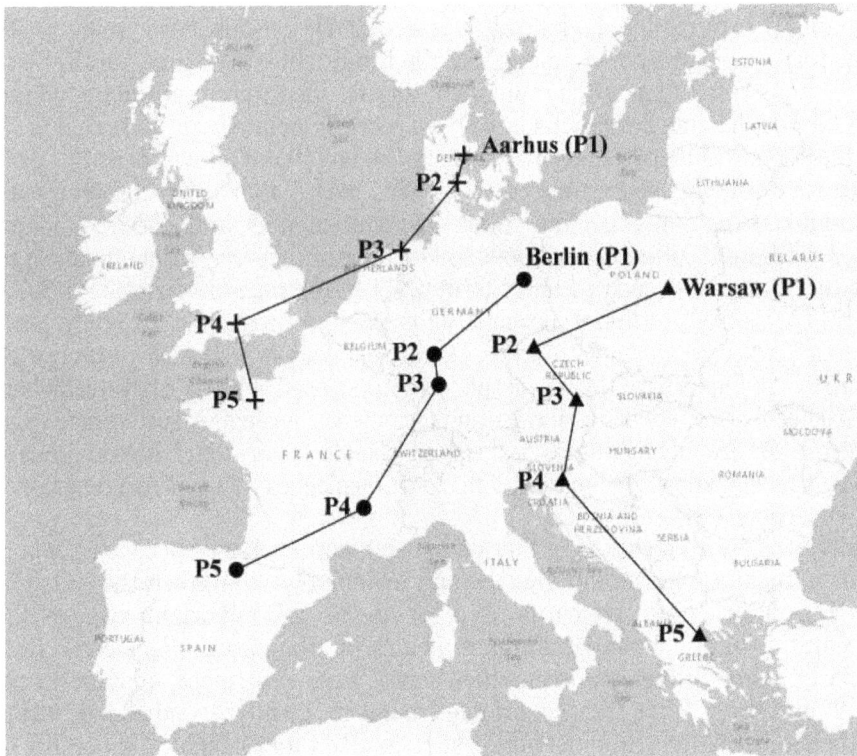

Figure 6. Climate shift over the European continent for the cities of Aarhus (Denmark), Berlin (Germany) and Warsaw (Poland), for the four 30-year shift time periods, namely, P1-P2 (from 1951-1980 to 1981-2010), P2-P3 (from 1981- 2010 to 2011-2040), P3-P4 (from 2011- 2040 to 2041-2070) and P4-P5 (from 2041- 2070 to 2071-2100)

hundred kilometres southwards, they may immediately realize what climate change actually means in terms of the changing climatic conditions and what the magnitude of these changes is. As an example, Figure 6 shows that Berlin's climate is shifting throughout Europe to reach North Spain by the end of the twenty-first century (2071-2100). Knowing that the climate in North Spain is much hotter and drier, with more frequent and intense heat waves, residents of Berlin could easily apprehend the magnitude of climate change and immediately envision the type of future climatic conditions they will have to cope with. Furthermore, displaying the different locations of Berlin's future climate – at different future time periods – might also raise awareness about the pace and dynamics of climate change, emphasizing on the fact that the speed of change is greatly increasing throughout the twenty-first century.

In addition of being a potentially efficient tool to raise awareness and communicate about climate change to a lay audience, the approach described in this paper might also be of great use for decision-makers and urban practitioners in charge of designing and implementing adaptation strategies in urban areas. Indeed, by closely looking at the current climatic conditions of the cities located southwards – along a given transect – decision-makers can readily envision the future climate impacts and vulnerabilities that their respective city will face. In the same line, by looking at the adaptation options that are currently implemented in the cities located southwards, urban practitioners can immediately and easily identify the ones that would have to be implemented in their own city to be well-adapted to the future climatic conditions. Such use of the climate analogues approach as decision-support tool shows great potential (Rohat *et al.*, 2016) but remains poorly explored. Hallegatte *et al.* (2007) showed that climate analogues could help assessing economic impacts of the future climate change and Kellett *et al.* (2011) demonstrated that such approach allows identifying adequate adaptation policies, although some limitations have been recently pointed out (Kellett *et al.*, 2015). One of the added values of our approach lies in the fact that it allows identifying climate analogues for several future time periods. This means that urban practitioners and decision-makers can identify the future climate impacts and efficient adaptation options for different future time periods, by looking at their city's climate analogue for these different periods, such as short-term (2011-2040), medium-term (2041-2070) and long-term (2071-2100) future. To exemplify this point, we computed the return periods of daily maximum temperatures for Aarhus, Berlin and Warsaw and their respective climate analogues for the four future 30-year time periods (Appendix Table AIV). Besides highlighting the great climate proximity between cities and their climate analogues, such results may be of valuable use for identifying future heat-related climate impacts and potential adaptation strategies – through knowledge sharing with climate analogues – for different time horizons of climatic challenges.

Finally, it is worth mentioning that using such approach as a decision-support tool for climate adaptation in urban areas would undeniably strengthen the collaboration among European cities, which is thought to be an important trigger for the implementation of efficient adaptation strategies in urban areas (Reckien *et al.*, 2015). Particularly along a given transect, cities could share experience, adaptation options and best practices. Such transect-oriented network could be embedded within the existing climate-related cities' networks, such as C40-Cities Climate Leadership Group or Covenant of Mayors.

While these potential uses appear promising and beneficial, limitations must be indicated at this point:

- On the one hand, certain limitations are inherent to the climate analogues approach. By taking into account only a limited number of climate statistics (five in this study), such method simplifies the definition of what climate really is. For instance, these five climate parameters do not integrate wind speed and humidity (which both play a role in the perceived temperature) and do not account for the local parameters of urban areas – e.g. the urban heat island – which can largely influence the climate and its impacts. Moreover, climate models' outputs that represent historical climate (1951-1980 and 1981-2010) are subject to spatially uneven biases that may lead to differences with weather stations data.

- On the other hand, certain factors can hinder the exchange of adaptation strategies between cities sharing similar climate at different time periods. For example, two cities of a similar transect might be too different – in terms of size, population's characteristics, functioning, shape, etc. – to efficiently share their adaptation options and best practices, which are often tightly linked to local characteristics. Moreover, the climate-related policies and infrastructures of a city might stem from political choices rather than from climatic conditions (Kellett *et al.*, 2015), hence limiting the utility of sharing adaptation measures between cities. Finally, cities are not always well-adapted to their current climate, meaning that their adaptation strategies should not be taken as an example of good practices. In this case, knowledge sharing would only allow identifying future impacts and vulnerabilities.

5. Conclusion

This paper has introduced and described a climate-matching method that can potentially be applied worldwide. As long as reliable climatic projections are available, it allows identifying and ranking climate analogues of any cities. This method also enables assessing the velocity of their climate shift, for any future time periods.

We exemplified this approach for the European continent and applied it on 90 different European cities. We successfully determined (84 per cent of success rate) their climate analogues for four time periods, namely 1981-2010, 2011-2040, 2041-2070 and 2071-2100.

On the basis of the spatial analysis of these climate-matching results, we have shown that European cities' climate will strictly shift southwards in the future. More specifically, this equator-ward climate shift appears to follow particular north-to-south transects, such as the three transects (Maritime, Continental and Eastern Europe) that we predetermined. Using these findings, we computed the southward velocity of this climate shift. Despite the heterogeneity of the results, the averages analysis has highlighted a significant rise (doubling) of the southward climate shift velocity throughout the twenty-first century. It reaches not less than 13.4 km year^{-1} in average for the shift from 2041-2070 to 2071-2100 (P4-P5). This finding is in line with other studies (Burrows *et al.*, 2011; Chen *et al.*, 2011; Diffenbaugh and Field, 2013). Such climate shift and increase of velocity have direct implications for urban inhabitants, who will have to adapt to this rapid climate shift and to the wide range of new climate-related issues that will occur in cities at a constantly growing pace.

One of the major added values of this study is that the findings are both scientifically sound and easily understandable by a lay audience. As a result, these can be used to raise awareness of both stakeholders and urban dwellers about the existence, the magnitude, the pace, and the dynamics of climate change. Knowing that their city's future climate will be similar to the current climate of other cities located southwards and that such shift is expected to double throughout the twenty-first century, people may immediately envision what the future climate will look like in their city and at what pace these changes will occur. Furthermore, we have shown that such approach may also be used as a decision-support tool. It enables cities to learn from each other, in terms of future impacts and vulnerabilities as well as in terms of adaptation options, policies and best practices. Because of the southward climatic shift, knowledge transfer between European cities will be made from southern cities to northern cities, along the same transect. Such practical application of climate analogues might strengthen collaboration between cities and enhance their involvement in climate-related networks. Despite some limitations, mainly associated with the climate matching method and with the differences between cities' characteristics, this exploratory study shows a great potential for future development, particularly regarding its applications as both a communication and decision-support tool in urban areas. On the basis of this exploratory study, further research could integrate different IPCC scenarios (such as the new set of representative concentration pathways) to assess the influence that different radiative forcing has on the southward climate velocity of cities. Further studies could also empirically test the efficiency of such approach as an awareness-raising tool. Finally, on the basis of specific case studies and with the help of cities' stakeholders and policymakers, further research could also demonstrate the applicability of such approach as an efficient decision-support tool for designing and strengthening adaptation strategies in urban areas, at different time horizons.

References

Anderson, A.S., Storlie, C.J., Shoo, L.P., Pearson, R.G. and Williams, S.E. (2013), "Current analogues of future climate indicate the likely response of a sensitive montane tropical avifauna to a warming world", *PLoS ONE*, Vol. 8 No. 7, p. e69393, doi: 10.1371/journal.pone.0069393.

Archie, K.M., Dilling, L., Milford, J.B. and Pampel, F.C. (2014), "Unpacking the 'information barrier': comparing perspectives on information as a barrier to climate change adaptation in the interior Mountain", *Journal of Environmental Management*, Vol. 133, pp. 397-410, doi: 10.1016/j.jenvman.2013.12.015.

Ash, J.D., Givnish, T.J. and Waller, D.M. (2016), "Tracking lags in historical plant species' shift in relation to regional climate change", *Global Change Biology*, Vol. 23 No. 3, pp. 1305-1315, doi: 10.1111/gcb.13429.

Aylett, A. (2015), "Institutionalizing the urban governance of climate change adaptation: results of an international survey", *Urban Climate*, Vol. 14, pp. 4-16, doi: 10.1016/j.uclim.2015.06.005.

Beniston, M. (2013), "Exploring the behaviour of atmospheric temperatures under dry conditions in Europe: evolution since the mid-20th century and projections for the end of the 21st century", *International Journal of Climatology*, Vol. 33 No. 2, pp. 457-462, doi: 10.1002/joc.3436.

Beniston, M. (2014), "European isotherms move northwards by up to 15 km year−1: using climate analogues for awareness-raising", *International Journal of Climatology*, Vol. 34 No. 6, pp. 1838-1844, doi: 10.1002/joc.3804.

Beniston, M. (2015), "Ratios of record high to record low temperatures in Europe exhibit sharp increases since 2000 despite a slowdown in the rise of mean temperatures", *Climatic Change*, Vol. 129 Nos 1/2, pp. 225-237, doi: 10.1007/s10584-015-1325-2.

Bicknell, J., Dodman, D. and Satterthwaite, D. (2009), *Adapting Cities to Climate Change: understanding and Addressing the Development Challenges*, Earthscan, London.

Böhm, U., Kücken, M., Ahrens, W., Block, A., Hauffe, D., Keuler, K., Rocker, B. and Will, A. (2006), "CLM – the climate version of LM: brief description and long-term applications", *COSMO Newsletter*, Vol 6, pp. 225-235.

Bulkeley, H. (2013), *Cities and Climate Change*, Routledge, London.

Bulkeley, H., Schroeder, H., Janda, K., Zhao, J., Armstrong, A., Yi Chu, S. and Ghosh, S. (2011), "The role of institutions, governance, and urban planning for mitigation and adaptation", in Hoornweg, D., Freire, M., Lee, M.J., Bhada-Tata, P. and Yuen, B. (Eds), *Cities and Climate Change: Responding to an Urgent Agenda*, World Bank, Washington, pp. 125-159.

Burrows, M.T., Schoeman, D.S., Buckley, L.B., Moore, P., Poloczanska, E.S., Brander, K.M., Brown, C., Bruno, J.F., Duarte, C.M., Halpern, B.S., Holding, J., Kappel, C.V., Kiessling, W., O'Connor, M.I., Pandolfi, J.M., Parmesan, C., Schwing, F.B., Sydeman, W.J. and Richardson, A.J. (2011), "The pace of shifting climate in marine and terrestrial ecosystems", *Science*, Vol. 334 No. 6056, pp. 652-655, doi: 10.1126/science.1210288.

Chen, I.C., Hill, J.K., Ohlemüller, R., Roy, D.B. and Thomas, C.D. (2011), "Rapid range shifts of species associated with high levels of climate warming", *Science*, Vol. 333 No. 6045, pp. 1021-1026, doi: 10.1126/science.1206432.

Christensen, O.B., Drews, M. and Christensen, J.H. (2007), "The HIRHAM regional climate model version 5", Technical Report No. 06-17, Danish Meteorological Institute, Copenhagen.

Climate Communication (2014), "Climate communication: science and outreach", available at: www.climatecommunication.org/ (accessed 15 March 2017).

CNRM (2008), "The scientific contents of ALADIN", available at: www.umr-cnrm.fr/aladinold/scientific/scientif.html (accessed 15 March 2017).

Collins, M., Booth, B.B., Harris, G.R., Murphy, J.M., Sexton, D.M.H. and Webb, M.J. (2006), "Towards quantifying uncertainty in transient climate change", *Climate Dynamics*, Vol. 27 No. 2, pp. 127-147, doi: 10.1007/s00382-006-0121-0.

CSIRO-Bureau of Meteorology (2016), "Climate analogues explorer tool", available at: www.climatechangeinaustralia.gov.au/en/climate-projections/climate-analogues/about-analogues/ (accessed 15 March 2017).

de Castro, M., Gallardo, C., Jylha, K. and Tuomenvirta, H. (2007), "The use of a climate-type classification for assessing climate change effects in Europe from an ensemble of nine regional climate models", *Climatic Change*, Vol. 81 No. S1, pp. 329-341, doi: 10.1007/s10584-006-9224-1.

Diffenbaugh, N.S. and Field, C.B. (2013), "Changes in ecologically critical terrestrial climate conditions", *Science*, Vol. 341 No. 341, pp. 486-492, doi: 10.1126/science.1237123.

ENSEMBLES (2009), "Data distribution Portal", available at: http://ensemblesrt3.dmi.dk/ (accessed 15 March 2017).

Eurostat (2012), "Eurostat regional yearbook 2012: focus on European cities", available at: http://ec.europa.eu/eurostat/en/web/products-statistical-books/-/KS-HA-12-001-12 (accessed 15 March 2017).

Gerstengarbe, F.W. and Werner, P.C. (2008), "A short update on Koppen climate shifts in Europe between 1901 and 2003", *Climatic Change*, Vol. 92 Nos 1/2, pp. 99-107, doi: 10.1007/s10584-008-9430-0.

Grenier, P., Parent, A.-C., Huard, D., Anctil, F. and Chaumont, D. (2013), "An assessment of six dissimilarity metrics for climate analogs", *Journal of Applied Meteorology and Climatology*, Vol. 52 No. 4, pp. 733-752, doi: 10.1175/JAMC-D-12-0170.1.

Hallegatte, S. (2009), "Strategies to adapt to an uncertain climate change", *Global Environmental Change*, Vol. 19 No. 2, pp. 240-247, doi: 10.1016/j.gloenvcha.2008.12.003.

Hallegatte, S., Hourcade, J.-C. and Ambrosi, P. (2007), "Using climate analogues for assessing climate change economic impacts in urban areas", *Climatic Change*, Vol. 82 Nos 1/2, pp. 47-60, doi: 10.1007/s10584-006-9161z.

Holdridge, L.R. (1947), "Determination of world plant formations from simple climatic data", *Science*, Vol. 108 No. 2727, pp. 367-368, doi: 10.1126/science.105.2727.367.

IPCC (2001), "Appendix I: glossary", in Houghton, J.T., Ding, Y., Griggs, D.J., Noguer, M., van der Linden, P.J., Dai, X., Maskell, K. and Johnson, C.A. (Eds), *Climate Change 2001: The Physical Science Basis, Contribution of Working Group I to the 3rd Assessment Report of the Intergovernmental Panel on Climate Change*, Cambridge University Press, Cambridge, MA, pp. 787-798.

Juhola, S. (2016), "Barriers to the implementation of climate change adaptation in land use planning: a multi level governance problem?", *International Journal of Climate Change Strategies and Management*, Vol. 8 No. 3, pp. 338-355, doi: 10.1108/IJCCSM-03-2014-0030.

Jylhä, K., Tuomenvirta, H., Ruosteenoja, K., Niemi-Hugaerts, H., Keisu, K. and Karhu, J.A. (2010), "Observed and projected future shifts of climatic zones in Europe and their use to visualize climate change information", *Weather, Climate, and Society*, Vol. 2 No. 2, pp. 148-167, doi: 10.1175/2010WCAS1010.1.

Kellett, J., Hamilton, C., Ness, D. and Pullen, S. (2015), "Testing the limits of regional climate analogues studies: an Australian example", *Land Use Policy*, Vol. 44, pp. 54-61, doi: 10.1016/j.landusepol.2014.11.022.

Kellett, J., Ness, D., Hamilton, C., Pullen, S. and Leditschke, A. (2011), "Learning from regional climate analogues", NCCARF Publication 1/12, ISBN: 978-1-921609-41-1, National Climate Change Adaptation Research Facility, Gold Coast.

Kjellström, E., Bärring, L., Gollvik, S., Hansson, U., Jones, C., Samuelsson, P., Rummukainen, M., Ullerstig, A., Willén, U. and Wyser, K. (2005), "A 140-year simulation of European climate with the new version of the Rossby Centre regional atmospheric climate model (RCA3)", SMHI Reports Meteorology and Climatology, No 108-SE-60176, Stockholm.

Kopf, S., Ha-Duong, M. and Hallegatte, S. (2008), "Using maps of city analogues to display and interpret climate change scenarios and their uncertainty", *Natural Hazards and Earth System Science Hazards*, Vol. 8 No. 4, pp. 905-918, doi: 10.5194/nhess-8-905-2008.

Lee, T. and Lee, T. (2016), "Evolutionary urban climate resilience: assessment of Seoul's policies", *International Journal of Climate Change Strategies and Management*, Vol. 8 No. 5, pp. 597-612, doi: 10.1108/IJCCSM-06.2015-0066.

Leibing, C., Signer, J., van Zonneveld, M., Jarvis, A. and Dvorak, W. (2013), "Selection of provenances to adapt tropical pine forestry to climate change on the basis of climate analogs", *Forests*, Vol. 4 No. 1, pp. 155-178, doi: 10.3390/f4010155.

Mycoo, M. (2015), "Communicating climate change in rural coastal communities", *International Journal of Climate Change Strategies and Management*, Vol. 7 No. 1, pp. 58-75, doi: 10.1108/IJCCSM-04-2013-0042.

Peacock, L. and Worner, S. (2006), "Using analogous climates and global insect distribution data to identify potential sources of new invasive insect pests in New Zealand", *New Zealand Journal of Zoology*, Vol. 33 No. 2, pp. 141-145, doi: 10.1080/03014223.2006.9518438.

Plummer, D., Caya, D., Côté, H., Frigon, A., Biner, S., Giguère, M., Paquin, D., Harvey, R. and de Elia, R. (2006), "Climate and climate change over North America as simulated by the Canadian Regional Climate Model", *Journal of Climate*, Vol. 19 No. 13, pp. 3112-3132, doi: 10.1175/JCLI3769.1.

Ramirez-Villegas, J., Lau, C., Köhler, A.K., Signer, J., Jarvis, A., Arnell, N., Osborne, T., Hooker, J. (2011), "Climate analogues: finding tomorrow's agriculture today", CGIAR Research Program on Climate Change, Agriculture and Food Security (CCAFS)", Working Paper No. 12, Copenhagen.

Reckien, D., Flacke, J., Olazabal, M. and Heidrich, O. (2015), "The influence of drivers and barriers on urban adaptation and mitigation plans-an empirical analysis of European cities", *PLoS One*, Vol. 10 No. 8, p. e0135597, doi: 10.1371/journal.pone.0135597.

Reckien, D., Flacke, J., Dawson, R.J., Heidrich, O., Olazabal, M., Foley, A., Hamann, J.J.-P., Orru, H., Salvia, M., De Gregorio Hurtado, S., Geneletti, D. and Pietrapertosa, F. (2013), "Climate change response in Europe: what's the reality? Analysis of adaptation and mitigation plans from 200 urban areas in 11 countries", *Climatic Change*, Vol. 122 Nos 1/2, pp. 331-340, doi: 10.1007/s10584-013-0989-8.

Rohat, G., Goyette, S. and Flacke, J. (2016), "Twin climate cities – an exploratory study of their potential use for awareness-raising and urban adaptation", *Mitigation and Adaptation Strategies for Global Change*, doi: 10.1007/s11027-016-9708-x.

Saxon, E., Baker, B., Hargrove, W., Hoffman, F. and Zganjar, C. (2005), "Mapping environments at risk under different global climate change scenarios", *Ecology Letters*, Vol. 8 No. 1, pp. 53-60, doi: 10.1111/j.1461-0248.2004.00694.x.

Schneider, S.H. (2006), "Climate change: do we know enough for policy action?", *Science and Engineering Ethics*, Vol. 12 No. 4, pp. 607-636, doi: 10.1007/s11948-006-0061-4.

Sinnott, R.W. (1984), "Virtues of the Haversine", *Sky Telescope*, Vol. 68 No. 2, pp. 158-159.

Ungar, J., Peters-Anders, J. and Loibl, W. (2011), "Climate twins – an attempt to quantify climatological similarities", *Environmental Software Systems, Frameworks of eEnvironment, 9th IFIP International Symposium, ISESS 2011, Brno, Czech Republic, Springer, Vienna*, pp. 428-436.

van der Linden, S.L., Leiserowitz, A.A., Feinberg, G.D. and Maibach, E.W. (2014), "How to communicate the scientific consensus on climate change: plain facts, pie charts or metaphors?", *Climatic Change*, Vol. 126 Nos 1/2, pp. 255-262, doi: 10.1007/s10584-014-1190-4.

van Meijgaard, E., van Ulft, L.H., van de Berg, W.J., Bosveld, F.C., van den Hurk, B.J.M., Lenderink, G. and Siebesma, A.P. (2008), "The KNMI regional atmospheric climate model RACMO version 2.1", KNMI Technical Report TR-302, Copenhagen.

Veloz, S.D., Williams, J.W., Blois, J.L., He, F., Otto-Bliesner, B. and Liu, Z. (2012a), "No-analog climates and shifting realized niches during the late quaternary: implications for 21st-century predictions by species distribution models", *Global Change Biology*, Vol. 18 No. 5, pp. 1698-1713, doi: 10.1111/j.1365-2486.2011.02635.x.

Veloz, S., Williams, J.W., Lorenz, D., Notaro, M., Vavrus, S. and Vimont, D.J. (2012b), "Identifying climatic analogs for Wisconsin under 21st-century climate-change scenarios", *Climatic Change*, Vol. 112 Nos 3/4, pp. 1037-1058, doi: 10.1007/s10584-011-0261-z.

Webb, L.B., Watterson, I., Bhend, J., Whetton, P.H. and Barlow, E.W.R. (2013), "Global climate analogues for winegrowing regions in future periods: projections of temperature and precipitation", *Australian Journal of Grape and Wine Research*, Vol. 19 No. 3, pp. 331-341, doi: 10.1111/ajgw.12045.

Williams, J.W. and Jackson, S.T. (2007), "Novel climates, no-analog communities, and ecological surprises", *Frontiers in Ecology and the Environment*, Vol. 5 No. 9, pp. 475-482 doi: 10.1890/070037.

Corresponding author

Guillaume Rohat can be contacted at: guillaume.rohat@unige.ch

Appendix. Supplementary Material – Equations

Equation for computing the averaged Euclidean distances (ED_{avg}) between the LOI's future climate and the current climate of a given grid point, for the five climate variables:

$$ED_{avg} = \sum_{m=1}^{n} abs\left(A_{(m)f}\right) - B_{(m)c}/n$$

where *abs* is the absolute value, A is the LOI's value for the month m for the future time period f, and B is the given grid point's value for the month m at the current time period c. For monthly mean precipitation and monthly mean temperature variables, number of months $n = 12$; for minimum winter temperature and maximum summer temperature variables, $n = 3$; for annual total precipitation variable, $n = 1$.

Equation for computing the similarity index of climate analogues, based on standardized Euclidean distances (*SED*):

$$SED = \sum_{v=1}^{8} \left[ED_{avg(v)} - X_S/\sigma_S \right]$$

where $ED_{avg(v)}$ is the averaged Euclidean distance for the climate variable v, Xv and σv are, respectively, the mean and the standard deviation of the set of ED_{avg} for all the grid points, for the climate variable v.

Table AI. List of the 90 investigated cities, with their coordinates, altitude and population

Transect	City	Latitude	Longitude	Altitude (m)	Population (inhab.)
Continental	Östersund	63.167	14.667	330	45,000
	Falun	60.6	15.616	110	37,000
	Örebro	59.26	15.22	40	100,000
	Linköping	58.4	15.61	60	98,000
	Jönköping	57.75	14.167	110	85,000
	Växjö	58.86	14.8	170	56,000
	Berlin	52.467	13.35	40	3,500,000
	Leipzig	51.333	12.417	110	523,000
	Dresden	51.05	13.73	120	524,000
	Erfurt	50.97	11.03	190	206,000
	Bayreuth	49.95	11.58	340	73,000
	Frankfurt	50.1	8.683	100	681,000
	Nuremberg	49.45	11.083	320	506,000
	Stuttgart	48.776	9.17	250	607,000
	Strasbourg	48.583	7.75	140	272,000
	Basel	47.55	7.6	260	165,000
	Zurich	47.378	8.54	410	379,000
	Besancon	47.24	6.03	250	117,000
	Geneva	46.217	6.15	370	189,000
	Lyon	45.767	4.833	170	485,000
	Clermont-Ferrand	45.78	3.08	410	140,000
	Grenoble	45.187	5.72	220	156,000
	Valence	44.93	4.89	130	62,000
	Rodez	44.35	2.567	630	25,000
	Toulouse	43.617	1.45	140	442,000
	Andorra la Vella	42.51	1.523	990	23,000
	Saragossa	41.65	−0.9	210	666,000
	Valladolid	41.63	−4.72	700	307,000
	Madrid	40.417	−3.717	650	3,165,000
	Cordoba	37.883	−4.767	130	329,000
Eastern Europe	St. Petersburg	59.917	30.417	20	5,222,000
	Pskov	57.8	28.433	50	205,000
	Velikiye Luki	56.33	30.533	100	99,000
	Rēzekne	56.5	27.34	160	33,000
	Daugavpils	55.875	26.53	110	90,000
	Polotsk	55.483	28.8	130	83,000
	Vitebsk	55.183	30.167	160	348,000
	Kaunas	54.89	23.89	30	309,000
	Vilnius	54.667	25.317	100	536,000
	Minsk	53.92	27.49	210	1,894,000
	Bialystok	53.34	23.166	130	295,000
	Pinsk	52.12	26.1	140	135,000
	Brest-Litovsk	52.08	23.7	170	323,700
	Warsaw	52.217	21.005	110	1,715,000
	Lodz	51.783	19.467	210	722,000
	Lublin	51.248	22.57	190	348,000
	Lutsk	50.75	25.335	190	212,000
	Cracow	50.067	19.933	220	759,000
	Lviv	49.85	24.02	290	724,000
	Kosice	48.716	21.25	200	241,000
	Budapest	47.498	19.04	110	1,732,000
	Debrecen	47.529	21.639	120	205,000

(continued)

Table AI.

Transect	City	Latitude	Longitude	Altitude (m)	Population (inhab.)
	Mukacheve	48.45	22.75	120	86,000
	Cluj-Napoca	46.76	23.583	340	304,000
	Timisoara	45.76	21.23	90	307,000
	Belgrade	44.817	20.467	170	1,352,000
	Prishtina	42.66	21.166	650	146,000
	Sofia	42.7	23.33	550	1,212,000
	Plovdiv	42.146	24.75	170	339,000
	Skopje	41.997	21.433	250	537,000
Maritime	Trondheim	63.6	10.383	10	179,000
	Bergen	60.383	5.333	10	266,000
	Haugesund	59.43	5.28	15	37,000
	Stavanger	58.967	5.75	10	129,000
	Kristiansand	58.15	8	10	88,000
	Aalborg	57.05	9.919	10	204,000
	Aarhus	56.15	10.22	20	320,000
	Herning	56.762	8.317	50	87,000
	Esbjerg	55.467	8.467	20	116,000
	Bremen	53.083	8.8	10	547,000
	Groningen	53.218	6.56	10	190,000
	Amsterdam	52.38	4.9	10	780,000
	Rotterdam	51.917	4.483	10	611,000
	Calais	50.95	1.85	10	73,000
	Lille	50.65	3.083	20	228,000
	Rouen	49.433	1.083	20	111,000
	Caen	49.18	−0.37	20	109,000
	Rennes	48.1	−1.667	40	208,000
	Brest	48.39	−4.48	50	140,000
	Vannes	47.65	−2.76	30	54,000
	Nantes	47.233	−1.538	10	293,000
	Bordeaux	44.833	−0.567	10	244,000
	San-Sebastián	43.32	−1.98	10	187,000
	Bilbao	43.25	−2.933	20	347,000
	Santander	43.46	−3.805	10	176,000
	Gijón	43.53	−5.7	20	276,000
	Vigo	42.23	−8.67	10	295,000
	Porto	41.15	−8.617	50	231,000
	Lisbon	38.733	−9.133	50	531,000
	Faro	37.03	−7.91	20	65,000

Source: http://ec.europa.eu/eurostat/statistics-explained/index.php/Statistics_on_European_cities

Table AII. Full results of the southward velocity computations for the 90 investigated cities and the seven shift time periods

Cities of each transect	1950-1980 to 1980-2010		1980-2010 to 2010-2040		2010-2040 to 2040-2070		2040-2070 to 2070-2100		1950-1980 to 2010-2040		2010-2040 to 2070-2100		1950-1980 to 2070-2100	
	Distance (km)	Velocity (km/year)	Distance	Velocity	Distance	Velocity	Distance	Velocity	Distance	Velocity	Distance	Velocity	Distance	Velocity
Continental														
Östersund	225.00	7.50	38.00	1.27	236.00	7.87	169.00	5.63	200.00	3.33	362.00	6.03	430.00	3.58
Falun	293.00	9.77	57.00	1.90	251.00	8.37	498.00	16.60	297.00	4.95	740.00	12.33	711.00	5.93
Örebro	89.00	2.97	235.00	7.83	125.00	4.17	257.00	8.57	294.00	4.90	308.00	5.13	600.00	5.00
Linköping	183.00	6.10	142.00	4.73	337.00	11.23	460.00	15.33	263.00	4.38	745.00	12.42	1007.00	8.39
Jönköping	88.00	2.93	28.00	0.93	729.00	24.30	225.00	7.50	104.00	1.73	930.00	15.50	1031.00	8.59
Växjö	191.00	6.37	127.00	4.23	167.00	5.57	195.00	6.50	242.00	4.03	352.00	5.87	594.00	4.95
Berlin	420.00	14.00	109.00	3.63	562.00	18.73	598.00	19.93	487.00	8.12	1102.00	18.37	1579.00	13.16
Leipzig	284.00	9.47	113.00	3.77	576.00	19.20	612.00	20.40	365.00	6.08	1181.00	19.68	1543.00	12.86
Dresden	256.00	8.53	137.00	4.57	389.00	12.97	531.00	17.70	393.00	6.55	893.00	14.88	1269.00	10.58
Erfurt	182.00	6.07	215.00	7.17	237.00	7.90	281.00	9.37	370.00	6.17	492.00	8.20	852.00	7.10
Bayreuth	115.00	3.83	183.00	6.10	441.00	14.70	203.00	6.77	296.00	4.93	496.00	8.27	779.00	6.49
Frankfurt	156.00	5.20	191.00	6.37	257.00	8.57	331.00	11.03	344.00	5.73	578.00	9.63	918.00	7.65
Nuremberg	112.00	3.73	223.00	7.43	293.00	9.77	408.00	13.60	326.00	5.43	699.00	11.65	1025.00	8.54
Stuttgart	199.00	6.63	239.00	7.97	460.00	15.33	708.00	23.60	438.00	7.30	1177.00	19.62	1536.00	12.80
Strasbourg	176.00	5.87	125.00	4.17	489.00	16.30	240.00	8.00	294.00	4.90	700.00	11.67	991.00	8.26
Basel	93.00	3.10	250.00	8.33	888.00	29.60	184.00	6.13	299.00	4.98	1071.00	17.85	1354.00	11.28
Zurich	343.00	11.43	86.00	2.87	59.00	1.97	1019.00	33.97	378.00	6.30	1060.00	17.67	1425.00	11.88
Besancon	177.00	5.90	263.00	8.77	161.00	5.37	679.00	22.63	440.00	7.33	794.00	13.23	1200.00	10.00
Geneva														
Lyon	450.00	15.00	35.00	1.17	118.00	3.93	690.00	23.00	467.00	7.78	793.00	13.22	1063.00	8.86
Clermont-Ferrand					383.00	12.77	516.00	17.20	240.00	4.00	898.00	14.97	975.00	8.13
Grenoble	224.00	7.47	360.00	12.00	270.00	9.00	318.00	10.60	581.00	9.68	568.00	9.47	1141.00	9.51
Valence					635.00	21.17	368.00	12.27	187.00	3.12	891.00	14.85	946.00	7.88
Rodez	92.00	3.07	191.00	6.37	184.00	6.13	490.00	16.33	193.00	3.22	668.00	11.13	839.00	6.99
Toulouse	194.00	6.47	93.00	3.10	385.00	12.83	478.00	15.93	241.00	4.02	857.00	14.28	1018.00	8.48
Andorra la Vella									52.00	0.87	324.00	5.40	354.00	2.95
Saragossa	451.00	15.03	35.00	1.17	265.00	8.83	169.00	5.63	482.00	8.03	388.00	6.47	794.00	6.62
Valladolid	58.00	1.93	187.00	6.23	103.00	3.43	507.00	16.90	234.00	3.90	583.00	9.72	816.00	6.80
Madrid	229.00	7.63	458.00	15.27	161.00	5.37	267.00	8.90	685.00	11.42	407.00	6.78	864.00	7.20
Cordoba	390.00	13.00	28.00	0.93					418.00	6.97				

(continued)

Table AII.

Cities of each transect	1950-1980 to 1980-2010		1980-2010 to 2010-2040		2010-2040 to 2040-2070		2040-2070 to 2070-2100		1950-1980 to 2010-2040		2010-2040 to 2070-2100		1950-1980 to 2070-2100	
	Distance (km)	Velocity (km/year)	Distance	Velocity	Distance	Velocity	Distance	Velocity	Distance	Velocity	Distance	Velocity	Distance	Velocity
Eastern Europe														
St. Petersburg	206.00	6.87	126.00	4.20	383.00	12.77	115.00	3.83	239.00	3.98	475.00	7.92	708.00	5.90
Pskov	251.00	8.37	180.00	6.00	270.00	9.00	318.00	10.60	393.00	6.55	559.00	9.32	896.00	7.47
Velikiye Luki	246.00	8.20	185.00	6.17	246.00	8.20	349.00	11.63	363.00	6.05	577.00	9.62	859.00	7.16
Rezekne	129.00	4.30	206.00	6.87	344.00	11.47	360.00	12.00	333.00	5.55	701.00	11.68	886.00	7.38
Daugavpils	204.00	6.80	129.00	4.30	257.00	8.57	353.00	11.77	319.00	5.32	609.00	10.15	859.00	7.16
Polotsk	179.00	5.97	193.00	6.43	295.00	9.83	370.00	12.33	349.00	5.82	661.00	11.02	922.00	7.68
Vitebsk	168.00	5.60	154.00	5.13	302.00	10.07	417.00	13.90	322.00	5.37	717.00	11.95	975.00	8.13
Kaunas	220.00	7.33	122.00	4.07	358.00	11.93	325.00	10.83	342.00	5.70	630.00	10.50	956.00	7.97
Vilnius	172.00	5.73	130.00	4.33	340.00	11.33	263.00	8.77	302.00	5.03	601.00	10.02	808.00	6.73
Minsk	61.00	2.03	116.00	3.87	313.00	10.43	342.00	11.40	177.00	2.95	644.00	10.73	754.00	6.28
Bialystok	106.00	3.53	175.00	5.83	473.00	15.77	557.00	18.57	275.00	4.58	802.00	13.37	1074.00	8.95
Pinsk	232.00	7.73	260.00	8.67	420.00	14.00	223.00	7.43	448.00	7.47	546.00	9.10	986.00	8.22
Brest-Litovsk	261.00	8.70	427.00	14.23	250.00	8.33	59.00	1.97	676.00	11.27	306.00	5.10	908.00	7.57
Warsaw	528.00	17.60	254.00	8.47	311.00	10.37	870.00	29.00	538.00	8.97	1076.00	17.93	1350.00	11.25
Lodz	438.00	14.60	198.00	6.60	300.00	10.00	931.00	31.03	573.00	9.55	1209.00	20.15	1318.00	10.98
Lublin	288.00	9.60	274.00	9.13	464.00	15.47	647.00	21.57	562.00	9.37	1086.00	18.10	1236.00	10.30
Lutsk	307.00	10.23	157.00	5.23	303.00	10.10	599.00	19.97	463.00	7.72	899.00	14.98	1002.00	8.35
Cracow	167.00	5.57	245.00	8.17	226.00	7.53	1143.00	38.10	412.00	6.87	1203.00	20.05	1087.00	9.06
Lviv	175.00	5.83	253.00	8.43	151.00	5.03	653.00	21.77	427.00	7.12	801.00	13.35	1226.00	10.22
Kosice					296.00	9.87	424.00	14.13	144.00	2.40	670.00	11.17	648.00	5.40
Budapest	82.00	2.73			474.00	15.80	197.00	6.57	265.00	4.42	654.00	10.90	872.00	7.27
Debrecen	144.00	4.80	139.00	4.63	277.00	9.23	346.00	11.53	215.00	3.58	619.00	10.32	821.00	6.84
Mukacheve	281.00	9.37	198.00	6.60	902.00	30.07	267.00	8.90	309.00	5.15	705.00	11.75	1265.00	10.54
Cluj-Napoca	77.00	2.57	316.00	10.53	369.00	12.30			487.00	8.12				
Timisoara			357.00	11.90	304.00	10.13	407.00	13.57	423.00	7.05	560.00	9.33	952.00	7.93
Belgrade	599.00	19.97	342.00	11.40	153.00	5.10	577.00	19.23	488.00	8.13	579.00	9.65	1029.00	8.58
Prishtina													842.00	7.02
Sofia							706.00	23.53						
Plovdiv	303.00	10.10												
Skopje													802.00	6.68

(continued)

Table AII.

Cities of each transect	1950-1980 to 1980-2010		1980-2010 to 2010-2040		2010-2040 to 2040-2070		2040-2070 to 2070-2100		1950-1980 to 2010-2040		2010-2040 to 2070-2100		1950-1980 to 2070-2100	
	Distance (km)	Velocity (km/year)	Distance	Velocity	Distance	Velocity	Distance	Velocity	Distance	Velocity	Distance	Velocity	Distance	Velocity
Maritime														
Trondheim	66.00	2.20	512.00	17.07	169.00	5.63	346.00	11.53	473.00	7.88	502.00	8.37	963.00	8.03
Bergen	112.00	3.73	119.00	3.97	749.00	24.97	517.00	17.23	228.00	3.80	1221.00	20.35	1311.00	10.93
Haugesund	133.00	4.43	734.00	24.47	115.00	3.83	317.00	10.57	728.00	12.13	390.00	6.50	1021.00	8.51
Stavanger	670.00	22.33	128.00	4.27	211.00	7.03	335.00	11.17	687.00	11.45	503.00	8.38	1114.00	9.28
Kristiansand	251.00	8.37	89.00	2.97	1029.00	34.30	481.00	16.03	311.00	5.18	1268.00	21.13	1186.00	9.88
Aalborg	38.00	1.27	365.00	12.17	548.00	18.27	320.00	10.67	390.00	6.50	866.00	14.43	1184.00	9.87
Aarhus	89.00	2.97	296.00	9.87	647.00	21.57	287.00	9.57	376.00	6.27	758.00	12.63	1125.00	9.38
Herning	281.00	9.37	192.00	6.40	232.00	7.73	117.00	3.90	461.00	7.68	281.00	4.68	738.00	6.15
Esbjerg	154.00	5.13	155.00	5.17	244.00	8.13	762.00	25.40	177.00	2.95	1005.00	16.75	1111.00	9.26
Bremen	206.00	6.87	302.00	10.07	281.00	9.37	207.00	6.90	493.00	8.22	480.00	8.00	972.00	8.10
Groningen	211.00	7.03	264.00	8.80	244.00	8.13	192.00	6.40	475.00	7.92	435.00	7.25	906.00	7.55
Amsterdam	263.00	8.77	395.00	13.17	140.00	4.67	203.00	6.77	596.00	9.93	336.00	5.60	851.00	7.09
Rotterdam	529.00	17.63	91.00	3.03	83.00	2.77	226.00	7.53	615.00	10.25	308.00	5.13	852.00	7.10
Calais	181.00	6.03	152.00	5.07	200.00	6.67	504.00	16.80	327.00	5.45	584.00	9.73	895.00	7.46
Lille	94.00	3.13	306.00	10.20	286.00	9.53	289.00	9.63	393.00	6.55	534.00	8.90	863.00	7.19
Rouen	174.00	5.80	85.00	2.83	139.00	4.63	269.00	8.97	259.00	4.32	401.00	6.68	659.00	5.49
Caen	96.00	3.20	154.00	5.13	192.00	6.40	306.00	10.20	250.00	4.17	471.00	7.85	593.00	4.94
Rennes	142.00	4.73	111.00	3.70	198.00	6.60	126.00	4.20	235.00	3.92	321.00	5.35	499.00	4.16
Brest	241.00	8.03	125.00	4.17	139.00	4.63	140.00	4.67	358.00	5.97	279.00	4.65	582.00	4.85
Vannes	91.00	3.03	338.00	11.27	180.00	6.00			393.00	6.55				
Nantes	64.00	2.13	171.00	5.70	198.00	6.60	767.00	25.57	225.00	3.75	911.00	15.18	997.00	8.31
Bordeaux	75.00	2.50	115.00	3.83	817.00	27.23	199.00	6.63	136.00	2.27	919.00	15.32	1033.00	8.61
San-Sebastián	76.00	2.53	86.00	2.87	345.00	11.50	344.00	11.47	154.00	2.57	610.00	10.17	724.00	6.03
Bilbao													623.00	5.19
Santander	23.00	0.77							117.00	1.95			565.00	4.71
Gijón			101.00	3.37					151.00	2.52				
Vigo	171.00	5.70												
Porto														
Lisbon					107.00	3.57	139.00	4.63	207.00	3.45	232.00	3.87		
Faro									351.00	5.85				

(continued)

Table AII.

Cities of each transect	1950-1980 to 1980-2010		1980-2010 to 2010-2040		2010-2040 to 2040-2070		2040-2070 to 2070-2100		1950-1980 to 2010-2040		2010-2040 to 2070-2100		1950-1980 to 2070-2100	
	Distance (km)	Velocity (km/year)	Distance	Velocity	Distance	Velocity	Distance	Velocity	Distance	Velocity	Distance	Velocity	Distance	Velocity
Average Continental	218.08	7.27	159.54	5.32	339.30	11.31	422.26	14.08	331.38	5.52	716.32	11.94	987.64	8.23
Average Eastern Europe	232.96	7.77	214.00	7.13	337.73	11.26	454.54	15.15	378.62	6.31	715.56	11.93	964.48	8.04
Average Maritime	177.24	5.91	224.42	7.48	312.21	10.41	321.43	10.71	354.30	5.90	591.96	9.87	873.96	7.28
Average all transects	209.54	6.98	198.24	6.61	330.32	11.01	402.79	13.43	353.90	5.90	678.43	11.31	944.30	7.87

Note: Blank cells represent combinations of cities and shift time periods for which no climate analogue was found

Table AIII. Sensitivity analysis showing the absolute difference of minimum, maximum and mean southward climate shift velocity (for the three transects and the seven shift time periods) when computations are based on the second best climate analogues (for each LOIs) rather than on the best climate analogues

Shift time period	Velocity (km year^{-1})	Continental	Transects E. Europe	Maritime	Average (km year^{-1})
P1-P2	Min-max	1.5-0.5	1.2-1.4	0.4-0.7	
	Mean	0.2	0.3	0.1	0.2
P2-P3	Min-max	1.1-3.2	0.5-2.1	2.0-1.3	
	Mean	0.2	0.3	0.4	0.4
P3-P4	Min-max	0.9-0.8	1.9-3.1	0.8-2.4	
	Mean	0.2	0.4	0.2	0.3
P4-P5	Min-max	0.3-1.8	1.1-2.4	0.1-2.8	
	Mean	0.3	0.3	0.2	0.3
P1-P3	Min-max	0.7-2.5	0.5-3.1	1.5-1.7	
	Mean	0.2	0.1	0.1	0.1
P3-P5	Min-max	0.6-4.7	0.8-3.3	1.2-3.7	
	Mean	0.1	0.4	0.4	0.4
P1-P5	Min-max	1.1-4.2	0.8-3.8	1.4-2.5	
	Mean	0.2	0.3	0.3	0.3

Table AIV. Return periods in years of mean daily maximum temperatures for Aarhus (Denmark), Berlin (Germany) and Warsaw (Poland) and their respective climate analogues for the four future 30-year time periods

Cities	Daily max temperature (°C)	P1 city	P2 city	P2 climate analogue	P3 city	P3 climate analogue	P4 city	P4 climate analogue	P5 city	P5 climate analogue
Aarhus	29	23	11	11	9	9	5	3	2	2
	31	78	33	30	26	22	13	8	4	4
	33	261	98	85	79	60	34	25	9	9
	35	884	290	239	240	161	86	74	21	23
Berlin	35	18	13	19	7	7	5	6	1	1.0
	37	48	40	52	17	18	13	14	4	4
	39	134	119	148	40	43	33	39	6	9
	41	377	359	419	99	106	85	108	12	17
Warsaw	33	10	6	7	4	4	3	3	2	2
	35	43	19	23	9	8	7	7	3	4
	37	180	65	77	23	20	22	20	9	10
	39	767	223	261	73	62	62	51	23	29

Robustness of geography as an instrument to assess impact of climate change on agriculture

Muhammad Iftikhar Ul Husnain
Management Sciences, COMSATS Institute of Information Technology, Islamabad, Pakistan, and University of Glasgow, Adam Smith Business School, Glasgow, UK

Arjunan Subramanian
University of Glasgow, Adam Smith Business School, Glasgow, UK, and

Azad Haider
Economics, Saint Mary's University, Halifax, Canada, and Management Sciences, COMSATS Institute of Information Technology, Islamabad, Pakistan

Abstract

Purpose – The empirical literature on climate change and agriculture does not adequately address the issue of potential endogeneity between climatic variables and agriculture, which makes their estimates unreliable. This paper aims to investigate the relationships between climate change and agriculture and test the potential reverse causality and endogeneity of climatic variables to agriculture.

Design/methodology/approach – This study introduces a geographical instrument, longitude and latitude, for temperature to assess the impact of climate change on agriculture by estimating regression using IV-two-stage least squares method over annual panel data for 60 countries for the period of 1999-2011. The identification and *F*-statistic tests are used to choose and exclude the instrument. The inclusion of some control variables is supposed to reduce the omitted variable bias.

Findings – The study finds a negative relationship between temperature and agriculture. Surprisingly, the magnitude of the coefficient on temperature is mild, at least 20 per cent, as compared to previous studies, which may be because of the use of the instrumental variable (IV), which is also supported by an alternative robust measure when estimated across different regions.

Practical implications – The study provides strong implications for policymakers to confront climate change, which is an impending danger to agriculture. In designing effective policies and strategies, policymakers should focus not only on crop production but also on other agricultural activities such as livestock production and fisheries, in addition to national and international socio-economic and geopolitical dynamics.

Originality/value – This paper contributes to the growing literature in at least four aspects. First, empirical settings introduce an innovative geographical instrument, Second, it includes a wider set of control variables in the analysis. Third, it extends previous studies by involving agriculture value addition. Finally, the effects of temperature and precipitation on a single aggregate measure, agriculture value addition, are separately investigated.

Keywords Causality, Agriculture, Endogeneity, Climate change

1. Introduction

The relationships between climate change and agriculture are complex and manifold (Bosello and Zhang, 2005). A growing body of economic research has analysed the impact of climate change on agriculture (Parry *et al.*, 1999; Mendelsohn and Dinar, 1999; Mathauda *et al.*, 2000; Aggarwal and Mall, 2002; Jones and Thornton, 2003; Kumar *et al.*, 2004; Seo *et al.*, 2005; Deschenes and Greenstone, 2007; Schlenker and Roberts, 2008; Deressa and Hassan, 2009; Zhai and Zhaung, 2009; Schlenker and Lobell, 2010; Fisher *et al.*, 2012; Lee *et al.*, 2012; Bezabih *et al.*, 2014; Dasgupta *et al.*, 2014; Babar *et al.*, 2014; Javed *et al.*, 2014), and predicted that climatic variations will inflict wide range of economic losses across different regions and sectors (Hunt and Watkiss, 2011), yet the results of these studies have remained empirically elusive and controversial across time, countries and methodologies. This difference of results warrants attention.

It is well known that agriculture-related activities significantly emit greenhouse gases that lead to global warming. However, most of the existing literature does not adequately address the potential reverse causality and endogeneity of climatic variables to agriculture, which makes their estimates of the relevant causal effects unreliable (Kevane and Hirani, 2012). Furthermore, majority of these studies do not include control variables, such as fertilizers, population, technological changes and agriculture land area, and suffer omitted variable bias that may lead to misleading cross-country estimates, as pointed out by Auffhammer *et al.* (2012).

Some of the previous studies which identified potential endogeneity problem have used traditional instruments to address this issue by using lag values of the independent variables in estimating the impact of climate change on agriculture. However, these studies make a very strong assumption that no bidirectional relationship exists between previous year temperature and precipitation and current agriculture, which does not convincingly address the endogeneity problem. Further, these studies do not adequately discuss about the plausibility of their instruments used for temperature. Another limitation of this lag independent variable instrument technique is that it does not exclude the possibility that these instruments can directly affect the agriculture, which downscales the usefulness of these instruments. Javed *et al.* (2014) uses the lag of dependent variable to address the issue of endogeneity in analysing the impact of climate change on agriculture yield in Pakistan. However, the rationale behind the choice of the lagged dependent variable as an instrument is missing in their analysis. They also do not provide any discussion about controlling the econometric issues arising from the inclusion of lagged dependent variable as a regressor.

This study introduces geographical instruments, longitude and latitude, for temperature to assess the impact of climate change on agriculture because instrumental variables method makes it possible to assert the relationship between agriculture and climate change as a causal relationship rather than merely a correlation. Geographical location is a plausible instrument for weather conditions as these coordinates, longitude and latitude, are significant predictors of climate variables being closely linked to each other. It is believed that novel instruments can affect agriculture only through the channel of temperature and not directly that endorses the validity of these instruments. The validity of these instruments is probed by tests of weak identification and over identification, the results of which support the choice of the instruments. The F-statistic of joint significance is greater than 10 for the excluded instruments, thus passing the test for weak identification. In the case of over identification tests, the null hypothesis that the selected instrument is not correlated with the error term is not rejected.

A further strength of adopted empirical methodology is that it helps to overcome the problem of measurement error that emerges from the available data on climate variables, which are thought to be less reliable because of the limitations of the methods used to measure climatic variations. An instrumental variable approach handles the attenuation bias that may arise from measurement errors in independent variables, which produce biased estimates of the coefficients (Miguel *et al.*, 2004). The inclusion of some of the control variables such as area under cultivation may reduce the omitted variable bias as fluctuations in production may depend also on the area under cultivation and climate change. Likewise, technological improvements, fertilizers and other inputs also play their due role in unearthing the relationship between climate change and agriculture.

No study has considered the reverse causality between the dependent and independent variables by using geographical instruments. This paper aims to address contributes to the growing literature in at least four aspects. First, empirical settings introduce an innovative geographical instrument to capture impact of climate on agriculture, which establishes a causal relationship between climate change and agriculture rather than the simply correlation. Second, this analysis highlights the importance of including a wider set of control variables in examining the impact of climate change on agriculture. Third, this research improves previous research by using agriculture value addition, one of the most comprehensive, reliable and comparable measure of agriculture activity, as it accounts for all agriculture-related activities such as livestock and fisheries, which are also vulnerable to climate change. Previous empirical literature uses crop production, economic growth or revenues from crops as a measure for agriculture (Mendelsohn and Dinar, 2003; Deressa and Hassan, 2009; Kavikumar, 2009), and these are not comprehensive measures of agriculture. For example, economic growth is an outcome of the economic activity of different sectors in the economy, and agriculture is only a part of this measure. Finally, the effects of temperature and precipitation on a single aggregate measure, agriculture value addition, are separately investigated. Specifically, this paper constructs temperature and precipitation data for panel of 60 countries for the period of 1999-2011 and combines this data set with agriculture data. The sample includes countries from all habitant continents of the world, which makes the results of this study more generalisable and comparable to previous studies. The study uses data from World Development Indicators and Climate Research Unit of the University of East Anglia. The main identification strategy uses year-to-year fluctuations in temperature and precipitation to estimate the impact of temperature and precipitation on agriculture for each country included in the sample. The effects of climate fluctuations are measured using relatively few assumptions. It examines aggregated outcomes directly, rather than relying on *a priori* assumptions about what mechanisms to include and how they might operate, interact and aggregate.

The plan of the paper is as follows. Section 2 discusses the theoretical background of the impacts of climate change on agriculture and the model. Section 3 elaborates on variables, data and empirical methodology. Section 4 presents the results, and Section 5 provides a detail discussion of the results. Conclusions are presented in Section 6.

2. Materials and methods
2.1 The conceptual underpinnings
This section focusses on the meaning and the impact of climate change on agriculture. The fourth assessment report of the Intergovernmental Panel on Climate Change unequivocally states that the climate system is warming, and if necessary policy actions are not taken in time, the increase in greenhouse gases emission will continue and affect the climate. These unabated changes will inflict wide range of economic losses across different regions and sectors (Hunt and Watkiss, 2011). How these large effects are captured has been a contentious yet very important debate.

Robustness of geography as an instrument to assess impact of climate change...

131

A variety of models are in use to assess the potential impact of climate change on agriculture, yet the impact is not fully understood (Mendelsohn *et al.*, 1996) as environmental indicators are not included in the impact assessment (Antle, 1996). The most famous models in literature are the crop simulation models, production function models, Ricardian approach, general equilibrium models (GEMs), integrated assessment models (IAMs) and panel data models. A brief description of these models is available in Table I.

Table I shows that different approaches and techniques have been in use to assess the complex relationship between cropland and climatic variations across the world. These techniques range from simply mean averages to quantitative crop simulation models, panel and statistical time series models (Jones and Thornton, 2003; Lobell *et al.*, 2008). The choice of a model that better suits to unearth the link between climate change and agriculture is a complex area of research, because of factors such as data deficiencies, interactive behaviour and role of economic and agriculture policies across regions, lack of competency in understanding, applying and handling the models and role of uncertainties that are hard to anticipate in projection of the agriculture yield responses to future climate variability. Therefore, a few broad concepts such as clarity of objective, knowledge of agriculture policies, characteristics of the population and obtaining a reliable high frequency data could help in choosing a better model for the analysis.

Table I. Different models used to measure impact of climate change on agriculture

Method	Description	Strengths	Weaknesses
Crop simulation models	Crops are grown under controlled experiments and predictions are made about climate effects (Hebbar, 2008)	These models can predict and forecast impact of climate on crop production under different scenarios (Geethalakshmi *et al.*, 2011)	These models are considered only agriculture oriented as they focus on plant physiology and compare productivity levels under different climate scenarios (Eitzinger *et al.*, 2003)
Production function approach	A mathematical function that links agriculture inputs to output	Explicitly measures macroeconomic effects of weather variability on agriculture (Adams, 1989)	Unable to capture adaptation behaviour of farmers popularly known as "dumb farmer" phenomenon
Ricardian approach	It is a cross-sectional, across climate method to measure how climate affects land values and net revenues (Mendelsohn and Dinar, 2003; Kavikumar, 2009)	Accounts for direct effect of climate on yield of different crops and indirect substitution of different inputs (Mendelsohn *et al.*, 1994)	It does not include transition costs and does not capture the impact of space invariant variable (Sohngen *et al.*, 2002). It also assumes prices as constant (Cline, 1996)
GEMs	These models link agriculture to climate change considering its link with other sectors of the economy (Calzadilla *et al.*, 2010a)	Asses complex system of relationship simultaneously (Calzadilla *et al.*, 2013)	Supress the special characteristics of variables (Mendelsohn and Dinar, 2009)
IAMs	Combine agriculture data and economic models (Mikiko *et al.*, 2003)	Incorporate information from other disciplines. Describe cause and effect of climate change (Mikiko *et al.*, 2003)	Complex in nature and take climate as exogenous variable (Dinar and Mendelsohn, 2011)
Panel data models	Used to see the impact of environment on agriculture yield (McCarl *et al.*, 2008)	Capture time and space specific characteristics of the variables (Saravanakumar, 2015)	Use deviation from country specific means that leads to large measurement errors (Schlenker and Lobell, 2010)

2.2 Model

The present study uses stochastic production function approach suggested by Just and Pope (1978)[1] to estimate the effect of temperature and precipitation on agriculture, controlling for fertilizers, agriculture input imports, agriculture land area and population. This function has the following basic form:

$$y = f(X, \beta) \tag{1}$$

where y is measure of agriculture value addition, $f()$ is a production function, X is vector of independent variables and β is vector of estimable parameters attached with X. The study uses the following estimable regression form of equation (1).

$$Y_{i,t} = B_0 + \beta_1 T_{i,t} + \beta_2 P_{i,t} + \beta_3 F_{i,t} + \beta_4 AII_{i,t} + \beta_5 POP_{i,t} + \beta_6 ALA_{i,t} + \alpha_i + \gamma_t$$
$$+ \in_{i,t} \dots$$

$$\tag{2}$$

where $Y_{i,t}$ is the agriculture value of ith country at time t, T is temperature, P is precipitation, F is fertilizer, AII is agriculture input imports, POP is population, ALA is arable land area and $\in$ is an error term.

3. Data and estimation scheme

3.1 Data

This study uses panel data on agriculture value added from 1999 to 2011 for 60 countries representing all six habitant continents corresponding to temperature, precipitation, fertilizer, population and agriculture input imports data during the same period. The detailed description of variables is given below.

3.1.1 Dependent variable. Agriculture value added is the dependent variable in the model, measured in current US dollars, which includes fishing, forestry and cultivation of crops and livestock production. It is a net output from all these activities that is obtained after subtracting intermediate inputs from all outputs. Degradation of natural resources and depreciation for assets is not considered. The origin of value added is determined by the International Standard Industrial Classification, revision 3. The data for the variable are taken from World Development Indicators.

3.1.2 Climate variables. A number of climate databases are used in empirical literature to assess the impact of climate on different social outcomes. These data sets use different methods to measure, some use blend of surface and satellite while the others solely rely on surface, precipitation and temperature across spatial scale.

3.1.2.1 Average annual temperature. The study obtained data on temperature developed by Climate Research Unit, University of East Anglia, in conjunction with the Hadley Centre (at the UK Met Office), collected at 5° latitude by 5° longitude resolution. The observations are added for each month for each node in each year, and then, all the nodes in a country are averaged, which ends up a unique annual observation.

3.1.2.2 Average annual precipitation. Precipitation series is constructed from global climate data set, widely used in climate-related studies, that provides data on different weather locations in a country on latitude/longitude/altitude bases. Then data are averaged in the same way as in the case of temperature.

3.1.3 Control variables. In the model, four control variables adapted from World Development Indicators are included. The details of these variables are given below.

3.1.3.1 Fertilizer consumption. Agriculture productivity and use of fertilizer are closely linked; therefore, we control for the impact of fertilizer by including fertilizer consumption in the model. It contains nitrogenous, potash and phosphate fertilizers used per unit of arable land. It excludes traditional nutrients such as animal and plant manure.

3.1.3.2 Agriculture land area. A country having a larger land area is expected to have higher agriculture value addition. To control for this potential bias, agriculture land area is included in the regression equation. According to FAO, agriculture land refers to the share of land area (square kilometres) that is arable and includes land under temporary crops (double-cropped areas are counted once), temporary meadows for mowing or for pasture, land under market or kitchen gardens and land temporarily fallow, but excludes land under trees grown for wood or timber.

3.1.3.3 Agricultural input imports. Agriculture input imports also contribute to the agriculture yield of a country and therefore is included in the estimation equation. It consists of crude materials but excludes crude fertilizers.

3.1.3.4 Total population. Population can affect the agriculture value added through different channels. It counts mid-year estimates of all residents, regardless of legal status or citizenship except for refugees, based on the definition in World Development Indicators.

3.1.4 Instrumental variables. Agriculture and temperature are endogenous, as two-way potential causality possibly exists between the variables. Latitude and longitude determine, to greater extent, the temperature and precipitation of a location, respectively. Therefore, these two coordinates are used as instrument in the model. The data are obtained on the capital city in each country and are available at https://en.wikipedia.org/wiki/List_of_capital

3.1.4.1 Longitude. These are lines/meridian that run between the North and South Poles and locate the position of a point East West. The Prime Meridian is assigned 0° which bisects the Earth into equal West and East halves. Both the East and West halves are measured from 0° to 180° with due East and West are called "90°" respectively.

3.1.4.2 Latitude. Latitude is an angle which ranges from 0° at the Equator to 90° (North or South) at the poles. Lines of constant latitude or parallels run East-West as circles parallel to the equator. Latitude is used together with longitude to specify the precise location that features on the surface of the Earth.

3.2 Estimation technique

The study estimates equation (2) and uses country fixed effect α_i to control country-specific time invariant characteristics, such as geographic location, soil quality and soil type. Year fixed effect γ_t is controlled for shocks such as changes in national agriculture policies, introduction of new crop seed and cost shocks such as fossil fuel and fertilizer price, which are common to all countries in the given year. $\epsilon_{i,t}$ is an unobservable error term with zero mean. In analysing panel data, the decision whether to use random or fixed effects model is made on the basis of Haussmann test. But the choice of fixed effects is supported on the basis of two arguments. First, for random effect, the sample should be a random selection from a larger population, but this assumption is violated, as this study choses 60 countries across the world on some specific criteria. Second, a random effects model assumes that a country-specific effect is independent of the covariates included in the regression (Poudel and Kotani, 2013), which is unlikely to be fulfilled, as climate variable is correlated with the country-specific effect of agriculture production. At the end, this paper introduces instrumental variables to control for potential endogeneity of the climate variables with the agriculture production. To avoid the risk of spurious regression, it seems plausible to test time series properties of the variables included in the model. To achieve this purpose, Levin, Lin and Chu (LLC) panel unit root test (Levin *et al.*, 2002) is applied.

3.2.1 Panel unit root test. This study uses specification presented by Kula *et al.* (2009) for panel unit root test:

$$\Delta y_{i,t} = \alpha_i + \beta_i y_{i,t-1} + \sum_{l=1}^{Li} \gamma_{l,} \Delta y_{i,t-l} + \epsilon_{i,t}$$

where subscripts i and t stand for country and year, respectively, while $\Delta y_{i,t} = y_{i,t} - y_{i,t-1}$ and α_i, β_i and γ_l are intercept and slope coefficients to be estimated, respectively; li is the lag length to be determined using Schwartz or Akaike information criterion and $\epsilon_{i,t}$ is an error term. The null hypothesis for LLC panel unit root test is *H0:* $\beta = 0$ for all i against the alternative *H1:* $\beta < 0$ for all i. The test assumes that data are independent and identically distributed (i.i.d) across individuals, and the null hypothesis postulates that each individual time series is non-stationary against the alternative hypothesis that each time series has no unit root.

4. Results
4.1 Descriptive statistics
Descriptive statistics of all the variables used are reported in Table II. They show a large variation in precipitation while small changes in temperature.

It is found that the volatility of temperature is the highest in Asia and lowest in Africa. Also, highest volatility in precipitation is found in Asia and least is found in North America. Substantial differences in the variation of fertilizer use are found across continents. Arable land area, agriculture value added, population and agriculture input imports also reveal huge variation depending upon the differences in the geo-economic characteristics of the continents.

4.2 Result of panel unit root tests
The results of unit root tests are reported in Table III, which show that all the variables included in the model are stationary at level with constant, level with constant and trend and level with no constant and no trend.

Table II. Descriptive statistics

Variables	Des. Stat	Overall	Africa	Asia	Europe	North America	South America	Oceania
Agriculture value	Mean	22.56	21.61	23.75	22.18	24.85	22.89	21.89
Added#	SD	1.66	1.17	1.69	1.44	0.81	1.05	1.76
Temperature	Mean	17.01	22.64	19.79	10.79	17.95	19.55	19.60
	SD	7.49	3.98	8.81	4.81	4.78	3.93	4.79
Precipitation	Mean	868	701	1133	739	817	871	1423
	SD	599	557	947	237	180	300	698
Fertilizer	Mean	230	142	323	181	94	223	669
	SD	352	262	451	102	26	193	934
Agriculture input Imports#	Mean	1.75	1.96	1.25	1.57	1.30	1.70	0.69
	SD	1.01	1.26	1.57	0.61	0.20	0.76	0.29
Population#	Mean	16.75	16.50	1.47	15.93	19.01	17.14	15.23
	SD	1.67	1.09	15.93	1.49	0.51	1.12	1.33
Agriculture land area#	Mean	11.63	11.78	1.83	10.33	14.55	12.92	11.79
	SD	2.08	1.72	10.33	1.77	0.69	1.06	2.86
N		780	169	169	286	26	91	39

Note: # shows that variables are measured in log form because of their large values

Table III. Unit root tests results

Variables	Individual effect[a]	Individual effect	None[c]
Individual linear trend[b]			
Agriculture value added	−20.86***	−20.62***	−15.13***
Temperature	−163.84***	−163.98***	0.82
Precipitation	−57.41***	−66.16***	−1.50*
Fertilizer	−11.51***	−14.69***	−1.79**
Agriculture input imports	−9.02***	−7.55***	−10.30***
Population	−4.69***	−42.68***	3.19
Agriculture land area	−4.88***	−100.03***	−35.53***

Notes: ***, ** and *show that the null hypothesis of the presence of unit root is rejected at 99, 95 and 90% confidence level, respectively; [a]results estimated using the equation with only constant included; [b]Results estimated using the equation with constant term and deterministic trend included; [c]results estimated using the equation without constant term and trend

It can be observed that the *t*-statistics slightly, but not systematically, improves as the authors include trend in the model; therefore, there is no violation of the assumptions of classical linear regression model. So, it is not necessary to include the deterministic trend in the estimated equation (2), as suggested by Wooldridge (2008) to handle the problem of trending variables.

4.3 Identifying the mechanism
In this section, the authors identify the mechanism behind the bias caused by the omission of control variables by probing the sensitiveness of the estimated temperature, precipitation and agriculture nexus by adding each of the additional control variables one at a time. In Table IV, the first column reports estimates from the restricted model that include only temperature and precipitation. Then, each additional control variable is included separately. The last column shows the full model.

Overall, the estimated results support the full model as the preferred specification as the additional control variables are jointly significant with sufficiently large value of adjusted R^2. The coefficients on temperature and precipitation increase systematically when the

Table IV. Regression results after controlling for additional control variables

Variables	Temperature and precipitation (1)	Added (fertilizer) (2)	Added (agriculture inputs imports) (3)	Added (population) (4)	Added (agriculture land area) (5)
Temperature	−0.016*	−0.016*	−0.023*	−0.047***	−0.048***
Precipitation	0.001	0.001	0.003*	0.001*	0.002***
Fertilizer		−0.000	−0.000	0.000	0.003***
Agriculture Inputs imports			0.474***	−0.014	0.028*
Population				0.993***	0.827***
Agriculture Land area					0.092***
Adjusted R^2	0.003	0.004	0.009	0.488	0.489
Observation	780	780	780	780	780

Note: ***, ** and *show that the null hypothesis of the presence of unit root is rejected at 99, 95 and 90% confidence levels, respectively

estimated model moves from restricted to full model. This suggests that after controlling for control variables indeed, the temperature and precipitation are relatively high. In addition, all the signs of the variables are according to the prior expectations. Temperature has a negative effect while precipitation and all control variables have positive effects on agriculture.

4.4 IV-two-stage least squares results
In climate- and agriculture-related literature, temperature is used as an exogenous variable that can affect agriculture. However, it is interesting to note that agriculture and temperature are endogenous and causality may run both ways, i.e. from temperature to agriculture and vice versa. Agricultural activities produce greenhouse gases that lead to an increase in temperature, which makes temperature endogenous variable with agriculture. To control for this endogeneity, the authors use a geographical instrument, longitude and latitude, which is substantially correlated with temperature and can affect agriculture through temperature, as two coordinates (longitude and latitude) can provide information about the temperature and its corresponding changes. No instrument is used for precipitation, as it is assumed to be exogenous to agriculture, and no theoretical evidence is available, which suggests a bi-directional causality between agriculture and precipitation.

Therefore, the study estimates regression by using IV-2SLS method to overcome the potential threat of endogeneity and reports the results of both stages (Stages 1 and 2). The authors test whether, through endogeneity test, variables are endogenous or exogenous. On the basis of Durban score ($p < 1$ per cent) and Wu-Hausman F-statistics ($p < 1$ per cent), the authors reject the null hypothesis of "variables are exogenous" in all alternative specifications, which means that 2SLS is a plausible technique as it caters for endogeneity. The method of 2SLS introduces instruments for endogenous regressors.

Results reported in Table V show that the instrumental variables have right signs and are statistically significant, which shows that they can be substituted as an instrument for temperature. Results of IV (Stage 2) reported in Table V are discussed in detail, as these are the conclusive results and have importance from research and policy perspectives.

4.5 Robustness
Panel A of Table V presents the estimated impact of climate change on agriculture in the panel of 60 countries selected from different continents while controlling for continent fixed and year fixed effects. To check the robustness of findings, the authors did the same exercise for different continents, and the results are reported in the same Panel of Table V. Results show, broadly speaking, that temperature has a negative significant effect on agriculture across all continents, except South America and North America (results not reported), where it shows a positive yet insignificant association between temperature and agriculture. The reason for this positive impact could be the small number of observations that makes the statistical analysis less reliable. It is interesting to note that the magnitude of the temperature coefficient is relatively small, which is just continuity of our previous findings.

At the end, some diagnostic tests are conducted to check the validity of instruments used in the analysis, as in instrumental variables technique, it is mandatory to test whether the instruments are valid and strong. For this purpose, F-test is used. The significantly high value of the F-statistics ($F = 137.19$ for full sample) is greater than all the critical values obtained under 2SLS (Stage 1), which rejects the null hypothesis of "instruments are weak". The critical F-value ranges from 50 in Africa to 1,064 in Oceania continent, supporting that the instrument used is not weak. One of the other characteristic of a good instrument is that it should be highly correlated with the variable for which it is being used. The partial

Table V. Regression results for 2SLS

2SLS	Full sample (1)	Africa (2)	Asia (3)	Europe (4)	South America (5)	Oceania (6)
Panel A						
Longitude	−0.077* (0.004)	−0.145*** (0.012)	−0.126*** (0.009)	−0.039** (0.021)	0.051 (0.031)	−0.399 (0.433)
Latitude	−0.142** (0.009)	−0.020*** (0.015)	−0.459* (0.28)	−0.480*** (0.028)	0.219*** (0.012)	0.080 (0.412)
Constant	6.083*** (2.13)	36.644*** (2.956)	23.222*** (6.712)	29.081*** (4.106)	24.286*** (7.877)	117.26 (127.34)
Year fixed effects	Yes	Yes	Yes	Yes	Yes	Yes
Country fixed effects	Yes	Yes	Yes	Yes	Yes	Yes
Clustered SE	Yes	Yes	Yes	Yes	Yes	
Adjusted R^2	0.346	0.507	0.748	0.639	0.807	0.99
F-statistics	70.84	50.23	107	174.08	147.82	1,064
Observation	780	169	169	286	91	39
Panel B						
IV Stage two						
Temperature	−0.076 (0.007)	−0.049*** (0.019)	0.15*** (0.009)	−0.022*** (0.008)	0.043*** (0.031)	−0.112*** (0.024)
Constant	8.162*** (0.222)	3.677*** (1.046)	7.394*** (0.82)	9.440*** (0.404)	12.780*** (0.742)	−12.39** (7.09)
Year fixed effects	Yes	Yes	Yes	Yes	Yes	Yes
Country fixed effects	Yes	Yes	Yes	Yes	Yes	Yes
Clustered SE	Yes	Yes	Yes	Yes	Yes	Yes
R^2	0.84	0.828	0.832	0.89	0.910	0.990
Wald χ^2	4,919	857	948	2,620	933	5,001
Observation	780	169	169	286	91	39

Notes: Regressions control for fertilizer, agriculture input imports, population and agriculture land area. Temperature is instrumented with geographical coordinates (longitude and latitude). ***, ** and *show that the null hypothesis of the presence of unit root is rejected at 99, 95 and 90% confidence levels, respectively. Robust standard errors are reported in parenthesis. North America was excluded from the analysis because of less number of observations; SE: standard errors

correlation coefficients (0.62) obtained under 2SLS (Stage 1) is sufficiently high which conclude that temperature and the instrument used are closely related. The range of partial correlation coefficients across different continents is from 0.36 in Africa to 0.88 in Oceania. In the last step, over identification test is applied to test over identification restrictions. The very small values of the Sargan score (0.03) and Basman χ^2 (0.02) do not provide sufficient evidence against the null hypothesis of "no over identification" to reject. Throughout the analysis, the authors allow the effects of country-specific characteristics and time variant properties to be absorbed using the standard procedure.

5. Discussion
5.1 Climate variables
The two climate variables, temperature and precipitation, are the main focus of this paper. The negative sign of the coefficient of temperature shows that temperature affects agriculture negatively. This negative effect can work through a wide array of channels (Dell *et al.*, 2012). For example, higher temperature could cause lower agriculture yields, reduce livestock and affect labour productivity. The negative relationship between temperature and agriculture uncovered by this study is consistent with the findings of previous studies such as Li *et al.* (2015), who report that climate change will cause a decrease in rice production in most areas of the world. Lehmann *et al.* (2015) also found that irrigated crops will face water shortages as a result of increased temperature, which will negatively affect yields. Aggarwal and Mall (2002) show that an increase of 1°C-2°C will lead to 3-17 per cent fall in rice production across different regions. Bezabih *et al.* (2014) conclude that in general, agriculture in Ethiopia is highly responsive to variation in temperature. Deressa and Hassan (2009) examine the effects of annual temperature on net farm incomes and show that marginal increase in temperature significantly and negatively affects net crop revenue per hectare in summer and winter in Ethiopia. Mathauda *et al.* (2000) discovered a reduction in rice yield from 3.2 to 8.4 per cent as a consequence of slight-to-extreme increases in temperature. Mendelsohn and Dinar (1999) use three different scenarios to show that crop yields are negatively affected by rise in temperature in developing countries.

Many studies report positive effects of climate change on productivity. Lee *et al.* (2012) found that in summer, an increase in temperature increases agriculture yield in tropical countries. Babar *et al.* (2014) unearthed that increased temperature in the season of Rabi, from November to April, increases crop yield in Pakistan as higher temperature helps crops to mature in time. The rising temperature in mountain terrain increases the crop area and helps winter crops to mature in time, leading to increase in yields (Hussain and Mudasser, 2007).

As stated above, the main empirical finding is that the temperature generates a statistically significant negative effect on agriculture. However, this study estimate is relatively smaller than the estimates of several previous studies. For example, Adams *et al.* (1998) predict decrease in crop productivity from minimum 45 per cent in northeast states to maximum 66 per cent in lake states in USA under different climate change scenarios. Likewise, Parry *et al.* (2004) find negative impact of climate change even after realising the direct beneficial effect of CO_2 on plant growth and farm-level adaptation up to 22 per cent on world crop. Seo *et al.* (2005) found that rise in temperature is harmful, and the damage could lead to 50 per cent decrease in current agriculture productivity in Sri Lanka. According to this study's results, the elasticities of agriculture with respect to temperature range from 8 per cent in Africa to 30 per cent in Asia, which are at least 20 per cent smaller than those of the previous studies. Furthermore, the negative effects of temperature further scale down as the authors introduce geographical instruments to control for potential endogeneity. This decrease ranges from 38 per cent in Africa to 92 per cent in Europe. The inclusions of some important exogenous variables as control variables and introduction of strong instruments for climate in the estimated model may partly explain the difference of the estimates of the impacts of temperature on agriculture compared to previous studies. The estimated alternative robust measures show that magnitude of the estimated coefficients on temperature remains mild when estimated across different regions.

The results show that a positive association exists between precipitation and agriculture, which is in line with findings of a number of previous studies. For example, Seo *et al.* (2005) show that increase in rainfall is beneficial to crops, and the net revenues from the crops could increase from 11 to 122 per cent in Sri Lanka. Malik *et al.* (2012) report that average seasonal precipitation shift towards south of Pakistan improves the availability of water in normally dry winter season for agriculture lands, which increases the crop yield in the region. However, results of many studies do not support a positive association between precipitation and agriculture. Lehmann *et al.* (2015) conclude that despite record breaking precipitation around the world in the past decade, huge agriculture losses have occurred through increased infestation of pests and fungi that required additional efforts for pest control and treatment. Byjesh *et al.* (2010) show that the pattern of monsoon rain in Himalayan range reduces production of maize. Interestingly, some studies (Deschenes and Greenstone, 2007) report that rise in temperature and precipitation is not going to affect yield of major crops.

5.2 Control variables
Results reported in Table IV show that fertilizers have a strong positive effect on agriculture value added, as expected. Javed *et al.* (2014) also found a strong positive and statistically significant impact of fertilizer on agriculture production in Pakistan. Researchers hold different views on the effect of population on agriculture. The positive relationship between population and agriculture found in this study is supported by Templeton and Scherr (1997) and Boserup (1965), who conclude that population pressure induces intensive use of labour and institutional changes and reduces fallow periods. However, this finding contradicts the Malthusian view on the relationship. Cuffaro (1997) states that population growth induces adjustments in agriculture in terms of technical progress and intensification, thus enhancing yield. However, this optimism may not be justified as there are serious and growing concerns about the impacts of rapid population growth on natural resources (Ehrlich and Ehrlich, 1990).

Results show that imports of agriculture inputs have a positive effect on agriculture. These imports include mechanical and non-mechanical imports used in agriculture and allied activities such as livestock and fisheries. So, the growth of agriculture increases the demand for imported inputs for agriculture, which in turn boosts agricultural output in the country. Several studies of green revolution of the twentieth century showed that state interventions were important in supporting critical stages of agricultural market development, as reported by Dorward *et al.* (2004). Arable land area is one of the confounding factors in analysing the impact of climate change on agriculture. An increase in arable land area may generate a negative effect on climate but a positive effect on agricultural output. The authors find that arable land area is positively related to agriculture, as expected. This result is also in line with the findings of Javed *et al.* (2014), who document a significant positive relation for cultivated area and agriculture production.

6. Conclusions

In this study, the impacts of temperature and precipitation on agriculture value addition are investigated using the method of instrumental variables. Two geographical measures, longitude and latitude, are used as instruments for temperature, but no instrument is used for precipitation. The estimated model also controls for potential confounding factors that could affect agriculture value added across 60 countries, sampled from all habitant continents, for the period of 1999-2011. The study findings indicate that temperature and precipitation are negatively and positively related to agriculture value addition, respectively. However, the magnitudes of the estimated effects of climate variables are relatively smaller (at least by 20 per cent) than those reported in previous empirical studies for different parts of the world. These impacts decrease with the introduction of geographical instruments in the model. The results of the previous studies overstate the effects of temperature on agriculture. The difference in these results may be due to the inclusion of instrumental variables and control variables and the use of a larger sample that consists of 60 countries. Most of the countries in the sample are European countries, which are less vulnerable to climate change according to previous literature. As expected, all the control variables, agriculture inputs imports, fertilizers, population and arable land area are positively related to agriculture value added.

The results of this study highlight the importance and the urgency of implementing effective policies to mitigate the adverse effects of current climate change on agriculture on a global scale. In designing effective policies and strategies, policymakers should focus not only on crop production but also on other agricultural activities such as livestock production and fisheries, in addition to national and international socio-economic and geo-political dynamics. They should also consider the possible long-term effects of agricultural activities on arable land area and precipitation. The agriculture imports including fertilizers needs to be encouraged, and cultivation area expansion can increase agriculture yield and help mitigate adverse effects of climate change. Results of the study also indicate that population policies have implications for growth of agriculture. To avoid possible future shortages of food due to adverse impact of climate on agriculture, policymakers should focus on feed storage, livestock species diversification and introduction of new weather resilient crop varieties along with an increase in cultivated areas, as pointed out by Olesen and Bindi (2002).

Note

1. This approach is also used by Poudel and Kotani (2013).

References

Adams, R.M. (1989), "Global climate change and agriculture: an economic perspective", *American Journal of Agricultural Economics*, Vol. 71 No. 5, pp. 1272-1279.

Adams, R.M., McCarl, B.A., Segerson, K., Rosenzweig, C., Bryant, K.J., Dixon, B.L., Conner, R., Evenson, R.E. and Ojima, D. (1998), "The economic effects of climate change on US agriculture", in Mendelsohn, R.O. and Neumann, J.E. (Eds), *The Impact of Climate Change on the United States Economy*, Cambridge University Press, Cambridge, pp. 18-54.

Auffhammer, M., Ramanathan, V. and Vincent, J.R. (2012), "Climate change, the monsoon, and rice yield in India", *Climatic Change*, Vol. 111 No. 2, pp. 411-424.

Aggarwal, P.K. and Mall, R.K. (2002), "Climate change and rice yields in diverse agro-environments of India: effect of uncertainties in scenarios and crop models on impact assessment", *Climatic Change*, Vol. 52 No. 3, pp. 331-343.

Antle, M.J. (1996), "Methodological issues in assessing potential impacts of climate change on agriculture", *Agricultural and Forest Meteorology*, Vol. 80 No. 1, pp. 67-85.

Babar, S., Rehman, N. and Amin, A. (2014), "Climate change impact on rabi and kharif crops of Khyber-Pakhtunkhwa", *Humanities and Social Sciences*, Vol. 21 No. 1, pp. 49-56.

Bezabih, M., Di Falco, S. and Mekonnen, A. (2014), "On the impact of weather variability and climate change on agriculture: evidence from Ethiopia", working paper series 14-15, *Environment for Development*, 18 July.

Bosello, F. and Zhang, J. (2005), "Assessing climate change impacts: agriculture", working paper No. 02, Climate Impacts and Policy Division, 18 July.

Boserup, E. (1965), *The Conditions of Agricultural Growth*, Aldine Publishing, New York, NY.

Byjesh, K., Kumar, S.N. and Aggarwal, P.K. (2010), "Simulating impacts, potential adaptation and vulnerability of maize to climate change in India, mitigation and adaptation strategies", *Global Change*, Vol. 15 No. 5, pp. 413-431.

Calzadilla, A., Rehdanz, K. and Tol, R.S.J. (2010a), "The economic impact of more sustainable water use in agriculture: a computable general equilibrium analysis", *Journal of Hydrology*, Vol. 384 Nos 3/4, pp. 292-305.

Calzadilla, A., Rehdanz, K., Betts, R., Falloon, P., Wiltshire, A. and Tol, R.S.J. (2013), "Climate change impacts on global agriculture", *Climatic Change*, Vol. 120 Nos 1/2, pp. 357-374.

Cline, W.R. (1996), "The impact of global warming on agriculture: comment", *American Economic Review*, Vol. 86 No. 5, pp. pp. 1309-1311.

Cuffaro, N. (1997), "Population growth and agriculture in poor countries: a review of theoretical issues and empirical evidence", *World Development*, Vol. 25 No. 7, pp. 1151-1163.

Dasgupta, S., Hossain, M.M., Huq, M. and Wheeler, D. (2014), "Facing the hungry tide climate change livelihood threats, and household responses in coastal Bangladesh", World Bank Working Paper 7148.

Dell, M., Jones, F.B. and Olken, A.B. (2012), "Temperature shocks and economic growth: evidence from the last half century", *American Economic Journal: Macroeconomics*, Vol. 4 No. 3, pp. 66-95.

Deressa, T.T. and Hassan, M.R. (2009), "Economic impact of climate change on crop production in Ethiopia: evidence from cross-section measures", *Journal of African Economies*, Vol. 18 No. 4, pp. 529-554.

Deschenes, O. and Greenstone, M. (2007), "The economic impacts of climate change: evidence from agricultural output and random fluctuations in weather", *American Economic Review*, Vol. 97 No. 1, pp. 354-385.

Dinar, A. and Mendelsohn, R. (2011), *Handbook on Climate Change and Agriculture*, Edward Elgar, Cheltenham Glos.

Dorward, A., Kydd, J., Morrison, J. and Urey, I. (2004), "A policy agenda for pro-poor agricultural growth", *World Development*, Vol. 32 No. 1, pp. 73-89.

Eitzinger, J., Stastna, M., Zalud, Z. and Dubrovski, M. (2003), "A simulation study of the effect of soil water balance and water stress on winter wheat production under different climate change scenarios", *Agricultural Water Management*, Vol. 61 No. 3, pp. 195-217.

Ehrlich, P.R. and Ehrlich, A.H. (1990), *The Population Explosion*, Simon & Schuster, New York, NY.

Fisher, C.A., Hanemann, N.W., Roberts, J.M. and Schlenker, W. (2012), "The economic impacts of climate change: evidence from agricultural output and random fluctuations in weather: comment", *American Economic Review*, Vol. 102 No. 7, pp. 3749-3760.

Geethalakshmi, V., Lakshmanan, A., Rajalakshmi, D., Jagannathan, R., Gummidi, S., Ramaraj, A.P., Bhuvaneswari, K.L., Gurusamy, L. and Anbhazhagan, R. (2011), "Climate change impact assessment and adaptation strategies to sustain rice production in cauvery basin of Tamil Nadu", *Current Science*, Vol. 101 No. 3, pp. 342-347.

Hebbar, K.B. (2008), "Predicting cotton production using infocrop-cotton simulation model, remote sensing and spatial agro-climatic data", *Current Science*, Vol. 95 No. 11, pp. 1570-1579.

Hunt, A. and Watkiss, P. (2011), "Climate change impacts and adaptation in cities: a review of the literature", *Climatic Change*, Vol. 104 No. 1, pp. 13-49.

Hussain, S.S. and Mudasser, M. (2007), "Prospects for wheat production under changing climate in mountain areas of Pakistan-an econometric analysis", *Agricultural Systems*, Vol. 94 No. 2, pp. 494-501.

Javed, A.S., Ahmad, M. and Iqbal, M. (2014), "Impact of climate change on agriculture in Pakistan: a district level analysis", Climate change working paper No. 3, Pakistan Institute of Development Economics, 28 January.

Jones, P.G. and Thornton, P.K. (2003), "The potential impacts of climate change on maize production in Africa and Latin America in 2055", *Global Environmental Change*, Vol. 13 No. 1, pp. 51-59.

Just, R.E. and Pope, R.D. (1978), "Stochastic specification of production functions and econometric implications", *Journal of Econometrics*, Vol. 7 No. 1, pp. 67-86.

Kavikumar, K.S. (2009), "Climate sensitivity of Indian agriculture: do spatial effects matter?", Working paper No. 45, South Asian Network for Development and Environmental Economics, IUCN, Kathmandu, 25 November.

Kevane, M. and Hirani, R. (2012), "Robustness of climate as an instrumental variable to estimate effect of GDP declines on political change in Africa", available at: http://bellarmine2.lmu.edu/economics/papers/Kevane_Hirani_climate_and_outcomes_in_SSA_v14.pdf (accessed 26 May 2016).

Kula, F., Aslan, A. and Gozbasi, O. (2009), "Random walk or mean reversion of balance of payment in OECD countries: evidence from panel data", *Journal of Applied Sciences*, Vol. 9 No. 19, pp. 3606-3608.

Kumar, K.K., Kumar, R.K., Ashrit, G.R., Deshpande, R.N. and Hanseen, W.J. (2004), "Climate impacts on indian agriculture", *International Journal of Climatology*, Vol. 24 No. 11, pp. 1375-1393.

Lee, J., Nadolnyak, D. and Hartarska, V. (2012), "Impact of climate change on agricultural production in Asian countries: evidence from panel study", Prepared for presentation at the Southern Agricultural Economics Association Annual Meeting, Birmingham, February 4-7.

Lehmann, J., Dim, C. and Katja, F. (2015), "Increased record-breaking precipitation events under global warming", *Climatic Change*, Vol. 132 No. 4, pp. 501-515.

Levin, A., Lin, C.F. and Chu, C.S.J. (2002), "Unit root tests in panel data: asymptotic and finite-sample properties", *Journal of Econometrics*, Vol. 108 No. 1, pp. 1-24.

Li, T., Angeles, O., Radanielson, A., Marcaida, M. and Manalo, E. (2015), "Drought stress impacts of climate change on rainfed rice in South Asia", *Climatic Change*, Vol. 133 No. 4, pp. 709-720.

Lobell, D.B., Burke, B.M., Tebaldi, C., Mastrandrea, M.D., Falcon, W.P. and Nylore, R.S. (2008), "Prioritizing climate change adaptation needs for food security in 2030", *Science*, Vol. 319 No. 5863, pp. 607-610.

McCarl, B.A., Villacencio, X. and Wu, X. (2008), "Climate change and future analysis: is stationarity dying?", *American Journal of Agricultural Economics*, Vol. 90 No. 5, pp. 1241-1247.

Malik, M.K., Mahmood, A., Kazmi, H.D. and Khan, M.J. (2012), "Impact of climate change on agriculture during winter season over Pakistan", *Agricultural Sciences*, Vol. 03 No. 8, pp. 1007-1018.

Mathauda, S.S., Mavi, H.S., Bhangoo, B.S. and Dhaliwal, B.K. (2000), "Impact of projected climate change on rice production in Punjab (India)", *Tropical Ecology*, Vol. 41 No. 1, pp. 95-98.

Mendelsohn, R. and Dinar, A. (1999), "Climate change, agriculture, and developing countries: does adaptation matter?", *The World Bank Research Observer*, Vol. 14 No. 2, pp. 277-293.

Mendelsohn, R. and Dinar, A. (2003), "Climate, water and agriculture", *Land Economics*, Vol. 79 No. 3, pp. 328-341.

Mendelsohn, R. and Dinar, A. (2009), *Climate Change and Agriculture-an Economic Analysis of Global Impacts, Adaptation and Distributional Effect*, Edward Elgar, Cheltenham Glos.

Mendelsohn, R., Nordhaus, W. and Shaw, D. (1994), "Measuring the impact of global warming on agriculture", *American Economic Review*, Vol. 84 No. 4, pp. 753-771.

Mendelsohn, R., Nordhaus, W. and Shaw, D. (1996), "Climate impacts on aggregate farm value: accounting for adaptation", *Agricultural and Forest Meteorology*, Vol. 80 No. 1, pp. 55-66.

Miguel, E., Satyanath, S. and Sergenti, E. (2004), "Economic shocks and civil conflict: an instrumental variables approach", *Journal of Political Economy*, Vol. 112 No. 4, pp. 725-753.

Mikiko, M., Yuzuru, M. and Tsuneyuk, M. (2003), *Climate Policy Assessment Asia-Pacific Integrated Modeling*, Springer-Verlag, Tokyo.

Olesen, E.J. and Bindi, M. (2002), "Consequences of climate change for european agricultural productivity, land use and policy", *European Journal of Agronomy*, Vol. 16 No. 4, pp. 239-262.

Parry, M.L., Rosenzweig, C., Iglesias, A., Fischer, G. and Livermore, M. (1999), "Climate change and world food security: a new assessment", *Global Environmental Change*, Vol. 9, pp. 51-67.

Parry, M.L., Rosenzweigb, L.M., Iglesiasc, C., Livermored, A. and Fischere, G. (2004), "Effects of climate change on global food production under SRES emissions and socio-economic scenarios", *Global Environmental Change*, Vol. 14 No. 1, pp. 53-67.

Poudel, S. and Kotani, K. (2013), "Climatic impacts on crop yield and its variability in Nepal: do they vary across seasons and altitudes?", *Climatic Change*, Vol. 116 No. 2, pp. 327-355.

Saravanakumar, V. (2015), "Impact of climate change on yield of major food crops in Tamil Nadu, India", Working paper No. 91-15, South Asian Network for Development and Environmental Economics, January.

Schlenker, W. and Roberts, J.M. (2008), "Estimating the impact of climate change on crop yields: the importance of nonlinear temperature effects", NBER Working Paper, 13799, February.

Schlenker, W. and Lobell, B.D. (2010), "Robust negative impacts of climate change on African agriculture", *Environmental Research Letters*, Vol. 5 No. 1, pp. 8-15.

Seo, N.S., Mendelosn, R. and Munasinghe, M. (2005), "Climate change and agriculture in Sri Lanka: a ricardian valuation", *Environment and Development Economics*, Vol. 10 No. 5, pp. 581-596.

Sohngen, B.R., Mendelsohn, R. and Sedjo, R. (2002), "A global market of climate change impacts on timber markets", *Journal of Agricultural and Resource Economics*, Vol. 26 No. 2, pp. 326-343.

Templeton, S. and Scherr, J.S. (1997), "Population pressure and the microeconomy of land management in hills and mountains of developing countries", EPTD Discussion Paper No. 26, Environment and Production Technology Division, Washington, DC.

Wooldridge, J.M. (2008), *Introductory Econometrics*, 4th ed., South-Western College Publishing.

Zhai, F. and Zhaung, J. (2009), "Agricultural impact of climate change: a general equilibrium analysis with special reference to Southeast Asia", ADBI Working Paper Series No. 131, 15 February.

Corresponding author

Azad Haider can be contacted at: azad.haider@smu.ca

9

Climate impact assessment and "islandness": challenges and opportunities of knowledge production and decision-making for Small Island Developing States

Aideen Maria Foley

Department of Geography, Birkbeck University of London, London, UK

Abstract

Purpose – Climate data, including historical climate observations and climate model outputs, are often used in climate impact assessments, to explore potential climate futures. However, characteristics often associated with "islandness", such as smallness, land boundedness and isolation, may mean that climate impact assessment methods applied at broader scales cannot simply be downscaled to island settings. This paper aims to discuss information needs and the limitations of climate models and datasets in the context of small islands and explores how such challenges might be addressed.

Design/methodology/approach – Reviewing existing literature, this paper explores challenges of islandness in top-down, model-led climate impact assessment and bottom-up, vulnerability-led approaches. It examines how alternative forms of knowledge production can play a role in validating models and in guiding adaptation actions at the local level and highlights decision-making techniques that can support adaptation even when data is uncertain.

Findings – Small island topography is often too detailed for global or even regional climate models to resolve, but equally, local meteorological station data may be absent or uncertain, particularly in island peripheries. However, rather than viewing the issue as decision-making with big data at the regional/global scale versus with little or no data at the small island scale, a more productive discourse can emerge by conceptualising strategies of decision-making with unconventional types of data.

Originality/value – This paper provides a critical overview and synthesis of issues relating to climate models, data sets and impact assessment methods as they pertain to islands, which can benefit decision makers and other end-users of climate data in island communities.

Keywords Decision-making, Climate change, Uncertainty, Islands, Climate models

1. Introduction

It is widely recognised that Small Island Developing States (SIDS) are among the most vulnerable to the effects of climate change (Betzold, 2015; Wang *et al.*, 2016). The Alliance of Small Island States (AOSIS) has argued that climate change negotiations should aim to hold global warming below 1.5°C (Wong, 2011). The Paris Agreement reflects growing support for this limit, with governments agreeing to hold the increase "well below" the 2°C threshold while "pursuing efforts" to keep within 1.5°C of warming relative to pre-industrial levels. An IPCC special report on global warming of 1.5°C is in preparation, which will address the vulnerabilities of islands and coastal areas in particular. Limiting warming to 1.5°C would reduce risks to fishery sustainability (Cheung *et al.*, 2016) and coral reefs (Schleussner *et al.*, 2016) and modelling suggests that the 1.5°C target is feasible if a temperature overshoot is allowed and large, early reductions in emissions are made (Su *et al.*, 2017). However, under Intended Nationally Determined Contributions (INDCs) as of 2016, a median warming of at least 2.6°C is anticipated by 2100 (Rogelj *et al.*, 2016). Furthermore, even under an agreement of zero emissions, inertia in the climate system commits us to further sea level rise, as much as 2.3 m per degree of warming within the next 2,000 years (Levermann *et al.*, 2013).

Methods used in global and regional climate impact assessments to assess climate change risks are poorly suited to SIDS and, especially, atoll countries, given the fundamental mismatches in the spatial scales of knowledge creation and decision-making/action. Equally, reliance on global-scale models combined with uncertainties associated with local climate impacts may obscure opportunities for adaptation as relevance and credibility of information can act as a barrier to decision-making (Moser and Ekstrom, 2010). Understanding the capacities and limitations of typical data sources used in impacts assessment is, therefore, paramount if we are to ensure that suitable information and decision-making techniques are available to support adaptation and minimize maladaptation, in island contexts.

A single, coherent definition of "islandness" is elusive. Characteristics such as isolation and peripherality are often cited; yet, island communities can be highly integrated with the mainland and the rest of the world (Grydehøj and Hayward, 2014), making them at once both open and closed (Pugh, 2016). Hay (2013) noted that characteristics such as isolation, remoteness and containment could also apply to continental locations and argued more specifically the importance of the sea in defining islandness, particularly as the source of boundedness. It has also been contended that islandness is a metaphysical sensation that arises from physical isolation (Conkling, 2007), with Hay (2006) noting the enhanced sense of place that is often associated with islands. Similarly, Taglioni (2011) distinguished between insularity as a physical feature of certain spaces and islandness as the aggregate experiences of islanders. As such, while this paper is concerned predominantly with the confluence of assumed particular physical characteristics of islands, such as smallness, land boundedness, isolation and fragmentation (Fernandes and Pinho, 2017) and the challenges these pose to climate impact assessment, it recognises that these characteristics are neither exclusive to islands nor do they amount to a comprehensive definition of "islandness".

Arguably, while smallness and land boundedness pose the technical challenge to climate modelling, the isolation and fragmentation of islands are associated with further knowledge gaps relating to observed environmental and vulnerability data, leading to a sub-optimal decision-making basis for managing climate risk, if we rely on conventional data sources alone. This paper explores these challenges of islandness in both top-down, model-led forms of climate impact assessment and bottom-up, vulnerability-led approaches. It examines the role that alternative forms of knowledge production can play in validating models, where they can match the scale of island decision-making and in guiding adaptation actions at the local level and highlights decision-making techniques that can support adaptation even when data is uncertain.

2. Perceptions of climate risk in Small Island Developing States

Risks posed to SIDS by climate change include food insecurity (Barnett, 2011), water resource issues (Dore and Singh, 2013) and a range of human health impacts (McIver *et al.*, 2016). The majority of recent extinctions have occurred on islands (Courchamp *et al.*, 2014) and the threats posed by climate change to biodiversity take on a particular urgency in light of the high levels of endemism among island species (Wetzel *et al.*, 2012). In the Republic of Kiribati and in Tuvalu, migration from rural peripheral islands to urban central islands is already being attributed, in part, to climate change, coupled with socio-economic factors (Locke, 2009). In the short-term, as rural island populations relocate to urban areas, access to housing, employment and services needs to be considered to ensure positive outcomes for those migrating (Birk and Rasmussen, 2014). In the longer-term, islanders may be faced with losing their entire territory to the sea, which would raise highly complex issues relating to sovereignty and citizenship status (Skillington, 2017; Yamamoto and Esteban, 2010) and how to preserve the "lived values" of islands (Graham *et al.*, 2013).

Community-based adaptation projects have highlighted the potential for people to be highly perceptive of and attuned to their local environment yet be unaware of the threats climate change poses within that environment (Dumaru, 2010). In part, this stems from the reality that climate change is but one issue facing SIDS (Kelman, 2014). For example, in a case study of Funafuti, Tuvalu, McCubbin *et al.* (2015) found that people were more concerned about food, water and overcrowding than climate change. Yet, there is clear potential for climate change to influence these issues, potentially exacerbating or diminishing them.

Perception and awareness of climate change and its impacts can, thus, act as a barrier to climate adaptation (Betzold, 2015). For example, in a survey of students at the University of the South Pacific, Scott-Parker *et al.* (2017) found that a majority of the respondents believed climate change risks to be overstated, which may reflect a lack of trust in scientific sources of information among a cohort that may include future regional leaders.

Empirical data have a role to play in raising awareness of impacts among communities that underestimate the effects of climate change and in defining the local scale and scope of impacts, to assist where decision-making is impeded by the perceived magnitude of the issue. Co-learning approaches, which use both local and external scientific knowledge, may help to expand community understanding of climate change, but the scope to implement such approaches is limited where fine-grained, long-term data relating to the full range of impact-relevant climate parameters is unavailable or uncertain. As noted by McCubbin *et al.* (2015), the information provided by climate models generally refers to temperature and sea level change across broad regions and, as such, provides little insight into the specific concerns of a small island community. The following section will explore the technical limitations that give rise to this information gap.

3. "Islandness" in climate impact assessment

3.1 Scale and downscaling in climate modelling

Many climate impact assessments typically follow a sequential, top-down, model-led approach (Moss *et al.*, 2010), which could also be referred to as a "predict then act" approach (Dessai *et al.*, 2009). Socio-economic scenarios inform emissions scenarios, which in turn are used to produce radiative forcing scenarios. These radiative forcing scenarios are used to run climate models and climate model outputs such as temperature, precipitation and soil moisture are ultimately used as the inputs to impact models used to assess risks and vulnerabilities, often for a specific sector. For instance, the Inter-Sectoral Impact Model Intercomparison Project (ISI–MIP) involved global impact models for water, agriculture, biomes, coastal infrastructure and malaria (Warszawski *et al.*, 2014), but within most sectors, there are multiple impacts models to consider also (e.g. water resources; Schewe *et al.*, 2014).

Given the central role of climate models in these types of impact assessment, it is important to consider their suitability and limitations as they pertain to islands. The state-of-the-art atmosphere–ocean General Circulation Models (GCMs) used in the Coupled Model Intercomparison Project 5 (CMIP5; Taylor *et al.*, 2011) varies in resolution, but the average is ~1.85 × 2.25° [1]. For reference, at the Equator, 1° corresponds to ~111 km. The small surface area of many islands, together with their land boundedness, means that the grid cells corresponding to their location are likely to be classed as ocean in the model. Furthermore, as noted by Fernandes and Pinho (2017), island geomorphology may vary greatly across a small surface area, with volcanic islands such as Hawaii showing particularly large variations in altitude and, correspondingly, in bioclimatic conditions, little, if any, of which can be captured in a GCM.

Downscaling approaches can be used to address, in part, the scale mismatch. Regional climate models (RCMs) are a form of dynamical downscaling, in which GCM outputs are "regionalised" by a model operating at a higher resolution over a limited area. Figure 1 illustrates the effect that enhancing resolution can have on land surface representation.

The Providing Regional Climates for Impacts Studies (PRECIS)-Caribbean initiative has led to significant modelling capacity and localised climate information in the Caribbean region (Taylor *et al.*, 2013). The coordinated regional downscaling experiment (CORDEX) South Asia RCM simulations use an ensemble of different models and have a spatial resolution of 0.44° (~50 km) (Ghimire *et al.*, 2015), which is high resolution relative to GCMs. Statistical downscaling, a less computationally expensive method, in which local climate parameters are related to large-scale modelled variables, can also be applied. In the USA, statistical downscaling has been used to generate climate scenarios at 1/8° (~14 km) (Ahmed *et al.*, 2013).

However, in the case of SIDS, these methods can only bridge part of the gap and may come with additional uncertainty. Enhanced resolution is not a guarantee of reliable information. Centella-Artola *et al.* (2015) analysed PRECIS RCM simulations over the Caribbean with a resolution of 50 km and found that the default configuration of the model does not capture many of the smaller islands. In configurations that include the islands, by marking the nearest or covering grid boxes as land, a difference in simulated cloud cover is noted over the eastern Caribbean, highlighting how the absence or inclusion of small islands in models has ramifications for the quality of the regional projection.

Similarly, Cantet *et al.* (2014) generated climate scenarios for the Lesser Antilles using the ALADIN – Climate RCM nested within ARPEGE GCM and noted the islands are considered as land by the RCM model, but are not resolved at all by the driving GCM, which raises an important point relating to the credibility of RCM outputs for islands. RCMs are intended to add regional detail to a global scenario and so are highly dependent on the driving conditions received from the parent GCM (Foley *et al.*, 2013; Karmalkar *et al.*, 2013). As such, the inability of a GCM to resolve island topography has the potential to significantly impact the simulative skill of the RCM.

3.2 Validating models in island contexts

Given these uncertainties, it is important to validate any type of downscaling approach against observed data. Where such data are limited, however, it may not be possible to establish an adequate record of past climate variability against which to assess the performance of models in the present. Nunn *et al.* (2014) note the tendency of impact studies

(a)

(b)

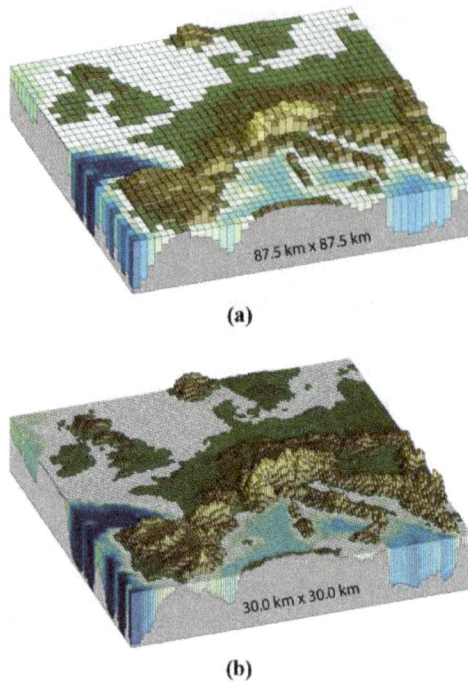

Notes: (a) Illustration of the European topography at
a resolution of 87.5 × 87.5 km; (b) same as (a) but
for a resolution of 30.0 × 30.0 km
Source: Reproduced from Cubasch *et al.* (2013,
Fig.1.14)

Figure 1. Horizontal resolutions considered in today's higher resolution models and in
very high resolution models now being tested

to focus on the most densely populated areas of islands in contrast to rural communities, which is attributable to the urban bias of governance and decision-making structures (Connell, 2010). In terms of data landscapes, there is equivalence in that peripheral areas tend to be most affected by data sparsity.

For instance, in a study of temperature trends in Fiji, Kumar *et al.* (2013) note that all meteorological stations analysed were located on or near the coast, with a scarcity of data in the island interior. However, differences in topography or land use may result in significant differences in climatic conditions even over short distances. For example, where data collection is limited to densely populated areas, urban heat effects may need to be taken into account. This effect is illustrated in Figure 2, using data from Koror, the largest city in Palau and Nekken Forestry, just 12 km away.

While data sparsity can also be an issue in continental locations, it is especially important in island contexts as the spatial gaps between observations can be very large due to the isolation and fragmentation of islands (Wright *et al.*, 2016). In their study using the PRECIS RCM in the Caribbean, Campbell *et al.* (2011) noted that the sparsity of Caribbean meteorological station data hampers validation of the model; data sparsity is also cited as a limitation in Whan *et al.*'s (2014) study of temperature extremes in the Western Pacific. In this way, challenges to validation can complicate the use of climate models, even for islands that are large enough to be adequately resolved within a model. For recent decades, satellite-based data sets such as NASA's Modern-Era Retrospective Analysis for Research and Applications (MERRA; Rienecker *et al.*, 2011) are a useful resource. However, for the purposes of establishing baselines for climate model evaluation, data are required over a longer timeframe, to sample the full range of natural climate variability.

3.3 Informing bottom-up vulnerability assessment

In the absence of specific data relating to future risks, current vulnerabilities and adaptations are often discussed as a means of exploring the potential for adapting to climate change (McCubbin *et al.*, 2015). Such an approach could be described as bottom-up, beginning with the identification of vulnerabilities, sensitivities and factors that increase resilience to climate-related threats (Falloon *et al.*, 2014; Wilby and Dessai, 2010). But climate change also creates new challenges, such as sea level rise and ocean acidification, which are unprecedented on human timescales and for which there are no traditional adaptations, or for which the limits of existing adaptations may be unknown (Weir *et al.*, 2017). Hence, the absence of appropriate data about future climate, against which to test adaptation strategies, may lead towards underestimation of future vulnerability.

Furthermore, while bottom-up approaches may not rely on climate data, they must use other kinds of data to gauge the impact of and vulnerability of communities to, past hazards. Here, similar themes emerge around the mismatch in spatial scales of knowledge and action on SIDS. For instance, in the aftermath of natural hazards, community level impacts can go unnoticed when local data (e.g. relating to numbers dead, injured and homeless) is combined and analysed at the national scale, potentially resulting in missed opportunities for effective intervention (Méheux *et al.*, 2007). Preston *et al.* (2011) finds that reliance on secondary data is common in vulnerability assessments, with none of the 45 studies reviewed incorporating primary data regarding biophysical factors and only nine per cent including primary data regarding socio-economic factors. Turvey (2007) also notes the challenges that data quality and availability issues pose in the creation of the composite vulnerability index (CVI) and particularly highlights the need for more information on the vulnerability of coastal environments to seasonal and interannual climate variability.

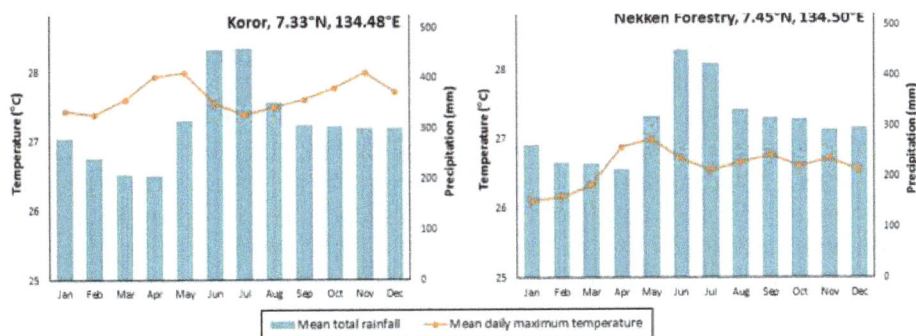

Source: NOAA's Climate Normals (Arguez *et al.*, 2010)

Figure 2. 1981-2010 climatological averages for rainfall and average temperature at two sites in Palau

4. Addressing the scale mismatch

4.1 Applications of alternative data sources

Developing a robust understanding of past climate change in islands is key, so that models can be evaluated appropriately and so that current sensitivities to change can be more clearly apprehended. Combining modern and historical data has the potential to produce novel insights into the links between local conditions and the larger, changing climate system.

Arguably, the strong sense of place associated with islands privileges and aids in the preservation of local knowledge on environmental risk. Local knowledge cannot provide the same types of information to decision makers such as GCMs and so cannot directly compensate for the issues identified around small island representation at model scales. However, it can provide important, additional information, which may relate not only to environmental change but also to potential human responses. For example, Fritz and Kalligeris (2008) highlight the case of the 1 April 2007 Solomon Islands tsunami, during which islanders knew to flee to higher ground after an earthquake based on ancestral knowledge of past events. Similarly, it may be informative to make use of narrative accounts and oral traditions (e.g. poetry and songs) of past climate and environmental changes (Janif *et al.*, 2016) to develop a long-term view of local contexts. In this manner, the cultural capital of island communities can connect with modern scientific knowledge.

Adger *et al.* (2013) argue that cultural aspects are infrequently incorporated into climate change analyses due to the challenge of merging the qualitative approaches more often used to study culture and the quantitative methods common in natural science. However, there are some examples. In a study of shoreline recession in the Solomon Islands, Albert *et al.* (2016) used historical aerial photography, satellite imagery and local historical insight to explore the interaction between sea level rise and other factors contributing to coastal recession, such as sea walls and extreme events. The merging of science and indigenous knowledge can also directly support adaptation in creative and unconventional methods. Hirsch (2015) described how Maldivians are embracing inter-island mobility and reimagining ancestral practices, giving the example of the smartphone app Nakaiy Nevi, which combines the indigenous Maldivian calendar system of *nakaiy* with weather observations. Weather-related traditional knowledge, relating plant and animal behaviours to meteorological phenomena, such as tropical cyclones, can also be incorporated into scientific forecasting tools (Magee *et al.*, 2016), although Chand *et al.* (2014) noted that the verification of traditional knowledge as a forecasting method is hindered by the lack of historical records for phenological responses (e.g. flowering times of mango trees as a predictor of cyclone activity).

Data rescue activities, including digitisation of historical meteorological records, facilitate such research. The Climate Data for the Environment (CliDE) system, a web-based climate data management tool, supports such activities and has been deployed in 14 Pacific island states (Martin *et al.*, 2015). Workshops and peer-to-peer data exchange have also been successfully used in the Western Pacific to extend existing data sets and build research capacity in the region, although the length and quality of some of the datasets identified remain an issue (McGree *et al.*, 2014). Citizen science strategies can also play a part here, building two-way collaborations between local communities and research teams, to not only disseminate research findings and inform decision makers but also harness local knowledge and experience and build synergies with existing priorities (Petridis *et al.*, 2017).

4.2 Robust decision-making strategies

Advances in computing capacity and climate modelling will inevitably enhance the quality of future climate projections that can be made for SIDS, but decisions about climate adaptation strategies will still have to be made against a backdrop of uncertainty. At each stage of the sequential process described previously, uncertainties and inconsistencies accumulate, widening the envelope of uncertainty associated with specific impacts. Even variation in methods and interpretations of uncertainty assessment can lead to different policy outcomes (Wesselink *et al.*, 2015). Uncertainty may also impact how adaptation actions are perceived; for example, if a community decided to migrate in response to potential, but as yet unobserved, effects of climate change, would the available evidence base impact the level of international support they might receive (Kelman, 2015)?

Uncertainty can be perceived as a barrier to decision-making or used as a rationale to avoid decision-making, but in reality, it should not preclude action as uncertainties and unknowns are inevitable in any decision-making scenario. While computation strategies, using conventional decision-support tools such as multi-criteria analysis and cost–benefit analysis, would have to be based on an uncertain knowledge base, compromise, judgement or inspiration strategies could be more appropriate choices (de Boer *et al.*, 2010) and may better reflect the important socio-political factors that shape the SIDS vulnerability. Within such strategies, robust decision-making could be applied to help decision makers understand the conditions under which a particular proposed policy would fail and ultimately identify policies that will endure under a range of scenarios (Lempert, 2013). A similar concept is decision scaling, introduced by Brown *et al.* (2012), which seeks to identify broadly the climate states that favour a particular decision and then establish the probability of occurrence using GCM data, thus lessening the specificity of information required from the models while enhancing the relevance of the data to the decision.

4.3 Visualising and communicating future climates

Given the inevitability of uncertainty, it is also worth considering how best to convey information about possible future climates. Communication of climate data relies heavily on visualisation; yet, there has been limited research into the effectiveness of different approaches, particularly in the context of understanding deep uncertainty (Spiegelhalter *et al.*, 2011). Kaye *et al.* (2012) noted that Web-based interfaces could offer innovative ways to explore uncertainty, compared to static approaches, proposing visualisations in which the user can specify an acceptable level of uncertainty, generating a map that only displays regions meeting that criteria. Wesselink *et al.* (2015) advised referring to processes and trade-offs when communicating results rather than relying on numerical ranges alone.

Links between spatial scales of data and risk perception may also benefit from further study. In a study of volcanic hazard mapping methods on Montserrat, Haynes *et al.* (2007) found that perspective photographs were significantly more effective than other visualisations, as people could better identify features and their orientation. Daly *et al.* (2010) subsequently used large-scale aerial photographs, with landmarks identified, in participatory research on coastal vulnerability in Samoa, allowing islanders to make links between the visualisations and their own perceptions and experiences of hazards. It has been suggested that map users may associate a high level of visual realism in geospatial images, with higher confidence in the quality of the data underlying those image (Kettunen *et al.*, 2012). How, then, might the inability of a model to adequately resolve an island, or indeed, resolve it at all, impact the perceived trustworthiness of data and communication of risk? Is a coarse map better or worse than no map at all? Questions such as these are particularly relevant in the context of islands, considering the strong sense of place shared by many island communities (Baxter *et al.*, 2015; Coulthard *et al.*, 2017).

5. Conclusions

This paper has outlined some of the ways in which characteristics commonly allied with "islandness", such as smallness, land boundedness and isolation, may limit the applicability of global and regional climate impact assessment methods to SIDS and, particularly, to atoll countries. Small island topography can be too detailed for current GCMs to resolve, but equally, local meteorological station data can only represent a point and records may be sparse or short in island peripheries. As well as the technical limitations of climate models and data sets, the absence or unreliability of local-level vulnerability data may be perceived as a barrier to adaptation actions.

Yet, there is a need for a robust decision-making basis in light of the kinds of climate adaptation policies under consideration. Climate science is increasingly harnessing big data (Schnase *et al.*, 2017) and by comparison, the decision-making basis for SIDS may seem limited on account of the constraints discussed, but this would be an oversimplification. Rather than viewing the issue as decision-making with big data versus with little or no data, a more productive discourse can emerge by conceptualising strategies of decision-making with *different* data. While climate models can still provide scenarios of future change, decision-making that uses those scenarios must recognise the inherent uncertainties. Continued recovery of historical observations, both of climate parameters and associated environmental phenomena identified through traditional knowledge, can aid in formulating baselines against which to assess models and evaluate long-term change. Looking forward, there is potential for crowdsourcing of climate data using citizen weather stations (Meier *et al.*, 2017) and smartphones (Mass and Madaus, 2014), which could play an important role in monitoring and understanding the onset of climate impacts. Alternative forms of knowledge production and robust decision-making can aid in raising awareness and enhancing the perception of climate change in island communities. The hybridization of science and indigenous knowledge offers much potential provided that the strengths and limitations of all forms of information are suitably acknowledged (Lebel, 2013) and that the merging of knowledge takes place in a culturally compatible manner (Mercer *et al.*, 2007).

Note

1. https://portal.enes.org/data/enes-model-data/cmip5/resolution

References

Adger, W.N., Barnett, J., Brown, K., Marshall, N. and O'Brien, K. (2013), "Cultural dimensions of climate change impacts and adaptation", *Nature Climate Change*, Vol. 3 No. 2, pp. 112-117.

Ahmed, K.F., Wang, G., Silander, J., Wilson, A.M., Allen, J.M., Horton, R. and Anyah, R. (2013), "Statistical downscaling and bias correction of climate model outputs for climate change impact assessment in the US northeast", *Global and Planetary Change*, Vol. 100, pp. 320-332.

Albert, S., Leon, J.X., Grinham, A.R., Church, J.A., Gibbes, B.R. and Woodroffe, C.D. (2016), "Interactions between sea-level rise and wave exposure on reef island dynamics in the Solomon Islands, *Environmental Research Letters*, Vol. 11 No. 5, pp. 54011.

Arguez, A., Durre, I., Applequist, S., Squires, M., Vose, R., Yin, X. and Bilotta, R. (2010), *NOAA's U.S. Climate Normals (1981-2010)*, NOAA National Centers for Environmental Information, doi: 10.7289/V5PN93JP (accessed 17 June 2017).

Barnett, J. (2011), "Dangerous climate change in the Pacific Islands: food production and food security", *Regional Environmental Change*, Vol. 11 No. 1, pp. 229-237.

Baxter, G. Cooper, K. Gray, D. Reid, P.H. Vergunst, J. and Williams, D. (2015), "The use of photo elicitation to explore the role of the main street in Kirkwall in sustaining cultural identity, community and a sense of place", available at: https://openair.rgu.ac.uk/handle/10059/1185 (accessed 20 June 2017).

Betzold, C. (2015), "Adapting to climate change in small Island developing states", *Climatic Change*, Vol. 133 No. 3, pp. 481-489.

Birk, T. and Rasmussen, K. (2014), "Migration from atolls as climate change adaptation: current practices, barriers and options in Solomon Islands", *Natural Resources Forum*, Vol. 38 No. 1, pp. 1-13.

Brown, C., Ghile, Y., Laverty, M. and Li, K. (2012), "Decision scaling: linking bottom-up vulnerability analysis with climate projections in the water sector", *Water Resources Research*, Vol. 48 No. 9, pp. W09537.

Campbell, J.D., Taylor, M.A., Stephenson, T.S., Watson, R.A. and Whyte, F.S. (2011), "Future climate of the Caribbean from a regional climate model", *International Journal of Climatology*, Vol. 31 No. 12, pp. 1866-1878.

Cantet, P., Déqué, M., Palany, P. and Maridet, J.-L. (2014), "The importance of using a high-resolution model to study the climate change on small islands: the Lesser Antilles case", *Tellus A*, Vol. 66 No. 1, pp. 24065.

Centella-Artola, A., Taylor, M.A., Bezanilla-Morlot, A., Martinez-Castro, D., Campbell, J.D., Stephenson, T.S. and Vichot, A. (2015), "Assessing the effect of domain size over the Caribbean region using the PRECIS regional climate model", *Climate Dynamics*, Vol. 44 Nos 7/8, pp. 1901-1918.

Chand, S.S., Chambers, L.E., Waiwai, M., Malsale, P. and Thompson, E. (2014), "Indigenous knowledge for environmental prediction in the pacific island countries", *Weather, Climate and Society*, Vol. 6 No. 4, pp. 445-450.

Cheung, W.W.L., Reygondeau, G. and Frölicher, T.L. (2016), "Large benefits to marine fisheries of meeting the 1.5°C global warming target", *Science*, Vol. 354 No. 6319, pp. 1591-1594.

Conkling, P. (2007), "On Islanders and Islandness", *Geographical Review*, Vol. 97 No. 2, pp. 191-201.

Connell, J. (2010), "Pacific islands in the global economy: paradoxes of migration and culture", *Singapore Journal of Tropical Geography*, Vol. 31 No. 1, pp. 115-129.

Coulthard, S., Evans, L., Turner, R., Mills, D., Foale, S., Abernethy, K., Hicks, C. and Monnereau, I. (2017), "Exploring 'islandness' and the impacts of nature conservation through the lens of wellbeing", *Environmental Conservation*, Vol. 44 No. 3, pp. 1-12.

Courchamp, F., Hoffmann, B.D., Russell, J.C., Leclerc, C. and Bellard, C. (2014), "Climate change, sea-level rise and conservation: keeping island biodiversity afloat", *Trends in Ecology and Evolution*, Vol. 29 No. 3, pp. 127-130.

Cubasch, U., Wuebbles, D., Chen, D., Facchini, M.C., Frame, D., Mahowald, N. and Winther, J.-G. (2013), "Introduction", in Stocker, T.F., Qin, D., Plattner, G.-K., Tignor, M., Allen, S.K., Boschung, J., Nauels, A., Xia, Y., Bex, V. and Midgley, P.M. (Eds), *Climate Change 2013: The Physical Science Basis. Contribution of Working Group I to the Fifth Assessment Report of the Intergovernmental Panel on Climate Change*, Cambridge University Press, Cambridge and New York, NY, pp. 119-158, doi: 10.1017/CBO9781107415324.007.

Daly, M., Poutasi, N., Nelson, F. and Kohlhase, J. (2010), "Reducing the climate vulnerability of coastal communities in Samoa", *Journal of International Development*, Vol. 22 No. 2, pp. 265-281.

Dessai, S., Hulme, M., Lempert, R. and Pielke, R. (2009), "Do we need better predictions to adapt to a changing climate?", *Eos, Transactions American Geophysical Union*, Vol. 90 No. 13, pp. 111-112.

de Boer, J., Wardekker, J.A. and van der Sluijs, J.P. (2010), "Frame-based guide to situated decision-making on climate change", *Global Environmental Change*, Vol. 20 No. 3, pp. 502-510.

Dore, M.H.I. and Singh, R.G. (2013), "Projected future precipitation scenarios for a small island state: the case of Mauritius", in Younos, T. and Grady, C.A. (Eds), *Climate Change and Water Resources*, Springer, Berlin Heidelberg, pp. 47-66.

Dumaru, P. (2010), "Community-based adaptation: enhancing community adaptive capacity in Druadrua Island, Fiji", *Wiley Interdisciplinary Reviews: Climate Change*, Vol. 1 No. 5, pp. 751-763.

Falloon, P., Challinor, A., Dessai, S., Hoang, L., Johnson, J. and Koehler, A.-K. (2014), "Ensembles and uncertainty in climate change impacts", *Frontiers in Environmental Science*, Vol. 2, pp. 33.

Fernandes, R. and Koehler, A.-K. (2017), "The distinctive nature of spatial development on small islands", *Progress in Planning*, Vol. 112, pp. 1-18.

Foley, A., Fealy, R. and Sweeney, J. (2013), "Model skill measures in probabilistic regional climate projections for Ireland", *Climate Research*, Vol. 56 No. 1, pp. 33-49.

Fritz, H.M. and Kalligeris, N. (2008), "Ancestral heritage saves tribes during 1 April 2007 Solomon Islands tsunami", *Geophysical Research Letters*, Vol. 35 No. 1, p. L01607.

Ghimire, S., Choudhary, A. and Dimri, A.P. (2015), "Assessment of the performance of CORDEX-South Asia experiments for monsoonal precipitation over the Himalayan region during present climate: part I", *Climate Dynamics*, pp. 1-24, available at: https://doi.org/10.1007/s00382-015-2747-2

Graham, S., Barnett, J., Fincher, R., Hurlimann, A., Mortreux, C. and Waters, E. (2013), "The social values at risk from sea-level rise", *Environmental Impact Assessment Review*, Vol. 41, pp. 45-52.

Grydehøj, A. and Hayward, P. (2014), "Social and economic effects of spatial distribution in island communities: comparing the isles of Scilly and Isle of Wight, UK", *Journal of Marine and Island Cultures*, Vol. 3 No. 1, pp. 9-19.

Hay, P. (2013), "What the sea portends: a reconsideration of contested island tropes", *Island Studies Journal*, Vol. 8 No. 2, pp. 209-232.

Hay, P.R. (2006), "A phenomenology of islands", *Island Studies Journal*, Vol. 1 No. 1, pp. 19-42.

Haynes, K., Barclay, J. and Pidgeon, N. (2007), "Volcanic hazard communication using maps: an evaluation of their effectiveness", *Bulletin of Volcanology*, Vol. 70 No. 2, pp. 123-138.

Hirsch, E. (2015), "It won't be any good to have democracy if we don't have a country: climate change and the politics of synecdoche in the Maldives", *Global Environmental Change*, Vol. 35, pp. 190-198.

Janif, S., Nunn, P., Geraghty, P., Aalbersberg, W., Thomas, F. and Camailakeba, M. (2016), "Value of traditional oral narratives in building climate-change resilience: insights from rural communities in Fiji", *Ecology and Society*, Vol. 21 No. 2, available at: https://doi.org/10.5751/ES-08100-210207

Karmalkar, A.V., Taylor, M.A., Campbell, J., Stephenson, T., New, M., Centella, A., Benzanilla, A. and Charlery, J. (2013), "A review of observed and projected changes in climate for the islands in the Caribbean", *Atmósfera*, Vol. 26 No. 2, pp. 283-309.

Kaye, N.R., Hartley, A. and Hemming, D. (2012), "Mapping the climate: guidance on appropriate techniques to map climate variables and their uncertainty", *Geoscientific Model Development*, Vol. 5 No. 1, pp. 245-256.

Kelman, I. (2014), "No change from climate change: vulnerability and small island developing states", *The Geographical Journal*, Vol. 180 No. 2, pp. 120-129.

Kelman, I. (2015), "Difficult decisions: migration from small island developing states under climate change", *Earth's Future*, Vol. 3 No. 4, 2014EF000278.

Kettunen, P., Irvankoski, K., Krause, C.M., Sarjakoski, T. and Sarjakoski, L.T. (2012), "Geospatial images in the acquisition of spatial knowledge for wayfinding", *Journal of Spatial Information Science*, Vol. 2012 No. 5, pp. 75-106.

Kumar, R., Stephens, M. and Weir, T. (2013), "Temperature trends in Fiji: a clear signal of climate change", *The South Pacific Journal of Natural and Applied Sciences*, Vol. 31 No. 1, pp. 27-38.

Lebel, L. (2013), "Local knowledge and adaptation to climate change in natural resource-based societies of the Asia-Pacific", *Mitigation and Adaptation Strategies for Global Change*, Vol. 18 No. 7, pp. 1057-1076.

Lempert, R. (2013), "Scenarios that illuminate vulnerabilities and robust responses", *Climatic Change*, Vol. 117 No. 4, pp. 627-646.

Levermann, A., Clark, P.U., Marzeion, B., Milne, G.A., Pollard, D., Radic, V. and Robinson, A. (2013), "The multimillennial sea-level commitment of global warming", *Proceedings of the National Academy of Sciences*, Vol. 110 No. 34, pp. 13745-13750.

Locke, J.T. (2009), "Climate change-induced migration in the pacific region: sudden crisis and long-term developments1", *Geographical Journal*, Vol. 175 No. 3, pp. 171-180.

Magee, A.D., Verdon-Kidd, D.C., Kiem, A.S. and Royle, S.A. (2016), "Tropical cyclone perceptions, impacts and adaptation in the Southwest Pacific: an urban perspective from Fiji, Vanuatu and Tonga", *Natural Hazards and Earth System Sciences*, Vol. 16 No. 5, pp. 1091-1105.

Mass, C.F. and Madaus, L.E. (2014), "Surface pressure observations from smartphones: a potential revolution for high-resolution weather prediction?", *Bulletin of the American Meteorological Society*, Vol. 95 No. 9, pp. 1343-1349.

Martin, D.J., Howard, A., Hutchinson, R., McGree, S. and Jones, D.A. (2015), "Development and implementation of a climate data management system for western Pacific small island developing states", *Meteorological Applications*, Vol. 22 No. 2, pp. 273-287.

McCubbin, S., Smit, B. and Pearce, T. (2015), "Where does climate fit? Vulnerability to climate change in the context of multiple stressors in Funafuti, Tuvalu", *Global Environmental Change*, Vol. 30, pp. 43-55.

McGree, S., Whan, K., Jones, D., Alexander, L.V., Imielska, A., Diamond, H., Ene, E., Finauhali, S., Inape, K., Jacklick, L., Kumar, R., Laurent, V., Malala, H., Malsale, P., Moniz, T., Ngemaes, M., Peltier, A., Porteus, A., Pulehetoa-Mitiepo, R., Seuseu, S., Skilling, E., Tahani, L., Teimitsi, F., Toorua, F. and Vaiimene, M. (2014), "An updated assessment of trends and variability in total and extreme rainfall in the western Pacific", *International Journal of Climatology*, Vol. 34 No. 8, pp. 2775-2791.

McIver, L., Kim, R., Woodward, A., Hales, S., Spickett, J., Katscherian, D., Hashizume, M., Honda, Y., Kim, H., Iddings, S., Naickere, J., Bambrick, H., McMichael, A.J. and Ebi, K.L. (2016), "Health impacts of climate change in pacific island countries: a regional assessment of vulnerabilities and adaptation priorities", *Environmental Health Perspectives*, Vol. 124 No. 11, pp. 1707-1714.

Méheux, K., Dominey-Howes, D. and Lloyd, K. (2007), "Natural hazard impacts in small island developing states: a review of current knowledge and future research needs", *Natural Hazards*, Vol. 40 No. 2, pp. 429-446.

Meier, F., Fenner, D., Grassmann, T., Otto, M. and Scherer, D. (2017), "Crowdsourcing air temperature from citizen weather stations for urban climate research", *Urban Climate*, Vol. 19, pp. 170-191.

Mercer, J., Dominey-Howes, D., Kelman, I. and Lloyd, K. (2007), "The potential for combining indigenous and western knowledge in reducing vulnerability to environmental hazards in small island developing states", *Environmental Hazards*, Vol. 7 No. 4, pp. 245-256.

Moser, S.C. and Ekstrom, J.A. (2010), "A framework to diagnose barriers to climate change adaptation", *Proceedings of the National Academy of Sciences*, Vol. 107 No. 51, pp. 22026-22031.

Moss, R.H., Edmonds, J.A., Hibbard, K.A., Manning, M.R., Rose, S.K., van Vuuren, D.P., Carter, T.R., Emori, S., Kainuma, M., Kram, T., Meehl, G.A., Mitcheel, J.F.B., Nakicenovic, N., Riahi, K., Smith, S., Stouffer, R.J., Thomson, A., Weyant, J. and Wibanks, T.J. (2010), "The next generation of scenarios for climate change research and assessment", *Nature*, Vol. 463 No. 7282, pp. 747-756.

Nunn, P.D., Aalbersberg, W., Lata, S. and Gwilliam, M. (2014), "Beyond the core: community governance for climate-change adaptation in peripheral parts of Pacific Island Countries", *Regional Environmental Change*, Vol. 14 No. 1, pp. 221-235.

Petridis, P., Fischer-Kowalski, M., Singh, S.J. and Noll, D. (2017), "The role of science in sustainability transitions: citizen science, transformative research and experiences from Samothraki island, Greece", *Island Studies Journal*, Vol. 12 No. 1, pp. 115-134.

Preston, B.L., Yuen, E.J. and Westaway, R.M. (2011), "Putting vulnerability to climate change on the map: a review of approaches, benefits and risks", *Sustainability Science*, Vol. 6 No. 2, pp. 177-202.

Pugh, J. (2016), "The relational turn in island geographies: bringing together island, sea and ship relations and the case of the Landship", *Social & Cultural Geography*, Vol. 17 No. 8, pp. 1040-1059.

Rienecker, M.M., Suarez, M.J., Gelaro, R., Todling, R., Bacmeister, J., Liu, E., Bosilovich, M.G., Schubert, S., Takacs, L., Kim, G.K., Bloom, S., Chen, J., Collins, D., Conaty, A., Silva, A., Joanna Joiner, W., Koster, R.D., Lucchesi, R., Molod, A., Owens, T., Pawson, S., Pegion, P., Redder, C.R., Reichle1, R., Robertson, F., Ruddick, A.G., Sienkiewicz, M. and Woollen, J. (2011), "MERRA: NASA's modern-era retrospective analysis for research and applications", *Journal of Climate*, Vol. 24 No. 14, pp. 3624-3648.

Rogelj, J., den Elzen, M., Höhne, N., Fransen, T., Fekete, H., Winkler, H., Schaeffer, R., Sha, F., Riahi, K. and Meinshausen, M. (2016), "Paris agreement climate proposals need a boost to keep warming well below 2°C", *Nature*, Vol. 534 No. 7609, pp. 631-639.

Schewe, J., Heinke, J., Gerten, D., Haddeland, I., Arnell, N.W., Clark, D.B., Dankers, R., Eiser, S., Fekete, B., Colon-Gonzales, F., Gosling, S., Kim, H., Liu, X., Masaki, Y., Portman, F., Satoh, Y., Stacke, T., Tang, Q., Wada, Y., Wisser, D., Albrecht, T., Frieler, K., Piontek, F., Warszawski, L. and Kabat, P. (2014), "Multimodel assessment of water scarcity under climate change", *Proceedings of the National Academy of Sciences*, Vol. 111 No. 9, pp. 3245-3250.

Schleussner, C.-F., Lissner, T.K., Fischer, E.M., Wohland, J., Perrette, M., Golly, A., Rogelj, J., Childers, K., Schewe, J., Frieler, K., Mengel, M., Hare, W. and Schaeffer, M. (2016), "Differential climate impacts for policy-relevant limits to global warming: the case of 1.5°C and 2°C", *Earth System Dynamics*, Vol. 7 No. 2, pp. 327-351.

Schnase, J.L., Duffy, D.Q., Tamkin, G.S., Nadeau, D., Thompson, J.H., Grieg, C.M., McInerney, M.A. and Webster, W.P. (2017), "MERRA analytic services: meeting the big data challenges of climate science through cloud-enabled climate analytics-as-a-service", *Computers, Environment and Urban Systems*, Vol. 61 Part B, pp. 198-211.

Scott-Parker, B., Nunn, P.D., Mulgrew, K., Hine, D., Marks, A., Mahar, D. and Tiko, L. (2017), "Pacific islanders' understanding of climate change: where do they source information and to what extent do they trust it?", *Regional Environmental Change*, Vol. 17 No. 4, pp. 1005-1015.

Skillington, T. (2017), "On the rights of the peoples of disappearing states", *Climate Justice and Human Rights*, Palgrave Macmillan, New York, NY, pp. 177-206.

Spiegelhalter, D., Pearson, M. and Short, I. (2011), "Visualizing uncertainty about the future", *Science*, Vol. 333 No. 6048, pp. 1393-1400.

Su, X. Takahashi, K. Fujimori, S. Hasegawa, T. Tanaka, K. Kato, E. Shiogama, H. Masui, T. and Emori, S. (2017), "Emission pathways to achieve 2.0°C and 1.5°C climate targets", Earth's Future, available at: https://doi.org/10.1002/2016EF000492.

Taglioni, F. (2011), "Insularity, political status and small insular spaces", *The International Journal of Research into Island Cultures*, Vol. 5 No. 2, pp. 45-67.

Taylor, K.E., Stouffer, R.J. and Meehl, G.A. (2011), "An overview of CMIP5 and the experiment design", *Bulletin of the American Meteorological Society*, Vol. 93 No. 4, pp. 485-498.

Taylor, M.A., Centella, A., Charlery, J., Bezanilla, A., Campbell, J., Borrajero, I., Stephenson, T. and Nurmohamed, R. (2013), "The precis Caribbean Story: lessons and legacies", *Bulletin of the American Meteorological Society*, Vol. 94 No. 7, pp. 1065-1073.

Turvey, R. (2007), "Vulnerability assessment of developing countries: the case of small-island developing states", *Development Policy Review*, Vol. 25 No. 2, pp. 243-264.

Wang, G., Power, S.B. and McGree, S. (2016), "Unambiguous warming in the western tropical Pacific primarily caused by anthropogenic forcing", *International Journal of Climatology*, Vol. 36 No. 2, pp. 933-944.

Warszawski, L., Frieler, K., Huber, V., Piontek, F., Serdeczny, O. and Schewe, J. (2014), "The inter-sectoral impact model intercomparison project (ISI–MIP): project framework", *Proceedings of the National Academy of Sciences*, Vol. 111 No. 9, pp. 3228-3232.

Weir, T., Dovey, L. and Orcherton, D. (2017), "Social and cultural issues raised by climate change in pacific island countries: an overview", *Regional Environmental Change*, Vol. 17 No. 4, pp. 1017-1028.

Wesselink, A., Challinor, A.J., Watson, J., Beven, K., Allen, I., Hanlon, H., Lopez, A., Lorenz, S., Otto, F., Morse, A., Rye, C., Saux-Picard, S., Stainforth, D. and Suckling, E. (2015), "Equipped to deal with uncertainty in climate and impacts predictions: lessons from internal peer review", *Climatic Change*, Vol. 132 No. 1, pp. 1-14.

Wetzel, F.T., Kissling, W.D., Beissmann, H. and Penn, D.J. (2012), "Future climate change driven sea-level rise: secondary consequences from human displacement for island biodiversity", *Global Change Biology*, Vol. 18 No. 9, pp. 2707-2719.

Whan, K., Alexander, L.V., Imielska, A., McGree, S., Jones, D., Ene, E., Finaulahi, S., Inape, K., Jacklick, L., Kumar, R., Laurent, V., Mala, H., Mlasale, P., Pulehetoa-Mitiepo, R., Ngemaes, M., peltier, A., Porteous, A., Seuseu, S., Skilling, E., Tahani, L., Toorua, U. and Vaiimene, M. (2014), "Trends and variability of temperature extremes in the tropical Western Pacific", *International Journal of Climatology*, Vol. 34 No. 8, pp. 2585-2603.

Wilby, R.L. and Dessai, S. (2010), "Robust adaptation to climate change", *Weather*, Vol. 65 No. 7, pp. 180-185.

Wong, P.P. (2011), "Small island developing states", *Wiley Interdisciplinary Reviews: Climate Change*, Vol. 2 No. 1, pp. 1-6.

Wright, E., Sutton, J., Luchetti, N., Kruk, M. and Marra, J. (2016), "Closing the pacific rainfall data void", *Eos*, Vol. 97, available at: https://doi.org/10.1029/2016EO055053

Yamamoto, L. and Esteban, M. (2010), "Vanishing Island States and sovereignty", *Ocean & Coastal Management*, Vol. 53 No. 1, pp. 1-9.

Corresponding author
Aideen Maria Foley can be contacted at: a.foley@bbk.ac.uk

10

Resource extractivism, health and climate change in small islands

Hilary Bambrick

School of Public Health and Social Work, Queensland University of Technology, Queensland, Australia

Abstract

Purpose – The extraction of natural resources has long been part of economic development in small islands. The damage to environment and health is extensive, even rendering once productive islands virtually uninhabitable. Rather than providing long-term benefits to the population or to the environment, the culture of "extractivism" – a nonreciprocal approach where resources are removed and used with little care or regard to consequences – has instead left many in far more fragile circumstances, increasingly dependent on external income. The purpose of this paper is to show how continued extractivism in small islands is contributing to global climate change and increasing climate risks to the local communities.

Design/methodology/approach – Through a series of case studies, this paper examines the history of extractivism in small islands in Oceania, its contribution to environmental degradation locally and its impacts on health.

Findings – It examines how extractivism continues today, with local impacts on environment, health and wellbeing and its much more far-reaching consequences for global climate change and human health. At the same time, these island countries have heightened sensitivity to climate change due to their isolation, poverty and already variable climate, whereas the damage to natural resources, the disruption, economic dependence and adverse health impacts caused by extractivism impart reduced resilience to the new climate hazards in those communities.

Practical implications – This paper proposes alternatives to resource extractivism with options for climate compatible development in small islands that are health-promoting and build community resilience in the face of increasing threats from climate change.

Originality/value – Extractivism is a new concept that has not previously been applied to understanding health implications of resource exploitation thorough the conduit of climate change. Small-island countries are simultaneously exposed to widespread extractivism, including of materials contributing to global climate change, and are among the most vulnerable to the hazards that climate change brings.

Keywords Adaptation, Climate change, Coal mining, Health impacts, Resource depletion, Small islands

Introduction

The extraction and export of natural resources has long been part of economic development in the Pacific and nearby islands and is at its heart founded in colonialism. The damage to local environment and health as a result of resource extraction in a number of countries has been extensive and, in some cases extreme, even rendering once productive islands virtually uninhabitable. Rather than providing long-term benefits to the population or to the environment or to the local economy, the culture of "extractivism" (Klein, 2015) – a nonreciprocal approach where resources are removed and used for monetary gain with little care or regard to consequences – has instead left many in far more fragile circumstances, increasingly dependent on external income and support.

The intention of the paper is to demonstrate through a series of case studies how extractivism has not only wrought havoc on small islands historically but that it is a practice that continues to occur to the present day. In particular, this paper intends to show that the negative impacts of extractivism are neither short-term nor limited to those islands. Rather, due to the nature of the resources now being extracted, these negative effects have increasingly global and long-term consequences, which in turn especially threaten the lives and livelihoods of those in the very communities from where the extracted resources originate.

The geographic scope of this paper is small islands in Oceania. These islands share similar colonial histories and countries involved in the resource extraction and are subject to similar kinds of environmental and health threats arising from extractivist practices. This paper is not intended to provide an exhaustive list of resource types and locations where extractivism has or does occur. There are many more examples of extractivism in small islands that have not been included here, as similar practices are repeated around the world (Klein, 2015).

The approach taken is to review and synthesize a selection of case studies that highlight the variety of resources that have been removed from small islands over time and where the benefits flow to external parties at significant cost to local communities – in resource depletion, environmental damage and impacts on community health and well-being. As the practice of extractivism continues, these themes of colonial power, inequity, short-term remain evident in the present day, with the added potential for damage to environment, health and society now on unprecedented scale.

Phosphate

Perhaps the best-known example of extractivism in small islands and the consequent damage to environment, society and health is from the island of Nauru. Located in Micronesia, Nauru, is a small island made up of phosphate rock. Phosphate mining commenced at the turn of past century and continued until resources were exhausted in the 1990s. Phosphate was extracted first by Europe and Australia, with about 30 per cent of it removed before independence in 1968, when the Nauruan government continued to extract it and place the money into a trust fund to create income for Nauruans (Gale, 2016, D'Odorico and Rulli, 2014). Prior to this nationalization of phosphate mining, the financial returns to Nauruans were microscopic (Pollock, 2014). With nationalization, the island nation subsequently underwent rapid economic development, and for nearly two decades, per capita income in Nauru was the highest in the world. This money, along with the phosphate reserves, is now gone (D'Odorico and Rulli, 2014).

The environmental loss has been extreme. Most of the island is missing, leaving just a coastal ring, and pollution from mining has also devastated surrounding fisheries from polluted run-off. Australia paid AU$135m in 1993 in compensation to rehabilitate the island, but rehabilitation has not occurred (Gale, 2016). With limited arable land and damaged local fisheries, the 10,000 residents of Nauru are almost entirely dependent on imported food. Despite a decline in recent decades, the reported prevalence of diabetes remains high at 14 per cent (Khambalia et al., 2011), and an extremely high prevalence of obesity remains (71 per cent) [World Health Organization (WHO), 2014]. The heavy burden from the serious sequelae of cardiovascular disease, retinopathy and renal disease (Win Tin et al., 2014) means that "healthy" life expectancy is 12 years below life expectancy at birth for both women and men, and overall increases in life expectancy fall below those for its Pacific neighbors [World Health Organization (WHO), 2015]. Once rich in vegetation and marine life, Nauru is now crowded and largely barren, dependent on income from an Australian government offshore detention center for asylum seekers (Fleay and Hoffman, 2014), and on payment for the eventual resettlement of refugees within the already resource scarce and socially fragile nation.

Similarly, Banaba Island in Kiribati, a near neighbor to Nauru, has all but been destroyed by phosphate mining which took place from 1900 to 1979 (Taeiwa, 2015). Here phosphate mining by Britain also forced migration from the island; all of the 1,003 inhabitants were removed in 1945 and taken to remote Rabi Island in Fiji, more than 1,000 km away (Edwards, 2014). Since phosphate mining ceased, some Banaba people have returned to Banaba Island, but as on Nauru, decades of mining has removed all the soil from the center of the island, leaving only the coastal ring habitable and with any vegetation (Taeiwa, 2015). Similar to Nauru, Banaba island has variable annual rainfall and is prone to drought, threatening food security further where there is so little remaining arable land.

The Banaba Islanders' new home of Rabi Island has an official population of around 3,000, based on the most recent (2007) Fiji census count (Fiji Bureau of Statistics, 2017) and estimated population growth rates (2007-2014) for the Cakuadrove province (Fiji Bureau of Statistics, 2015) to which Rabi Island belongs. Locals estimate population size to be around 5,000 (Bambrick and Moncada, 2015). The people of Rabi are food insecure, with added risk from climate change causing declining rainfall and more intense cyclones which damage food crops and fisheries, and the supply of clean water is tenuous. As a consequence, Rabi Island has high rates of diarrhea and other communicable disease while also being at risk from vector-borne diseases such as dengue which may intensify with changing climate (Bambrick and Moncada, 2015). These problems are compounded by being a small ethnic minority on a remote Fiji island, with limited infrastructure, services and health care.

The mining of phosphate to produce fertilizer in the wealthy west has not only devastated the islands of Nauru and Banaba, leaving dependence and declining health standards in its wake, but also disrupted the distant ecological systems where it has been applied in industrial scale agriculture. While phosphate is naturally stored in rock and slowly released to plants, the intense application of phosphate-rich fertilizer to increase productivity pollutes the surrounding lakes, streams and oceans, triggering eutrophication and algal blooms, depleting oxygen in the water (United Nations Environment Programme, 2001).

Fisheries

The Pacific Ocean supplies half of the world's tuna (Jin *et al.*, 2015). The region has long been subject to large-scale fishing of skipjack, yellowfin and bigeye tuna by international fleets, often using the purse-seine technique, where a school of fish is encircled with a net and everything within it is captured. While some large-scale international fishing is regulated, and taxes on foreign fishing provide substantial income to many Pacific Islands (Bell *et al.*, 2013), much of the fishing is unsupported by international agreements and is therefore unregulated and frequently illegal (Grewe *et al.*, 2015). The immense size of the Pacific Ocean and the limited capacity for local surveillance means that the illegal extractivism of the Pacific's ocean resources can occur largely unabated and with no royalties being paid to the countries whose waters are being exploited. Furthermore, purse-seining risks substantial unintended by-catch, whereby other non-target species are caught in the nets. The massive scale of the (especially unregulated) fishing industry in the Pacific and the use of purse seining threatens marine ecosystems and reduces the availability of fish for local use. Local catch rates are declining in the Pacific and are expected to continue to decline over coming decades (Albert *et al.*, 2015).

Climate change puts additional pressures on fisheries through its impacts on marine ecosystems and will have significant impacts on local food security where communities are dependent on coastal and offshore fisheries as a primary food source. There is less data available on the status of coastal fisheries (Albert *et al.*, 2015), but the consequences of climate change (in tandem with growing populations and therefore increased consumption pressures) are expected to be negative for fish availability and safety. This is particularly the case for coral reef fish for which climate change may cause a decline in abundance of 20 per cent by 2050 (Bell *et al.*, 2013). Rising temperatures may reduce local fish stocks, change species profiles and increase the risk of ciguatera contamination in reef fish by promoting the abundance of toxic algae (Llewellyn, 2010, Derne *et al.*, 2010).

Coral reef ecosystems, vital for sustaining coastal fisheries, may be further disrupted by coral fungal disease outbreaks from increased ocean acidification resulting from rising carbon dioxide levels and warmer ocean temperatures (Williams *et al.*, 2014). Ecological disturbance of reef systems can increase predominance of algal associated fish species (Feary *et al.*, 2007), and closed atoll lagoons with minimal flushing (such as in Kiribati and Marshall Islands) are especially at risk (Andrefouet *et al.*, 2015). Lack of fish affects not only food security in Pacific communities but also social cohesion and the capacity to continue local traditions, as fishing and activities around fisheries are frequently central to cultural and social events (Kittinger *et al.*, 2015). Perhaps more positively, the potential for warmer temperatures to increase tuna abundance in the eastern Pacific and possibly increase productivity of freshwater aquaculture (Bell *et al.*, 2013) could help offset overfishing and declines in reef and coastal fisheries and related threats to Pacific food security but would require a transition to oceanic fish becoming a more prominent local food source rather than where these fish are primarily extracted by foreign trawlers.

Logging

Illegal logging by foreign parties is another form of extractivism that takes place in the Oceania that is widespread throughout the region (Dinnan and Walton, 2016). While import and export bans have reduced the amount of forest loss that takes place, there remains a market for illegal and unregulated logging of high value timber (Felbab-Brown, 2011). In particular, foreign companies that enter under the umbrella of a local person or company as a means to avoid the logging and export bans that otherwise limit extractivism. Solomon Islands is a particular target because of their valuable hardwood forests and has experienced, for example, the smuggling of valuable unprocessed logs to international traders (Radio Australia, 2012) under the guise of nominating a local person as sole director of a foreign owned company to enter the country and commence business following barring by court order (Sanga, 2016). The rate of logging and export of primary forest in Solomon Islands is unsustainable and with no local processing there is little local economic benefit (Australian Centre for International Agircultural Research, 2012). This lack of manufacturing or value-adding to gain economic benefit from resource depletion is common in small islands with limited capacity for manufacture (including skilled labor, infrastructure, regulations), with only a few countries, such as Vanuatu and Fiji, having developed small-scale local industry associated with forestry (Australian Centre for International Agircultural Research, 2012). Unregulated logging creates local environmental problems such as erosion and increased landslide risk, and with loss of primary forest, loss of biodiversity (Franklin and Steadman, 2010) and invasion of pest species, including exotic grasses that make forest re-generation more difficult (Denslow *et al.*, 2006). Further, like with fossil fuels (below), forest loss also contributes to climate change through the removal of carbon sinks and directly to carbon emissions if the wood is then burned.

Fossil fuels

The non-reciprocal approach to resource extraction in small islands in Oceania now extends to coal, oil and gas. As with phosphate, fisheries and logging, the extraction of fossil fuels causes local damage to environment (e.g. soil degradation, biodiversity loss) and health (Morrice and Colagiuri, 2013). With coal, oil and gas, however, the potential scale of ecological disruption is also much greater, with global implications through the release of greenhouse gases. Climate change is disrupting the earth's life support systems on a grand scale as carbon from fossil fuels, that has been trapped in the material for millennia, is released when it is burned (IPCC, 2013). The deleterious consequences of extractivism in Oceania are no longer confined to the region but rather extend worldwide.

Oil and gas mining is subject to extractivist culture in Oceania. Timor-Leste is a small, least developed country situated in the Indonesian Archipelago. Although Timor-Leste gained independence from Indonesia in 2002, it remains subject to a temporary maritime border which favors Australia in the ownership of major oil and gas fields that lie between the two countries, rather than a border that sits halfway between the two, which would favor Timor-Leste. This arises from an historical agreement between Indonesia and Australia (the Timor Gap Treaty) which in 1991 gave Australia rights to the area (Commonwealth of Australia, 1995). Since 2007, the profits from the oil and gas have been shared evenly between Australia and Timor-Leste (Commonwealth of Australia, 2006) although more recently relations between the two countries became strained with allegations of espionage against Australia as Timor-Leste mounted its case for what it sees as a fairer distribution of resources (Belot and Stewart, 2017). While the shared profits provide some much-needed income to Timor-Leste, the burning of this oil and gas for energy contributes to greenhouse gas emissions, and Timor-Leste is one of the most vulnerable countries globally to the impacts of climate change. It is poor and reliant on subsistence agriculture, with highly variable rainfall already subject to both severe drought and severe flooding and with both periods of drought and rainfall intensity expected to increase with climate change. The extremely mountainous topography adds to Timor-Leste's risk, with erosion and landslides associated with intense rainfall.

Despite the apparent economic benefits to developing countries, mining fossil fuels is not a healthy option. Oil and gas mining contribute to environmental damage and social disruption at source, but it is coal mining especially that has well-established direct risks to the workers and surrounding communities.

Coal mining has become a principle source of income in the region. Indonesia has begun mining coal in the Province of West Papua for use in its coal-fired power stations in Sulawesi (Somba, 2008), whereas the Government of Papua New Guinea, another country already damaged by extractivism in gold and copper mining, is also turning to coal, having recently invested millions in developing its own local industry (Wilson, 2017). Echoing the sequelae of phosphate mining in Nauru and Banaba, the local communities themselves see little economic benefit as profits flow instead to largely overseas owned mining companies, with some taxes going to government (Somba, 2008).

Coal dust is an important contributor to particulate air pollution, which is a cause of death and illness from cardiovascular and respiratory disease. The miners themselves are not the only ones who are affected, but the surrounding community and the communities through which coal is transported (Environment Defenders Office NSW, 2010). While there are some data collected on the occupational health of miners (more so in wealthier countries with relatively functioning workplace safety regulations), there is very little data on the contribution of coal dust to respiratory disease in the broader population. Estimates from Australia do, however, give an indication of how substantial this might be. In the coal mining Hunter region in the state of New South Wales, it is estimated that 42 million kg of coal dust is distributed each year, largely through transport of the material to major ports (Environment Defenders Office NSW, 2010). Coal mining in West Papua may not be of this scale, but the environmental protections are even poorer. The mine produces poor health outcomes for local workers and residents; also the quality of coal is especially low and thus highly polluting, contributing to even more poor air quality and increased carbon emissions where it is burned for power in Sulawesi.

There are additional potential impacts from coal mining on the local communities. There may be increased income for workers (alongside the high occupational risks) and also increasing income disparity between those working in the mines and those who are not. Reliance on mineral exports for income is an established source of civil conflict (Ballard and Banks, 2003). The attraction of a higher income may trigger an exit from agriculture for creating livelihoods and a subsequent loss of local food production and a greater reliance on more expensive and possibly less nutritious imports.

Income flowing into communities from the presence of mining is also likely to be minimal with profits going to the private companies external to the communities. The inwards migration of workers for the mines may promote social disruption, causing local tensions and conflict and increased pressure on local resources of water, land and food.

Much of the world is now turning toward renewable energy (solar, wind and hydroelectric power). For some of these countries keen to develop economically, trying to extract what fossil fuels they can while it still has a dollar value has become a somewhat desperate undertaking. Some countries have relied on the money from mining to build their fragile economies, and the threat of economic collapse is very real if a mining company were to suddenly withdraw.

Health impacts of fossil fuel extractivism

The immediate health risks of coal mining to workers and surrounding communities are well-established (GBD 2013 Collaborators, 2015). Globally, around 25,000 people die each year from coal workers' pneumoconiosis (black lung) each year, and around 3.7 million people die from ultrafine particle outdoor air pollution each year (World Health Organization, 2016). A significant component of outdoor (ambient) air pollution which is black carbon originating from the processes of coal mining and transport, whereas the combustion of coal for electricity also adds nitrogen, sulfur compounds and heavy metals. Deaths from outdoor air pollution are mostly attributable to ischemic heart disease and stroke, chronic obstructive pulmonary disease or acute lower respiratory infections and lung cancer (World Health Organization, 2016). Even in relatively wealthy mining communities located in countries with established environmental and occupational protections, there are not inconsequential health risks to workers and communities. In Queensland, Australia for example, there have been 21 cases of coal workers' pneumoconiosis (black lung) since 2015 (Queensland Parliament, 2017) and pollution from a coal mine fire that burned for 45 days in the Australian State of Victoria was directly responsible for 10 or more deaths in the nearby town in 2014 (Victorian Government, 2015). The health consequences from the particulate pollution from coal dust distributed widely over communities during coal transportation is much more difficult to estimate but not insignificant; nationally the negative health impacts of coal mining, transport and combustion in Australia is estimated to cost around AU$2.6bn (US$2.05bn) every year (Australian

Academy of Technological Sciences and Engineering, 2009). The risk to community health is likely even greater in countries where there are fewer environmental protection regulations and less monitoring of particulate and gaseous emissions.

But importantly, the impacts of coal mining do not stop at the local or regional level; the health consequences of fossil fuel extractivism in the Pacific and elsewhere extend well beyond the proximate in both space and time. Coal's role in driving global climate change is unequivocal: burning coal is a chief villain in producing the emissions that are causing the planet to warm. The associated climatic changes have real health consequences, particularly in places which are already exposed to climate-associated health risks. Climate change, driven in large part by coal, will therefore increase the health burden in small islands. Exposure to increasing weather extremes including extreme heat, cyclones and storm surges augmented by sea-level rise are some direct consequences of global climate change of particular relevance to islands in Oceania, whereas less direct consequences include potential for increased transmission of vector-borne diseases, threats to water supply and food security through altered seasonal rainfall (McIver et al., 2015). Meanwhile, the damage to natural resources caused by extractivism of coal and other materials and the related health impacts and economic dependence, impart reduced resilience in these same communities and minimize their capacity to cope with the new health and environmental hazards caused by climate change.

Small islands in Oceania are especially vulnerable to climate variability because of their isolation, poverty, limited land area, topography (low lying and subject to flooding, or mountainous and subject to landslides) and frequent reliance on subsistence farming and imported foods (Nunn, 2007, Mimura et al., 2007). The poorest countries especially have few resources to adequately manage climate risks. Endemic poverty and minimal access to capital limit countries' capacity to respond and adapt to acute climate events and long-term changes (Boko et al., 2007), whereas coping actions are necessarily reactive and targeted at immediate threats; longer-term vulnerability may even be enhanced by unsustainable and ineffective practices and emergency responses (United Nations Development Programme, 2007). In coming years, for example, islands in the Pacific are expected to experience increasing climate extremes; periods of drought and heavy rainfall, cyclones and heat may become more frequent and more intense, oceans may become warmer and more acidic and sea-level rise may contaminate artesian water supplies and cause flooding (Australian Bureau of Meteorology and CSIRO, 2011).

The negative health consequences are many, especially for communities which are already poor and have a high burden of ill health and which are subject extreme or variable climate. So far these have been well documented for Pacific Island countries (McIver et al., 2015) and would be likely to be similar in other small islands in the region. The principal ways in which climate change affects health include:

- Direct trauma from extreme weather events, in particular cyclones and flooding, as events increase in intensity, and for flooding – frequency. Those especially at risk are people living in in exposed coastal and low-lying areas, or in mountainous areas (risk of landslide).

- Heat-related illness, as the frequency and intensity of hot days increases. Those especially at risk from the direct impacts of heat exposure are the elderly, people with underlying health conditions, young children and people who labor physically or outdoors.

- Vector-borne disease, in particular the mosquito-transmitted diseases such as dengue, chikungunya and zika viruses and malaria. As a general rule, mosquitoes thrive in warmer, wetter conditions and disease transmission intensifies. While malaria is not currently widespread in the region, climate change acts against eradication efforts in countries such as Timor-Leste and increases the potential for widening the geographic areas of transmission. As new areas become more climatically suitable for transmission, previously unexposed and highly susceptible populations are at risk.

- Increased food- and water-borne disease, especially in areas where clean water supply and sanitation is already inadequate and also where infrastructure is damaged by flooding. Prolonged drought makes water unavailable for hygiene purposes and can increase concentrations of pathogens in the water that is available, whereas warmer ambient temperatures increase multiplication of pathogenic organisms in food and on preparation surfaces, for example. Food security is decreased through more variable and less certain rainfall patterns, with drought and flood damaging crops and destroying livestock, severe events such as cyclones or extreme heat damaging coral ecosystems on which reef fish depend.

Other priority health outcomes under climate change in the small islands in the Pacific and elsewhere are zoonoses (diseases transmitted by animals such as leptospirosis), respiratory illnesses (including asthma and allergy), psychosocial ill-health (e.g. arising from loss of lives, resources and community) and non-communicable diseases (e.g. diabetes and obesity, associated with less nutritious diet, greater reliance on imported and processed foods). The potential for a far greater health burden into the future is exacerbated with concomitant population pressures (including rapid growth and urbanization), inadequate health systems and limited technical capacity (McIver et al., 2015).

Even these are all relatively simple and direct health consequences of climate change. But there are other, more complicated and less direct consequences that involve human systems and how they function.

Climate change is considered a "threat multiplier". Because of how it changes disease transmission (intensifying it in already exposed areas, expanding into fringe areas, and many of these diseases are related to poverty) and its impact on food production and water security, it makes poverty alleviation and attainment of other development goals more difficult. For example:

- As resources (food, water, land) become scarcer, conflict increases – especially in areas that are already more marginal or unstable (Bowles et al., 2015). Climate change potentially overwhelms the coping mechanisms of individuals and communities, lowers the threshold at which violence occurs and weakens states.

- More time spent sick (e.g. with diarrhea) means less time spent in education or working for those who are sick and their careers, limiting earning capacity now and in the future.

- Displacement of people due to both long-term pressures of climate change, and also to sudden disasters, leading to social upheaval and oftentimes violence. Women are especially vulnerable to violence under such circumstances (Klein, 2004). Displacement may occur more as a trickle, with people leaving areas as they become less habitable or productive, or in case of a severe weather event or of conflict, displacement may be sudden, en masse and unorganized. In best case scenarios, it will occur in a more organized way on well-negotiated terms, with island nations already planning for "migration with dignity" (Maclellan, 2011).

It is well known that these health outcomes are all climate-sensitive because there is insufficient collection of health and climate data, it is difficult to estimate the size of the effect of climate variables on health in small and under resourced at risk islands (Hales, 2013). It is also known broadly what those risks are likely to be because there have previously been demonstrated connections between weather and/or climate and disease. Transmission of vector-borne diseases such as dengue and filariasis is dependent, for example, on rainfall, humidity and temperature, as these set the parameters for mosquito survival and behavior and pathogen replication and that many islands are exposed already to these and similar diseases, including chikungunya, Ross River virus and leptospirosis (Bambrick and Hales, 2013). Similarly, food- and water-borne diseases such as typhoid and cholera occur when water supplies are unsafe or disrupted, which may be seasonally associated or follow an extreme event such as a cyclone, drought or flooding (Hales, 2013, Bambrick, 2013). Extreme events cause deaths and injuries, and it is expected that these might become more frequent and more extreme.

Climate also affects wellbeing more broadly – food security, livelihoods and social and economic development – and these in turn affect regional security by driving high unemployment, social disintegration and involuntary and economic migration. Climate change contributes to poor health outcomes by exacerbating difficult living conditions. A warmer, wetter, more hostile environment, driven in part by resource extractivism, will widen the health and economic inequalities between those from whom resources are taken and those who are profiting from them.

While all small islands in Oceania have heightened vulnerability to the health impacts of climate change, some communities are at greater risk than others, depending on their exposure and underlying characteristics; climate change will exacerbate these existing vulnerabilities. Communities already suffering food insecurity and other climate-sensitive health problems will, without adequate adaptation, be at even greater risk, thus also increasing the likelihood – and associated social and economic costs – of forced migration (Westphal et al., 2013, Mortreux and Barnett, 2009). Some population subgroups are at greater risk than others. Children are especially vulnerable to both increasing food insecurity and diarrheal disease, for example, because of demands of growth and development. The combination of poor nutrition and infectious disease lead to poor child growth and high rates of illness and mortality among children (Guerrant et al., 2008, Picot et al., 2012, Black et al., 2013).

Ending the extractivism era – opportunities for adaptation

There are clear links between resource extractivism in Oceania, health and major ecological disturbances, not the least of which is climate change. Extractivism-fueled climate change is threatening not only the lives and livelihoods of small islands but also their very existence, and therefore strong and urgent action to limit the extent of global warming is required. Anote Tong, then President of Kiribati, a country at risk of disappearing under rising sea-level caused by climate change, took his country's call for a global moratorium on all new coal mines at the Paris Conference of the Parties (COP21) in 2015.

Despite knowing all about the far-reaching consequences of resource extractivism for environment and health, and as communities face devastation, there is little sign of abatement. Notwithstanding the extremely clear causal links between coal and climate change, and between climate change and poor health outcomes for vulnerable island communities, coal mining in the region persists. In fact, the push to extract coal is accelerating as mining companies attempt to maximize sales before coal becomes entirely globally unmarketable (Denniss, 2016).

Resource extractivism is not always about the actions of a foreign power but has, in some cases, been internalized, becoming a tool by some island governments to fast-track understandably desirable economic development by boosting indicators of national income. As seen, however, any gains made are not universal and may not even trickle down to the local communities, nor are they long-lived, and they come at considerable costs to environment, community and health.

"Climate compatible" development, in contrast, builds resilient communities and is urgently required to protect the health of island communities. Even if a total coal mining (and combustion) moratorium (on both new and old mines) was enacted immediately, the climate system is slow to respond. Because of the momentum of warming that is already in the climate system, an immediate cessation of all fossil fuel burning would bring an average rise of another 0.5°C over the next two decades before it can plateau and start to decline (IPCC, 2013). No matter how strong the emissions reduction actions are taken today, the health impacts that are already becoming apparent in Pacific communities will continue and intensify for some time and, importantly, hinder development and resilience building.

Alternatives to extractivism must be found, including income-generating activities that are climate compatible and lead to inclusive economic development. Ideally these activities are community-led, are health promoting and at the same time build resilience to climate change that is already occurring.

But where to start? A significant barrier to understanding climate change adaptation needs in small islands in Oceania, and in less developed countries elsewhere, is insufficient data. Studies of climate and health and have concentrated in the more developed countries and often at broad (i.e. average) population level over large areas (Bambrick et al., 2015). The absence of suitable climate and health data, such as detailed watershed maps, national health surveys or hospital admissions, is a problem common to many developing countries including small islands (Bambrick and Hales, 2013). This absence of data, alongside minimal local capacity, has limited research in some of the world's most vulnerable communities. Data that does exist are frequently at large geographic scale, such as country or region (Nicholles et al., 2012, Bambrick and Hales, 2013). Averaging the data over large areas may miss important local problems and have limited relevance at the village, town or even whole island level as important local risks are overlooked. Lack of routine collection or unstandardized collection may mean that data are temporally and spatially sparse, and seasonal and geographic effects are missed, so that an association between one or more climate variables (such as rainfall and temperature) and a given health outcome (such as a vector borne disease) are unable to be reliably quantified. The sporadic, out of date statistics on even basic health indicators compounds poor understanding of climate-sensitive health risks and limits capacity to appropriately plan for and manage very real and present risks. Adaptation activities risk responding to the wrong type of threat, leading to potentially wasteful and even maladaptive practices.

Because of this frequent uncertainty, a key approach to adaptation is to concentrate on building resilience and adaptive capacity – the set of resources and the ability to use those resources, that are a prerequisite to adaptation (Nelson et al., 2007) – so that populations are better able to cope with the hazards of climate change, whatever they turn out to be. Population health and economic functioning are key determinants of adaptive capacity (Brooks et al., 2005) and interventions that improve a fundamental aspect of community health, such as children's nutritional status, could contribute directly to enhancing adaptive capacity by reducing social and economic burden regardless of the specific climate hazards in a given community. This "no regrets" approach, whereby actions are undertaken that benefit population health without full knowledge of the specific shape or size of the impacts that eventuate, is an essential strategy for taking meaningful and immediate action to bring substantial benefits to at-risk populations in both the short and longer term. Improving underlying population health could thus be considered a "first order" priority in climate change adaptation.

Work is currently underway, for example, by national governments and international development agencies to strengthen the health systems within small-island countries in Oceania. This includes implementing early warning systems for extreme events or for vector-borne disease which would benefit population health under current as well as future climate. Another, even more fundamental means of improving population health and increasing population resilience to climate change is to ensure clean water and adequate sanitation (Sheffield and Landrigen, 2011), and this has flow-on effects for economic and social functioning in families and communities. By reducing the burden of food- and water-borne disease, populations are healthier and more productive. In particular, the most vulnerable population groups benefit: Child mortality is reduced, and other key indicators of child health, such as height and weight status, are improved. Children spend more time in school and parents spend less time caring for sick children. Improving food security in areas already at risk from extreme weather is another pillar on which to build greater resilience. This might take the form of planting drought tolerant crops, changing the season in which to plant, changing locations of farms to more protected areas, storing water for irrigation and so on. Timing physical laboring to avoid the hottest parts of the day may be another relatively easy adaptive strategy that would bring immediate health benefits to workers' health and their productivity.

Beyond these basic yet essential first-order interventions that would contribute to health and consequently economic development are additional "second order" activities that would further help countries to reach their economic development goals. These not only need to be income generating but also should not damage health (or better yet, contribute directly to improved health), the local environment, nor contribute to climate change.

Rather than logging, for example, small-island communities and countries could look to grow their forest regions as global carbon sinks to offset emissions from elsewhere. With other countries seeking to offset their own emissions, there is increasing recognition of the potential market value of carbon capture, an essential ecosystem service. Other ways to maximize the value of their resource rich environment might include small-scale but high value production, such as niche agricultural and artisanal food production, and relatedly, building local business skills through enterprise training. In West Papua, for example, local communities would receive much greater income from sustainable fishing in unspoiled oceans than from the license fees paid by international mining companies, with development of coral reef ecotourism another option (Somba, 2008). Tourism that showcases unique and often spectacular natural environments is another potential avenue to develop, which again can be high value but small-scale so as not to overwhelm communities and environment. In Nauru, for example, the limestone pinnacles left by the removal of phosphate has produced an unusual landscape, whereas some unique native vegetation is beginning to return in some areas and if managed well could become a highly appealing attraction (Gale, 2016).

First-order and second-order adaptation priorities can counter some of the increased risk to climate change caused by local extractivism by building community resilience through improved heath and economic wellbeing. These relationships are illustrated in Figure 1.

Second-order priorities for inclusive development requires the right projects for the right communities that enhance rather than damage health. It is essential that these activities are community-driven and based on local needs and aspirations. Projects will be most effective if they are participatory and community managed and engage stakeholders throughout the process (Ebi, 2008b). This added potential for economic development from such activities will in turn enhance the adaptive capacity of small-island communities to respond to climate change.

Ideally, targeted adaptation activities should be also evaluated for their effectiveness; their impacts on community health and on other resources that determine resilience, such as economic and environmental benefits. A successful community intervention used in one place, and properly evaluated, may be more efficiently implemented in other communities with similar characteristics, with lessons learned considered in subsequent iterations. Evaluation would also assist with evidence-based policy formation aimed at reducing adverse climate impacts (Ebi, 2008a) and decisions on whether to upscale projects to national level, for instance. In the absence of such evaluation, the potential benefits to other communities of similar interventions cannot be estimated, whereas mistakes that trigger unintended and adverse consequences could go unreported and risk being repeated time and again. Evaluation of adaptive interventions is unfortunately rare, and so there remain substantial gaps about the potential for compatible development activities to reduce these adverse impacts in more vulnerable communities, in Oceania and elsewhere.

Conclusions

Extractivism of local resources in in small islands in Oceania has contributed to environmental degradation locally through pollution and loss of natural capital. It has reduced food security, forced the migration of whole populations, promoted cultural loss and social instability and created fragile economies.

The key feature of extractivism is that a more powerful, often external body, be it another country or a corporation, profits disproportionately from the arrangement while causing loss and damage to the local community from where the resource is taken. In the case of coal mining in Oceania, this damage is both relatively direct and immediate (local environmental damage and health impacts of particulate pollution, as well as broader disruptive sociocultural consequences) as well as complex and longer term, the result of climate change caused by the burning of fossil fuels causing poorer health outcomes for generations to come.

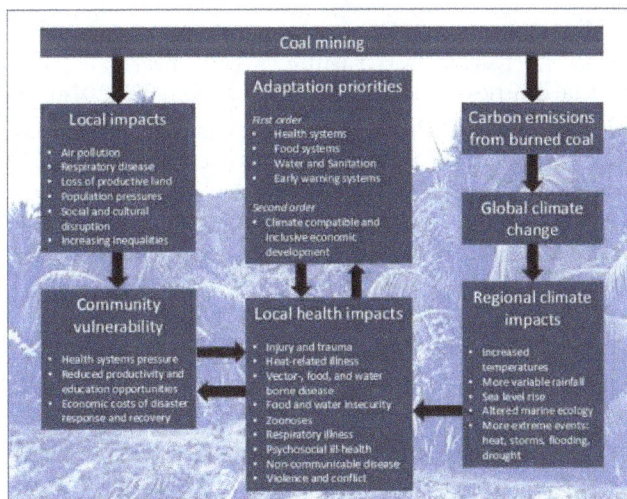

Figure 1. How coal mining affects local health, both directly and through its contribution to climate change, and the role of adaptation priorities in improving community resilience

Extractivism depletes natural resources and causes environmental degradation, leaving those exposed increasingly dependent on external income, such as from international aid, mining royalties, or – in the case of the island of Nauru – dependent on Australia's asylum seeker industry. Not all extractivism causes system-wide ecological damage, sometimes "merely" local resource depletion, pollution and cultural loss. At its more extreme end as demonstrated in this paper, extractivism forces whole communities to relocate and, in the case of the coal mining, contributes to the disruption of an entire planetary system.

The continuing practice of extractivism by a colonial or foreign corporate power is not restricted to small islands in Oceania, but this resource-rich region has been exploited in a number of ways for well over 100 years, with significant consequences for health. Coal mining in the region is a continuation of an established and oft repeated, historical narrative. What makes coal as the extracted resource particularly destructive is both its demonstrable health impacts on the local communities *and* its longer-term contribution to global warming, which in turn adversely affects the health of these same communities.

The harmful consequences of extractivism in small islands in Oceania throughout past century were seemingly unintended although one could argue potentially foreseeable. The local damage was visible even in its early days, but the drive to remove resources for monetary gain was overwhelming. Lessons from history have not been learned, with coal and other fossil fuel extraction continuing to expand and bringing even larger-scale consequences. There are clear links between resource extractivism, health and ecological disturbances, including global climate change. Yet, despite all that is now known about the far-reaching consequences of resource depletion for environment and health, and as communities face devastation, extractivism continues.

Alternatives to resource extractivism are urgently needed to ensure inclusive development and healthier futures, but these must be community-led and locally appropriate. Climate compatible development in vulnerable small islands can bring immediate health benefits and build community resilience in the face of increasing threats from climate change. "Migration with dignity" is an adaptation action of last resort but is increasingly becoming the only option left for some island communities.

References

Albert, S., Aswani, S., Fisher, P.L. and Albert, J. (2015), "Keeping food on the table: human responses and changing coastal fisheries in Solomon Islands", *PLoS One*, Vol. 10 No. 7, p. e0130800.

Andrefouet, S., Dutheil, C., Menkes, C.E., Bador, M. and Lengaigne, M. (2015), "Mass mortality events in atoll lagoons: environmental control and increased future vulnerability", *Global Change Biology* Vol. 21 No. 1, pp. 195-205.

Australian Academy of Technological Sciences and Engineering (2009), *The Hidden Costs of Electricity: Externalities of Power Generation in Australia*, Australian Academy of Technological Sciences and Engineering (ATSE), Victoria.

Australian Bureau of Meteorology and Csiro (2011), *Climate Change in the Pacific: Scientific Assessment and New Research*, Vol. 1, Regional Overview, BOM and CSIRO, Canberra.   fic

Australian Centre for International Agircultural Research (2012), ACIAR's Forestry Research in Paci Island Countries.

Ballard, C. and Banks, G. (2003), "Resource wars: the anthropology of mining", *Annual Review of Anthropology*, Vol. 32 No. 1, pp. 287-313.

Bambrick, H. (2013), *Assessment Report on Climate Change Risks to Health at District Hospitals: Climate Adaptation Strategy for Health*, Government of Samoa and United Nations Development Programme.

Bambrick, H. and Hales, S. (2013), *Climate Adaptation Strategy for Health and Action Plan*, Government of Samoa and United Nations Development Programme.

Bambrick, H. and Moncada, S. (2015), "Managing climate-sensitive health risks in vulnerable pacific island communities: lessons from Rabi Island", *Our Common Future Under Climate Change*, Paris.

Bambrick, H., Moncada, S. and Briguglio, M. (2015), "Climate change and health vulnerability in informal urban settlements in the Ethiopian rift valley", *Environmental Research Letters*, Vol. 10 No. 5.

Bell, J., Ganachaud, A., Gehrke, P., Griffiths, S., Hobday, A., Hoegh-Guldberg, O., Johnson, J., Le Borgne, R., Lehodey, P., Lough, J., Matear, R., Pickering, T., Pratchett, M., Gupta, A., Senina, I. and Waycott, M. (2013), "Mixed responses of tropical pacific fisheries and aquaculture to climate change", *Nature Climate Change*, Vol. 3, pp. 591-599.

Belot, H. and Stewart, E. (2017), "East Timor tears up oil and gas treaty with Australia after Hague dispute", ABC News [Online], available at: www.abc.net.au/news/2017-01-09/east-timor-tears-up-oil-and-gas-treaty-with-australia/8170476 [Accessed 15 March 2017].

Black, R.E., Victora, C.G., Walker, S.P., Bhutta, Z.A., Christian, P., De Onis, M., Ezzati, M., Grantham-Mcgregor, S., Katz, J., Martorell, R. and Uauy, R., Maternal and Child Nutrition Study Group (2013), "Maternal and child undernutrition and overweight in low-income and middle-income countries", *Lancet*, Vol. 382 No. 9890, pp. 427-451.

Boko, M., Niang, I., Nyong, A., Vogel, C., Githeko, A., Medany, M., Osman-Elasha, B., Tabo, R. and Yanda, P. (2007), "Africa. climate change impacts, adaptation and vulnerability. contribution of working group II", in Parry, M., Canziani, O., Palutikof, J., Van Der Linden, P. and Hanson, C. (Eds), *Fourth Assessment Report of the Intergovernmental Panel on Climate Change*, Cambridge University Press, Cambridge.

Bowles, D., Braidwood, M. and Butler, C. (2015), "Unholy Trinity: Climate change, conflict and health", in Butler, C. (Ed.), *Climate Change and Global Health*, CABI, Wallingford, Oxfordshire.

Brooks, N., Adger, N. and Kelly, P. (2005), "The determinants of vulnerability and adaptive capacity at the national level and the implications for adaptation", *Global Environmental Change*, Vol. 15 No. 2.

Commonwealth of Australia (1995), "Treaty between Australia and the Republic of Indonesia on the zone of coopoeration in an area between the Indonesian Province of East Timor and Northern Australia (Timor gap treaty)", in Australian Department of Foreign Affairs and Trade (Ed.), *Australian Treaty Series*, 11 December 1989, Australian Government Publishing Service, Timor Sea.

Commonwealth of Australia (2006), "Treaty between Australia and the Democratic Republic of Timor-Leste on certain maritime arrangements in the Timor Sea (CMATS)", in Australian Department of Foreign Affairs and Trade (Ed.), *Australian Treaty Series*, 12 January 2006: Australian Government Publishing Service, Sydney.

D'odorico, P. and Rulli, M. (2014), "The land and its people", *Nature Geoscience*, Vol. 7, pp. 324-325.

Denniss, R. (2016), "The great coal fire sale heats up", *Australian Financial Review*, 14 December 2015.

Denslow, J.S., Uowolo, A.L. and Hughes, R.F. (2006), "Limitations to seedling establishment in a mesic Hawaiian forest", *Oecologia*, Vol. 148 No. 1, pp. 118-128.

Derne, B., Fearnley, E., Goater, S., Carter, K. and Weinstein, P. (2010), "Ciguatera fish poisoning and environmental change: a case for strengthening health surveillance in the pacific?", *Pacific Health Dialog*, Vol. 16 No. 2, pp. 99-108.

Dinnan, S. and Walton, G. (2016), "Politics, organised crime and corruption in the Pacific", *State, Society and Governance in Melanesia*, 2016/24.

Ebi, K. (2008a), "Healthy people 2100: Modelling population health impacts of climate change", *Climatic Change*, Vol. 88 No. 1, pp. 5-19.

Ebi, K.L. (2008b), "Adaptation costs for climate change-related cases of diarrhoeal disease, malnutrition, and malaria in 2030", *Global Health*, Vol. 4 No. 9, p. 9.

Edwards, J.B. (2014), "Phosphate mining and the relocation of the banabans to northern Fiji in 1945: lessons for climate change-forced displacement", *Le Journal De La Société Des Océanistes*, Vol. 2014, pp. 138-139.

Environment Defenders Office NSW (2010), Technical Fact Sheet: Air Quality - Dust Monitoring.

Feary, D.A., Almany, G.R., Mccormick, M.I. and JONES, G.P. (2007), "Habitat choice, recruitment and the response of coral reef fishes to coral degradation", *Oecologia*, Vol. 153 No. 3, pp. 727-737.

Felbab-Brown, V. (2011), *Not as Easy as Falling off a Log: The Illegal Logging Trade in the Asia-Pacific Region and Possible Mitigation Strategies*, Brookings Institution, Washington, DC.

Fiji Bureau of Statistics (2015), "Population and labour force estimates of 2014", *Statistical News, 99*.

Fiji Bureau of Statistics (2017), Functional Age Groups - Total Population by enumeration area.

Fleay, C. and Hoffman, S. (2014), "Despair as a governing strategy – Australia and the offshore processing of asylum-seekers on Nauru", *Refugee Survey Quarterly*, Vol. 33 No. 2, pp. 1-19.

Franklin, J. and Steadman, D.W. (2010), "Forest plant and bird communities in the Lau group, Fiji", *PLoS One*, Vol. 5 No. 12, p. e15685.

Gale, S. (2016), "The mined-out phosphate lands of Nauru, equatorial western pacific", *Australian Journal of Earth Sciences*, Vol. 63 No. 3.

GBD 2013 Collaborators (2015), "Global, regional, and national age-sex specific all-cause and cause-specific mortality for 240 causes of death, 1990-2013: a systematic analysis for the global burden of disease study 2013", *Lancet*, Vol. 385, pp. 117-171.

Grewe, P.M., Feutry, P., Hill, P.L., Gunasekera, R.M., Schaefer, K.M., Itano, D.G., Fuller, D.W., Foster, S. D. and Davies, C.R. (2015), "Evidence of discrete yellowfin tuna (Thunnus albacares) populations demands rethink of management for this globally important resource", *Scientific Reports*, Vol. 5 No. 1, p. 16916.

Guerrant, R.L., Oria, R.B., Moore, S.R., Oria, M.O. and Lima, A.A. (2008), "Malnutrition as an enteric infectious disease with long-term effects on child development", *Nutrition Reviews*, Vol. 66 No. 9, pp. 487-505.

Hales, S. (2013), *Health Risk Analysis: Climate Adaptation Strategy for Health*, Integrating Climate Change Risks in the Agriculture and Health Sectors in Samoa, Samoa.

IPCC (2013), *Climate Change 2013: The Physical Science Basis: Contribution of Working Group I to the Fifth Assessment Report of the Intergovernmental Panel on Climate Change*, in Stocker, T., Qin, D., Plattner, G.K., Tignor, M., Allen, S., Boschung, J., Nauels, A., Xia, Y., Bex, V. and Midgley, P. (Eds), Intergovernmental Panel on Climate Change, Cambridge and New York, NY.

Jin, S., Yan, X., Zhang, H. and Fan, W. (2015), "Weight-length relationships and Fulton's condition factors of skipjack tuna (Katsuwonus pelamis) in the western and central pacific ocean", *PeerJ*, Vol. 3, p. e758.

Khambalia, A., Phongsavan, P., Smith, B.J., Keke, K., Dan, L., Fitzhardinge, A. and Bauman, A.E. (2011), "Prevalence and risk factors of diabetes and impaired fasting glucose in Nauru", *BMC Public Health*, Vol. 11, p. 719.

Kittinger, J.N., Teneva, L.T., Koike, H., Stamoulis, K.A., Kittinger, D.S., Oleson, K.L., Conklin, E., Gomes, M., Wilcox, B. and Friedlander, A.M. (2015), "From reef to table: social and ecological factors affecting coral reef fisheries, artisanal seafood supply chains, and seafood security", *PLoS One*, Vol. 10 No. 8, p. e0123856.

Klein, A. (2004), *Preventing and Responding to Sexual Violence in Disasters: A Planning Guide for Prevention and Response*, LaFASA & NSVRC, Louisiana.

Klein, N. (2015), *This Changes Everything: Capitalism Vs the Climate*, Simon and Schuster, London.

Llewellyn, L.E. (2010), "Revisiting the association between sea surface temperature and the epidemiology of fish poisoning in the South Pacific: reassessing the link between ciguatera and climate change", *Toxicon: official Journal of the International Society on Toxinology*, Vol. 56 No. 5, pp. 691-697.

Mciver, L., Kim, R., Hales, S., Honda, Y., Spickett, J., Woodward, A., Bambrick, H., Hashizume, M., Iddings, S., Katscherian, D., Kim, H. and Naicker, J. (2015), Human Health and Climate Change in Pacific Island Countries.

Maclellan, N. (2011), "Kiribati's policy for 'migration with dignity'", Inside Story.

Mimura, N., Nurse, L., Mclean, R.F., Agard, J., Briguglio, L., Lefale, P., Payet, R. and Sem, G. (2007), *Small islands. Climate Change 2007: Impacts, Adaptation and Vulnerability. Contribution of Working Group II to the Fourth Assessment Report of the Intergovernmental Panel on Climate Change*, in Parry, M.L., Canziani, O.F., Palutikof, J.P., Van Der Linden, P.J. and Hanson, C.E. (Eds), Cambridge University Press, Cambridge.

Morrice, E. and Colagiuri, R. (2013), "Coal mining, social injustice and health: a universal conflict of power and priorities", *Health & Place*, Vol. 19, pp. 74-79.

Mortreux, C. and Barnett, L. (2009), "Climate change, migration and adaptation in Funafuti, Tuvalu", *Global Environmental Change*, Vol. 19 No. 1, pp. 105-111.

Nelson, D., Adger, W. and Brown, K. (2007), "Adaptation to environmental change: contributions of a resilience framework", *Annual Review of Environment and Resources*, Vol. 32 No. 1, pp. 395-419.

Nicholles, N., Vardakoulias, O. and Johnson, V. (2012), *Counting on Uncertainty: The Economic Case for Community Based Adaptation in North-East Kenya*, New Economics Foundation and Care International.

Nunn, P. (2007), *Climate, Environment, and Society in the Pacific during the Last Millennium*, Elsevier, Amsterdam.

Picot, J., Hartwell, D., Harris, P., Mendes, D., Clegg, A.J. and Takeda, A. (2012), "The effectiveness of interventions to treat severe acute malnutrition in young children: a systematic review", *Health Technology Assessment*, Vol. 16 No. 19, pp. 1-316.

Pollock, N. (2014), "Nauru phosphate history and the resource curse narrative", *Le Journal De La Société Des Océanistes*, Vol. 2014 Nos 138/139, pp. 107-120.

Queensland Parliament (2017), *Black Lung, White Lies: Inquiry into the re-identification of coal workers' pneumoconiosis in Queensland Coal Workers' Pneumoconiosis Select Committee*, Brisbane.

Radio Australia (2012), *Asian Company Suspected of Smuggling Valuable Solomon Islands' Logs*, Australian Broadcasting Corporation (ABC).

Sanga, L. (2016), *Asian logging firm ignores Solomons court order, lands in West Guadalcanal. Pacific Islands Report*, East-West Centre Pacific Islands Development Program.

Sheffield, P. and Landrigen, P. (2011), "Global climate change and children's health: threats and strategies for prevention", *Environmental Health Perspectives*, Vol. 119 No. 3, pp. 291-298.

Somba, N. (2008), "Papua tribe questions govt over Issed mining licences", The Jakarta Post, 18 August 2008.

Taeiwa, K. (2015), *Consuming Ocean Island: Stories of People and Phosphate from Banaba*, Indiana University Press, Bloomington, Indiana.

United Nations Development Programme (2007), *Human Development Report*.

United Nations Environment Programme (2001), *Environmental Aspects of Phosphate and Potash Mining*, UNEP, France.

Victorian Government (2015), *Hazelwood Mine Fire Inquiry Report 2015/2016 Vollume II – Investigations into 2009-2014 deaths*, Hazelwood Mine Fire Inquiry, Melbourne.

Westphal, M., Hughes, G. and Brömmelhörster, J. (2013), *Economics of Climate Change in East Asia*, Asian Development Bank, Manila.

Williams, G.J., Price, N.N., Ushijima, B., Aeby, G.S., Callahan, S., Davy, S.K., Gove, J.M., Johnson, M.D., Knapp, I.S., Shore-Maggio, A., Smith, J.E., Videau, P. and Work, T.M. (2014), "Ocean warming and acidification have complex interactive effects on the dynamics of a marine fungal disease", *Proceedings Biological Sciences*, Vol. 281 No. 1778, p. 20133069.

Wilson, C. (2017), "Papua New Guinea moves to launch new coal mining industry", Mongabay, 16 May 2017.

Win Tin, S.T., Kenilorea, G., Gadabu, E., Tasserei, J. and Colagiuri, R. (2014), "The prevalence of diabetes complications and associated risk factors in Pacific Islands countries", *Diabetes Research and Clinical Practice*, Vol. 103 No. 1, pp. 114-118.

World Health Organization (2016), Ambient (outdoor) air quality and health. WHO.

World Health Organization (WHO) (2014), Noncommunicable Diseases (NCD) Country Profile: Nauru.

World Health Organization (WHO) (2015), Nauru: WHO Statistical Profile.

Corresponding author

Hilary Bambrick can be contacted at: h.bambrick@qut.edu.au

11

Understanding climate change vulnerability, adaptation and risk perceptions at household level in Khyber Pakhtunkhwa, Pakistan

Wahid Ullah and Takaaki Nihei
Department of Regional Geography, Graduate School of Letters, Hokkaido University, Sapporo, Japan, and

Muhammad Nafees, Rahman Zaman and Muhammad Ali
Department of Environmental Sciences, University of Peshawar, Peshawar, Pakistan

Abstract

Purpose – This study aims to investigate risks associated with climate change vulnerability and in response the adaptation methods used by farming communities to reduce its negative impacts on agriculture in Pakistan.

Design/methodology/approach – The study used household survey method of data collection in Charsadda district of Khyber Pakhtunkhwa province, involving 116 randomly selected respondents.

Findings – Prevalent crops diseases, water scarcity, soil fertility loss and poor socio-economic conditions were main contributing factors of climate change vulnerability. The results further showed that changing crops type and cultivation pattern, improved seed varieties, planting shaded trees and the provision of excessive fertilizers are the measures adapted to improve agricultural productivity, which may reduce the climate change vulnerability at a household level.

Research limitations/implications – The major limitation of this study was the exclusion of women from the survey due to religious and cultural barriers of in Pashtun society, wherein women and men do not mingle.

Practical implications – Reducing climate change vulnerability and developing more effective adaptation techniques require assistance from the government. This help can be in the form of providing basic resources, such as access to good quality agricultural inputs, access to information and extension services on climate change adaptation and modern technologies. Consultation with other key stakeholder is also required to create awareness and to build the capacity of the locals toward reducing climate change vulnerability and facilitating timely and effective adaptation.

Originality/value – This paper enriches existing knowledge of climate change vulnerability and adaptation in this resource-limited country so that effective measures can be taken to reduce vulnerability of farming communities, and enhance their adaptive capability.

Keywords Pakistan, Vulnerability, Climate change, Adaptation, Agriculture

1. Introduction

Recent changes in climate have confronted people all over the world – for some, it is a matter of changes in weather and for others a matter of survival. The real injustice of climate change is that although developing countries have contributed less to the annual global carbon dioxide emissions they are suffering most from its effects (Dazé, 2011; Van Aalst, 2006). It is expected that changing climatic conditions is likely to increase the frequency and magnitude of some extreme weather events and disasters like flood, droughts, storms and cyclones (Mirza, 2003; Greenough et al., 2001; Field, 2012). This is due to geographical locations of some of the most vulnerable regions of the world, their high exposure, limited assets, rapid and unmanaged population growth, as well as the likelihood of their mal-adaptation (Huq et al., 2004; Hay and Mamura, 2010; Atta-ur-Rahman and Khan, 2011).

South Asian countries including Pakistan are among those most affected by the risks associated with climate change (Ali and Erenstein, 2016). Pakistan's vulnerability to the impacts of climate change has been increasing with time despite its contribution to global warming being negligible. In 2012, Pakistan was at the 12th position, 8th in 2015 and 7th place among top countries of the world exposed to the vagaries of climate change and global warming (Kreft and Eckstein, 2013; IUCN, 2009). Recently, disasters related to climate change such as floods, droughts, cyclones and storms have hit Pakistan hard (Tingju et al., 2014; Mueller et al., 2014; Atta-ur-Rahman and Khan, 2013). These disasters have not only become more frequent but also caused more damage (Qasim et al., 2015). Even after having been consistently affected by climate exigencies year after year, the country's response to solve the issue has remained lackluster. The burden of natural disasters in Pakistan is underlined by the fact that they have affected millions and killed thousands of people countrywide (Atta-ur-Rahman and Khan, 2013). Among others, rapid population growth, uncontrolled development and unmanaged expansion of infrastructure are the most common factors resulting in more people being vulnerable to natural hazards than ever before (Cardona et al., 2003).

Studies suggest that poor people in rural areas of Pakistan are the most vulnerable to climate change (Ali and Erenstein, 2016; Deressa et al., 2009; Füssel, 2007). These communities are struck hard by those changes in climate identified by many studies conducted throughout the country (Tingju et al., 2014; Atta-ur-Rahman and Khan, 2013; Qasim et al., 2015; Abid et al., 2015). This is particularly because agriculture is climate sensitive and because of the huge number of rural populations predominantly dependent on agriculture as their mainstay of livelihood. Among others, the main challenges faced by farming communities in Pakistan include insufficient irrigation water; lack of technical knowledge, lack of education and limited number of extension facilities; widespread poverty among farmers and inadequate credit facilities; expensive farm inputs such as seeds, fertilizers and pesticides; lack of roads from field to market; low prices of agricultural output and the absence of agriculture-based industries (Abid *et al.*, 2015; Khan, 2013). Additionally, the inappropriate use of modern agricultural inputs such as fertilizers and pesticides has led to alarming environmental pollution (Khan *et al.*, 2013; Shahzada *et al.*, 2012; Yousaf and Naveed, 2013; Saif-ur-rehman and Shaukat, 2013).

A consistent major problem for Pakistan's authorities is that natural disasters occur regularly at all scales. Unfortunately, the authorities responsible for disaster risk reduction in Pakistan have not made adequate use of recent developments in scientific methodologies, methods and tools for cost-effective and sustainable interventions (Atta-ur-Rahman and Khan, 2013; Qasim *et al.*, 2015). Research aimed at identifying the main drivers of climate change vulnerability (CCV), adaptation and risk perceptions at household level is urgently needed to reduce the negative impacts of climate change on agriculture (Abid *et al.*, 2016).

Studies report that farmers use several techniques to adapt agriculture to CCV (Ali and Erenstein, 2016). Some of these techniques used at farm level include diversification in crop practices and changing the timing of operations (Deressa *et al.*, 2009); changing farm management practices such as type and amount of agricultural inputs applied (Abid *et al.*, 2016); livelihood diversification (Hussain and Mudasser, 2007); institutional changes, mainly government responses, such as subsidies/taxes and improvement in agricultural markets (Mendelsohn, 2001); and technological developments such as growing new and heat-tolerant crop varieties and advances in irrigation and water management techniques (Deressa *et al.*, 2009; Hussain and Mudasser, 2007). Climate change is generally detrimental to agriculture, but can partly be offset by deploying the various adaptation methods at farm level (Ali and Erenstein, 2016; Abid *et al.*, 2015). However, the degree to which a certain agriculture sector is exposed and vulnerable to climate change depends on the adaptive capacity of community or area to withstand or react to those changes (Adger *et al.*, 2003; Ullah *et al.*, 2015). In addition, some adaptation methods are highly localized and cannot be directly adopted and implemented in other regions or agriculture settings.

Knowing the importance of agriculture for rural communities in Khyber Pakhtunkhwa (KPK) province of Pakistan, the significance of identifying CCV, adaptation strategies and risk perceptions at farm level is crucial. Therefore, a growing number of agricultural experts have shifted their research interests toward the issue of climate change and its impacts on agriculture in Pakistan. The focus of these experts is on identifying perceptions of climate change related risks (Qasim *et al.*, 2015; Abid *et al.*, 2015; Khan *et al.*, 2013), vulnerability (Atta-ur-Rahman and Khan, 2013; Rasul *et al.*, 2012; Khan and Salman, 2012) and adaptations (Huq *et al.*, 2004; Abid *et al.*, 2016; Deressa *et al.*, 2009) particularly at farm level. Despite this, little research has addressed these issues in the case of KPK Province. Hence, this study provides an analysis of farmers' responses to farm risks, which has always been important issue and particularly so under changing climatic conditions. The paper also provides detailed farm-level evidence and discussion to highlight the actual situation of farmers and their decision environment. The specific objectives of the study include:

(1) to identify main factors of CCV;

(2) to investigate adaptation techniques deployed at farm level to reduce odd impacts of climate related risks; and

(3) to explore perceptions of rural farmers regarding their concerns on the impacts of climate change in the KPK province.

2. Research methods

This section presents methods, describing the main characteristics of the scoping study and the applied techniques of data collection and data analysis.

2.1 Study area

To assess how communities are vulnerable to climate change and in response deploy methods for adaptation, a comprehensive case study approach was chosen. KPK province was selected as a sample site because it was previously identified as vulnerable to climate change (Saif-ur-rehman and Shaukat, 2013; Ullah *et al.*, 2015; Rasul *et al.*, 2012; Malik, 2012), and agriculture contributes approximately 38 per cent to the provincial gross domestic product and provides employment for 44 per cent of the total population (Atta-ur-Rahman and Khan, 2013; Khan, 2012; Khan, 1994). The KPK province is divided into two parts for analysis purposes: northern and southern halves. The former is water sufficient and the latter is water deficient. The study area is the junction point situated along the bank of the Kabul River. This area faces two extreme conditions: drought and flood causing huge harm to humans, land and other property. Natural disasters, especially floods and droughts, have severely affected agriculture in the Charsadda district of KPK province. Two villages, Gulabad and Shabara, were selected from Charsadda district for this study (Figure 1). Gulabad lies on the main Peshawar – Charsadda Road; Shabara is located 3 km to the east of the main Peshawar – Charsadda Road and approximately 1 km to the east of the main

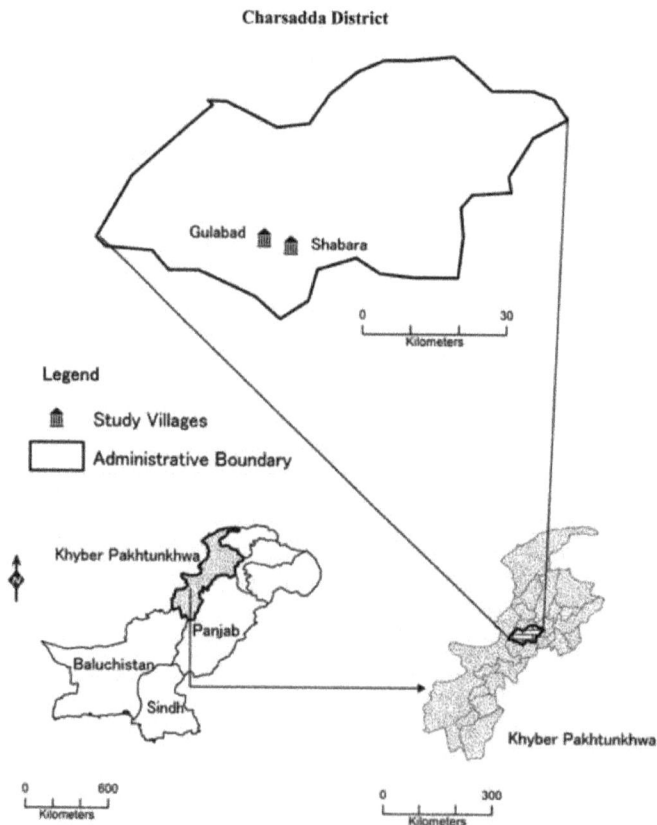

Figure 1. Map of the study area

Peshawar–Islamabad Express Motorway. Other facilities such as transportation, distance to district headquarters, access to internet and availability of mobile service and roads are equally available in both villages. However, access to water for irrigation is the main attribute differentiating the villages.

2.2 Data collection and sampling design

The study used the household survey (HHS) method for collecting data. A bottom-up approach was used to investigate actual farmers' experiences with climate and their responses to various climate conditions that might influence their decisions. Initially, a meeting was arranged with knowledgeable people such as village elders, experienced farmers, elected members of the village and school teachers in Gulabad to identify potential villages for the study. Before the meeting, the study objectives were explained to them, and they were asked to provide best-fit case study villages based on their knowledge of the area. Seven potential villages were identified for the research team to visit using criteria of flood damage, access to transportation, distance to district headquarters, internet access, availability of roads and water availability for irrigation. Due to the time and financial constraints, sample selection was reduced to three villages based on distance from rivers and irrigation techniques used for farming. Out of these three villages, the research team randomly selected two: Gulabad and Shabara.

Using a structured questionnaire, the survey targeted representatives of households that in most cases were household heads. However, in cases of unavailability, other adult members of the household were interviewed. It is important to mention that this study was performed in a region where people with Pashtun ethnicity reside. In Pashtun societies, women are not allowed to mingle with male members (Qasim *et al.*, 2015). Hence, this study only covers perceptions of male respondents as it was impossible to capture women's perceptions for cultural and religious reasons. Two field assistants were hired to help the first author of this paper collect primary data from the study area. A one-day intensive workshop was arranged to train field assistants prior to visit study villages. All questionnaires were administered personally to the respondents by the research team. In total, 116 households (45 from Gulabad and 71 from Shabara) were randomly selected for interviewing. The sampling frame included residents of both villages, which were male farmers. All interviews were conducted based on shared research principles and ethics (Bogner *et al.*, 2009).

Before starting the HHS, the purpose and objective of the study was clearly explained and respondents were asked for informal verbal consent. During the field survey, the research team did not come across any household who refused to be interviewed. This might be because the study team and respondents shared the same language and other cultural attributes, which made respondents less hesitant. Generally, the HHS lasted for approximately one hour. The survey included questions on household socio-economic characteristics – for example, age, gender, farming experience, occupation, education, assets for livelihood like transportation, electrical, mechanical assets, agricultural or farm equipment, climate-related vulnerability perceptions, knowledge on climate change and its impacts on agriculture, adaptation techniques used to reduce adverse effects of climate change and constraints to adaptation. Perceptions on concerns from climate-related risks were also part of the survey. Perceptions on concerns from weather related changes were grouped into excess rainfall, temperature change and droughts from the answers given during the survey while concerns regarding agricultural production were grouped into severe weather condition, crop diseases and lack of access to agricultural inputs. The reported concerns from changing climate (each measured on a five-point Likert scale) included decrease in income from agriculture (1), no compensation in case of human or agricultural losses (2), increased threats (3), no access to irrigation (4) and no access to alternative income (5).

3. Results and discussion

The study findings start with the analysis of social and demographic characteristics (Section 3.1) and availability of assets for livelihood and farm characteristics at household level (Section 3.2). Next, results on key factors contributing to the CCV of farmers are presented (Section 3.3). The analysis then moves on to consider adaptation techniques used to cope with climate change at farm level (Section 3.4) and finally the concerns of respondents interviewed on climate-related risks in the region are presented (Section 3.5).

3.1 Demographic and farm characteristics

Demographic and farm characteristics of sampled households are presented in Table I. In both villages, respondents were all men, married and performing farming as a primary job. In both villages, the majority of respondents were in the age range of 31-50 years. In Gulabad, 44 per cent of the sampled households had 6-10 years of education and in Shabara this proportion was 35 per cent. The household size was large in both villages, i.e. 8-9 persons per household, providing a labor force for farming. The majority of farmers were experienced. Slightly more than half of the respondents in Gulabad had 20-35 years of farming experience, but in Shabara the ratios were evenly distributed with almost 25 per cent, among the all categories (Table I). Some respondents (20 per cent and 25 per cent in Gulabad and Shabara, respectively) could not provide their experience in years and explained that they have been involved in farming since childhood. In both villages, 76-89 per cent of farmers used tractors for plowing land. More than 60 per cent of households in both villages plowed their land a minimum of twice per year. Inhabitants of Gulabad irrigated their agricultural land with canal water, but in Shabara, more than 73 per cent of farmers used tube-wells for irrigation and the rest waited for rainfall. The land in Shabara is drier compared with Gulabad; hence, it needs water more frequently to get desirable yields. Surveyed respondents mostly heard about climate change either through electronic media, friends or from elders. Farmers' unions are important media for sharing knowledge and experience but in our study sites, those unions did not exist. Respondents further reported that they had never been invited by local government for training programs on agriculture, which could have built their capacity and taught them advanced farming techniques.

The literature has widely covered the importance of understanding farmers' socioeconomic and demographic characteristics in the context of CCV and adaptation concerning it (Abid *et al.*, 2016; Bryan *et al.*, 2009). The results have shown how policymakers can best support poor farmers, who are most vulnerable to climate impacts given limited resources to make changes in their farming practices. Providing support to the poorest farmers is critical because they are the least equipped and the most vulnerable (Bryan *et al.*, 2009). Addressing these issues requires strong leadership and government involvement in planning for adaptation and implementing measures to facilitate adaptation at the farm level (Bryan *et al.*, 2009; Adger and Kelly, 1999).

Table I. Socioeconomic and farm characteristics of sampled households

Indicators	Category type	Village name	
		Gulabad (%)	Shabara (%)
Household size (persons)		7	9
Gender	Male	100	100
Marital status	Married	100	100
Occupation	Farming	100	100
	Shop keeping	7	4
	Others	7	1
Age (years)	11-30	4	17
	31-50	58	45
	>50	38	38
Education (years)	0	22	42
	1-5	33	23
	6-10	44	35
Farming experience (years)	Since childhood	20	25
	<20	11	24
	20-35	51	27
	>35	18	24
Land preparation	Use tractor	89	76
	Use both tractors and bullocks	11	24
Frequency of plowing per year	Once	18	7
	Twice	53	65
	Three and more	29	27
Means of irrigation	Canal	100	–
	Tube-well	–	73
	Rain fed	–	21
	Both tube-well and rain fed	–	6
Frequency of irrigation	Nil	–	15
	Weekly	22	37
	Twice a month	51	27
	Three times a month	24	13
	Monthly	2	8
Source of information on climate change	Media	36	10
	Village elders	44	70
	Own view	9	1
	Friends	7	11
	Do not know	4	4
Farmer's unions		Does not exist	Does not exist
Farmer training programs		Does not exist	Does not exist

Note: (–) means no responses were given
Source: Author's field survey

3.2 Assets for livelihood

Respondents of the HHS were asked about the assets they owned to support their livelihood. These assets included transportation, electrical, mechanical and agricultural or farm equipment (Table II). In both villages, most respondents owned televisions, cellphones and other common agricultural farm equipment such as spraying device, water pump and scale. Few respondents owned a tractor or post-harvest facilities. Very few respondents had access to advanced techniques of farming due to their poor socio-

Table II. Assets for livelihood in study area

Asset type	Gulabad (%)		Shabara (%)	
	Yes	No	Yes	No
Motorbike	38	62	32	68
Bicycle	27	73	28	72
Electric generator	31	69	10	90
Cellphone	76	24	69	31
Regular phone	7	93	–	100
Television	38	62	48	52
Radio	4	96	4	97
Camera	2	98	–	100
Washing machine	80	20	90	10
Other	78	20	94	6
(e.g. sewing machine)				
Tractor	2	98	–	100
Plow	80	20	90	10
Chemical spraying device	80	20	85	15
Water pump	73	27	73	27
Wooden cart	9	91	–	100
Grain/flourmill	4	96	–	100
Scale	82	18	83	17

Note: (–) means no responses were given
Source: Author's field survey

economic situation. The majority of them faced poor economic conditions, which expose them to the vagaries of climate change. Other researchers have consistently mentioned that having more agricultural assets and access to improved technology stimulates agricultural growth, expands food supply and so results in poverty alleviation (Ali and Erenstein, 2016; Abid *et al.*, 2016; Deressa *et al.*, 2009).

Previous studies suggest that analyzing vulnerability does not only involve identifying threats but also the resilience and recovery from the negative impacts of changing climatic conditions [39]. This includes individual and household characteristics, socioeconomic status, farm characteristics, distance from markets and access to extension and credit. As suggested, households that wish to reduce the risks associated with climate change and have the resources or access to resources needed to make the appropriate changes are generally more resilient and have greater capacity to adapt (Abid *et al.*, 2015; Deressa *et al.*, 2011). Knowledge of these assets helps in understanding how livelihoods work, and how people respond to climatic variability and adapt to change. Hence, livelihoods are built on these assets – individuals, households and groups depend on these assets for agricultural production (Jodha *et al.*, 2012). The general conception is that farmers with more capital better survive the negative results of climate change (Deressa *et al.*, 2011; Blaikie *et al.*, 2014). In the case of our study area in particular, and Pakistan in general, the livelihoods of poor farmers are particularly at risk from the ever-increasing exposure to natural disasters like floods, droughts, heavy monsoons and heat waves (Qasim *et al.*, 2015; Abid *et al.*, 2015).

3.3 Major factors contributing to climate change vulnerability
This section addresses question about what perceived factors are the primary contributors to CCV in the study area. After a detailed literature review, 16 variables were identified that are considered important in CCV analysis (Abid *et al.*, 2016). The scores given for each

indicator were measured and ranked on a five-point Likert-scale of very low (1), low (2), moderate (3), high (4) and very high (5), depending on how farmers perceived it in relation to the changing climatic conditions in the study area. Indicators that received a score 4 or 5 were considered primary contributing indicators to CCV, whereas those receiving 1 or 2 were perceived as low contributing factors.

There were mixed responses among respondents concerning gauging indicators of CCV. This might be due to the nature of different environmental risks people are exposed to in both Shabara and Gulabad. Consequently, the perceived threats from CCV were also seen differently (Figures 2 and 3). Indicators whose impacts were less threatening were perceived as low contributors to CCV in Gulabad and Shabara. Those indicators included drinking water, forest degradation, transportation system, grazing area, land resource, landslides, irrigation facility, animal diseases and minimum extreme temperature (Figure 3). However, soil problems (40 per cent), crop pests (42 per cent) and droughts (38 per cent) were perceived as low contributing factors to CCV. Flood was a major contributor to CCV in Gulabad and 38 per cent ranked it as a high and 47 per cent as a very high contributor.

In the case of Shabara, most of the indicators perceived as low, very low and moderately contributors to CCV were ranked similarly to Gulabad. A mix of responses was observed in ranking drinking water, with almost equal numbers of respondents in the different rankings concerning changes in availability of drinking water (Figure 3). Unlike in Gulabad, the agricultural land in Shabara is either irrigated with tube-wells or rain fed. Therefore, more than 20 per cent of respondents ranked irrigation with regard to CCV respectively as very low, low, moderate and highly vulnerable to negative impacts of climate change. In the case of flood, 27 per cent of respondents perceived that it was highly vulnerable to CCV. Flood is a major threat to CCV in both villages, possibly due to their proximity to the Kabul River

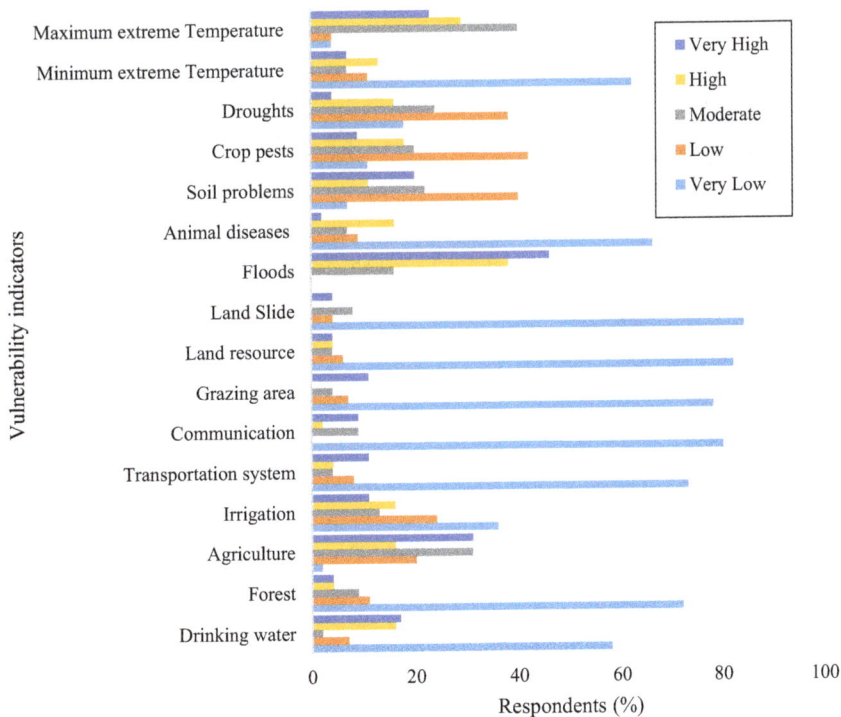

Figure 2. Perceptions on causes of CCV in Gulabad

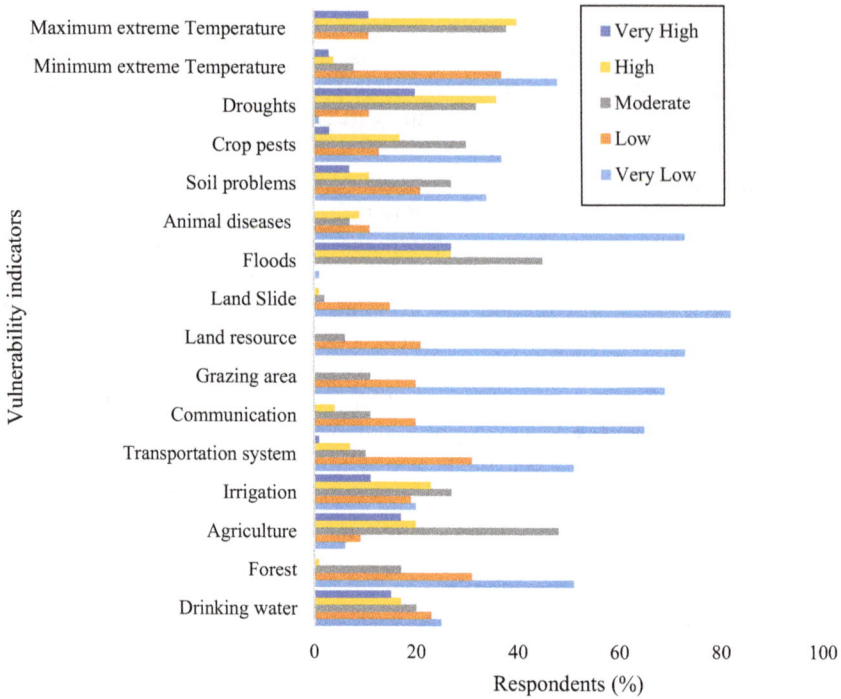

Figure 3. Perceptions on causes of CCV in Shabara

that floods almost every year in the monsoon season of July-September, resulting in serious problems for the socioeconomic and physical environment of the study area.

Comparing these results with other studies around the world, as well as some conducted in Pakistan, clearly indicates that many rural populations especially those involved in farming are severely affected by impacts of changes in climate (Mirza, 2003; Greenough *et al.*, 2001; Field, 2012; Ali and Erenstein, 2016). For instance, Abid *et al.* (2015) found similar conditions in Panjab province of Pakistan where longer summers, decrease in precipitation and changes in agricultural growing season were recorded by the farming communities. An increasing trend has been seen in the frequency and magnitude of natural disasters like extreme temperatures, floods and droughts (Maheen and Hoban, 2017). Temperatures are predicted to rise by 3°C by 2040 and up to 5-6°C by the end of the century. Monsoon rains will drastically reduce but have higher intensity. According to Abid *et al.* (2016), farmers' identification of various risks shows the importance of climate-related conditions for their farm-level operations. However, differences in how risk is perceived by farmers in different regions are common due to changes in the environmental setting, geographical location, availability of resources and economic status of an individual.

Disaster risk management experts believe that the main causes of vulnerability to hazards are not solely environmental but also result from ignorance of the people and destitution of the country (Adger and Kelly, 1999; Wisner *et al.*, 2012; Mustafa, 1998; Smit and Pilifosova, 2003). More than half of Pakistan's population lives in extreme poverty and many live in disaster-prone areas. This specific social segment cannot be expected to make disaster risk reduction a priority although they suffer severely from disasters when they occur. One of the physical vulnerabilities of the people living in highly vulnerable areas might be attributed to this social issue (Mustafa, 2002). In the 2010 Pakistan flood, in many areas people ignored warnings about impending disasters for various reasons including lack of awareness, education and lack of trust between locals and government officials. Another example of social and economic vulnerability is seen in the irrigation system of the country, where high demand for water has led to inappropriate irrigation resulting in worsening the flood and drought conditions (Mustafa, 2002).

3.4 Adaptation techniques used by farm households
This section borrows previous methodology used to identify how farmers adapt to climate-related risks at farm level in Panjab province of Pakistan (Deressa *et al.*, 2011; Abid *et al.*, 2016). Adaptation strategies included in this study were grouped into four major categories:

(1) changing cropping practices, e.g. crop type and variety or planting date;

(2) changing farm management techniques such as fertilizer, pesticide, seed quality or irrigation;

(3) advanced land use management measures, i.e. changing farm management techniques such as sowing and harvesting, planting shade trees, stopping cutting trees, using less water, storing water and soil conservation; and

(4) livelihood options, including shifting from single to multiple crops, shifting from farming to livestock keeping, migration and renting more crop land.

Farmers in Gulabad changed crop variety, type and quality of fertilizer, pesticide and seed quality; plant shade trees; and shift from single to multiple crops to cope with climate change (Figure 4). The prominent methods of adaptation to reduce the negative impacts of climate change in Shabara included changing crop type, changing seed quality, plant shade trees and stopping cutting trees. The methods of adaptation to climate change were mostly similar in both villages.

Changes in cropping practices implemented by respondents were dependent on the nature of problem. A change in crop type or variety was mostly adopted due to pest and insect attacks on crops that negatively affected agricultural production. To overcome this problem, households reported that they had tried new fertilizers and pesticides to ensure desired production. It was also mentioned that these adaptation strategies did not help

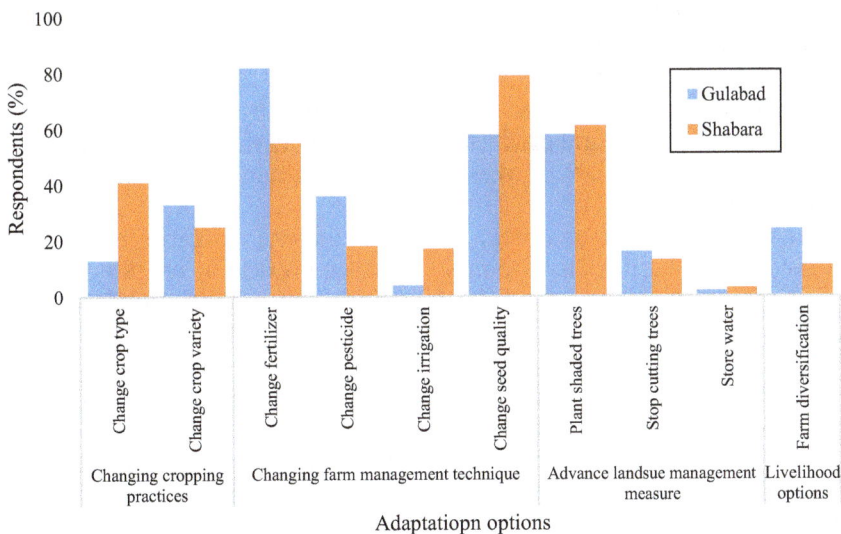

Figure 4. Adaptation strategies used to manage climate change and associated risks by farm households in Gulabad and Shabara

improve their yields. Farmers have started buying heat-tolerant wheat varieties to cope with extreme hot weather events, following farmers in Punjab province who usually get higher yields compared with farmers in the study area and in KPK province generally.

Changing farm management practices include changing fertilizer, pesticide, seed quality and frequency of irrigation were implemented by sampled households. For instance, in Shabara, during dry years (which happen often), farmers changed their irrigation frequency to ensure the desired production. Crops are exposed to pest attacks in cases of more rainfall. Farmers try different combinations of pesticides but mostly buy the cheapest available, partly due to availability in the village market – it is easier to buy it within their vicinity. The case of fertilizers is similar to that for pesticides. In response to loss of fertile soil layers by floods in 2005 and 2010, farmers used more fertilizers to balance nutrients in the soil and increase crop productivity. All farmers thought that their production had decreased in the last decade or so, and they try to add more fertilizer to their soil than before. It was reported that wheat and sugarcane production had reduced by more than half compared to 10 years ago. Most farmers believed that this was due to changes in climatic conditions, poverty and lack of support from local government.

Advanced land use management techniques were also adopted to protect livelihoods against negative impacts of climate-related risks. Although respondents understood the importance of water for agriculture, none used less water or stored it for winter (the season when there is insufficient water for irrigation). In Gulabad, availability of water for irrigation was considered a less important issue because they had good access to river water. Although a canal was built in Shabara, farmers get no benefit due to a lack of water channels to land. More than half of the respondents in Gulabad, and approximately 60 per cent in Shabara, have started planting Eucalyptus around their land especially near the river to reduce soil erosion from floods during July-September. However, farmers were unaware that Eucalyptus trees have high-water demand (Forrester *et al.*, 2010). Farmers in Shabara showed more concern after learning this fact because they already face water scarcity for irrigation.

All sampled respondents were primarily farmers and their whole household depended on it (Table I). Other livelihood options are currently rare. Farmers have tried shifting from single to multiple crops, planting Eucalyptus or replacing wheat with maize in some parcels of land. Due to the unavailability of grazing areas, few respondents were willing to depend on livestock for subsistence. One household migrated from a neighboring village to Shabara in response to loss of agricultural land due to floods in 2010 but was not satisfied with soil fertility in Shabara. Small numbers of farmers rented extra land within the village to increase their overall agricultural production and meet household food demand, which had been affected by infertile soil, lack of access to good agricultural inputs and high exposure to droughts and floods.

Due to the high exposure of agricultural communities to vagaries of climate change across Pakistan, many farmers tend to minimize these impacts by adapting. However, the impact of climate-related events strongly depends on the capacity to adapt to the risks. Although adaptation practices are potentially important, not all farmers use such practices due to lack of knowledge on what techniques are appropriate (Baig and Amjad, 2014; Ahmad *et al.*, 2013). This situation particularly applies to Pakistan where knowledge about the current process of adaptation and vulnerability aspects at farm level is still very limited due to lack of research on environmental vulnerability and local-level risk perceptions (Hussain and Mudasser, 2007). Abid *et al.* (2016) and Adger and Kelly (1999) reported that those who adapt in a timely manner may not only reduce the negative impacts of climate change but also profit compared with those who adapt late or not at all. However, the approaches to adaptation assume that people have access to the resources needed to put these strategies in place. For the most vulnerable people in many communities this is simply not the case. When people do not have secure access to these resources, their options are limited and they are less able to act on adaptation (Dazé, 2011; Mertz *et al.*, 2009).

3.4.1 Constraints to adaptation. During the HHS, farmers were asked to identify constraints that they perceived to be the most important barriers to changing their farming practices (Figure 5). Although farmers referred to several barriers to adaptation, the most important in both Gulabad and Shabara included poverty, lack of support from government and lack of assets. In Shabara, lack of water for irrigation, lack of information and knowledge on climate change weather and rainfall pattern, of an effective and timely early warning system and of market and price information were also among the dominant constraints to adaptation.

The capacity of a household to cope with climate risks depends to some degree on the enabling environment of the community, and is reflective of the resources and processes of the region (Smit and Pilifosova, 2003). At the local level, the ability to undertake adaptations can be influenced by such factors as managerial ability; access to financial, technological and information resources; infrastructure; the institutional environment within which adaptations occur; political influence; kinship networks and the socio-economic status of the household (Blaikie *et al.*, 2014; Watts and Bohle, 1993; Kelly and Adger, 2000; Roncoli *et al.*, 2002; Eakin 2003). However, to understand the importance of factors shaping farmers' decisions and responses to adapt, it is necessary to explore their perceptions regarding the barriers they face (Bryan *et al.*, 2009). Hence, in the next chapter, farmers' concerns about risks associated with climate change are described in detail.

3.5 Concerns regarding impacts of climate change on agriculture

This section explores perceptions of farm households regarding weather-related changes (Figure 6), primary risks to agricultural production (Figure 7) and finally their concerns from changes in climate in the region (Figure 8). The results showed that, in Gulabad, 69 per cent of the respondents perceived excess rainfall as a major weather threat, whereas in Shabara, 83 per cent of respondents perceived drought as the primary weather-related risk. The proximities of both villages to the Kabul River and the absence of any precautionary measures (e.g. high river boundary walls or effective

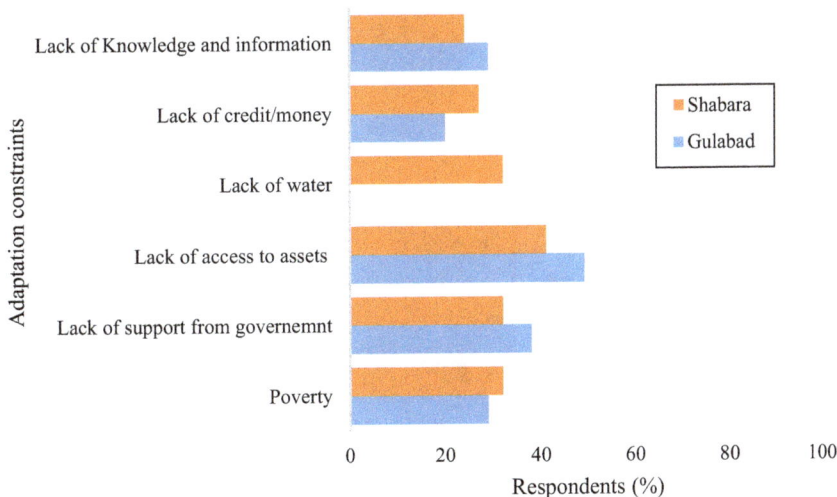

Figure 5. Perceived constraints to climate change adaptation at farm level in Gulabad and Shabara

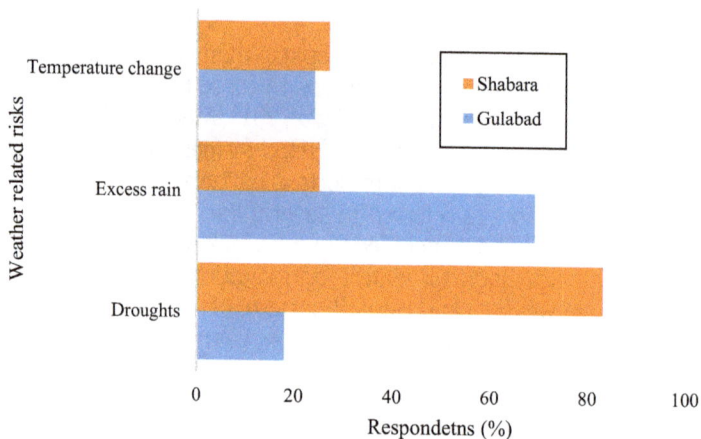

Figure 6. Perceptions on weather-related risks in Gulabad and Shabara

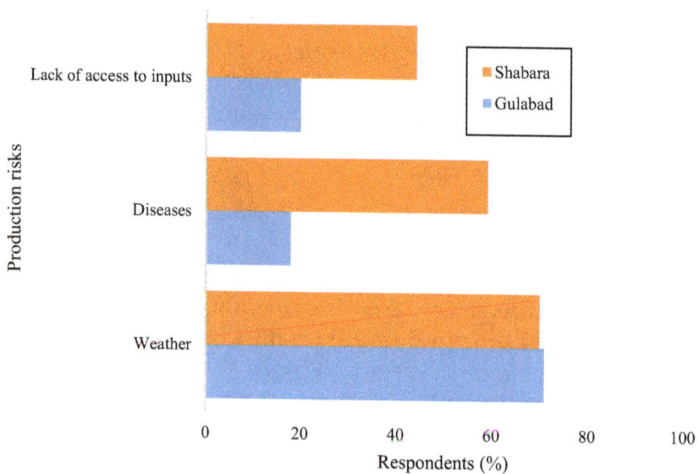

Figure 7. Perceived primary production risks in Gulabad and Shabara

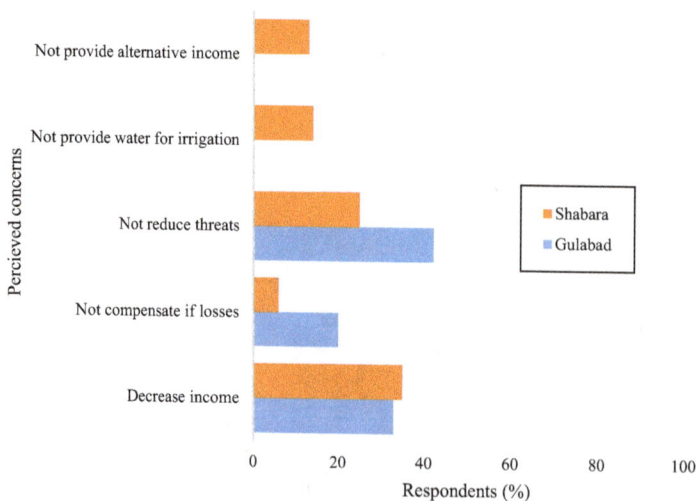

Figure 8. Perceived worries from changing climatic conditions in Gulabad and Shabara

early warning system) expose human lives, livestock, land and infrastructure to flood risk in the monsoon season.

Respondents were further asked about perceived primary risks for agricultural production. The majority of respondents, approximately 70 per cent in both villages, reported changing weather conditions especially hotter summers as the primary risk. Furthermore, crop diseases and lack of access to good quality agricultural inputs such as heat-tolerant seeds, fertilizers and pesticides; to improved soil conservation techniques and to post-harvest facilities were among other factors significantly affecting agricultural production in the study area.

Farmers were asked about their concerns regarding climate-related risks and the impacts on their household. Respondents in both villages acknowledged changes occurring in climate in the region (Figure 8). Respondents perceived that floods, droughts and warmer summers were due to climate change and were among the serious threats to their farming. As mentioned earlier, all surveyed households primarily depend on farming for subsistence. Respondents were of the view that these changes in climate would decrease farm income of their household. Some respondents (42 per cent and 25 per cent in Gulabad and Shabara respectively) reported that changing climatic situations would increase threats to agriculture, such as low yields, less fertile soil and more crop diseases.

Perceptions and response of farmers toward uncertain conditions are important as they can describe the decision-making behavior of farmers (Rasul et al., 2012). Hence, it is important to study risk perceptions of those farmers who live in a risky environment. In this regard, perceiving climate variability is the first step in the process of adapting agriculture to climate change, as discussed by Deressa et al. (2009) in Ethiopia and Abid et al. (2016) in Pakistan. In addition to the limited access to farm resources such as fertilizers, pesticides, good quality seed, water for irrigation, labor, land and infrastructure there are other environmental factors including floods, droughts and storms that increase CCV of farmers (Rafiq and Blaschke, 2012; Ali, 2013; Asif, 2013; Bukhari and Sayal, 2011).

Furthermore, the literature consistently emphasizes the concerns arising from the impacts of climate change and its variability on agricultural production worldwide (Ullah et al. 2015; Hay and Mamura, 2010; Ali and Erenstein, 2016). Continuous reduction and inconsistency in yield of major crops has been reported across Pakistan due to climate-related risks (Tingju et al., 2014; Ahmad et al., 2013). This suggests that in addition to analyzing risks to which people are exposed there is a need to investigate the quality of the options they have for coping and how they are ultimately managing risks. This understanding can facilitate identification of the most vulnerable groups and can also create opportunities to identify effective and sustainable adaptation strategies (Adger and Kelly, 1999; Rasul et al., 2012; Ahmad et al., 2013).

4. Conclusion

This study provides insights into the climate-related risk perceptions of farmers, including their vulnerability and adaptive responses, constraints that limit their adaptive capacity and concerns regarding negative impacts of climate change on agriculture at farm level. Identifying individual's risk perception is important as it determines their responses and helps in designing a context-specific policy. Farmers perceived that in the future, climate change would be an even greater threat to agriculture.

The study identified that climate change is negatively affecting agriculture, which is in most cases is the only primary subsistence activity of rural farmers all over Pakistan, particularly in the KPK Province. This situation is consistent with other climate-related studies conducted throughout the country and in our study area. Recent changes in climatic conditions have exposed rural farming communities to numerous risks. For instance, farmers mentioned that disastrous floods, severe droughts, storms, extreme maximum temperatures, changes in rainfall pattern, crop diseases and loss of farmland due to floods, are among the worst situations negatively affecting agricultural production. Major adaptation methods identified by households were changing fertilizer, changing seed quality, changing crop type or variety and planting shade trees. Lack of access to financial services and to information on agricultural training and lack of support from provincial and local governments were among the major constraints to adaptation. Farmers in the study area had no access to agricultural extension or farmer training that could build their capacity. Government and other relevant stakeholders should provide easy access to those services so that farmers can learn advanced farming techniques and how to effectively adapt.

Therefore, building capacity of the locals toward reducing CCV and facilitating effective adaptations are important. Future policies need to address barriers to the adoption of advanced adaptation techniques at the farm level. There is a dire need for research on identifying locally specific adaptation of agriculture to climate change so that farmers can decide the most suitable adaptation measure to apply. Support from agricultural extension bodies, research institutions and policy makers is also needed to provide updated information on weather and access to quality inputs used for improving yields. Cooperation among farmers is also key in improving their adaptive capacity and resolving other problems at the community level. The study also recommends that other researchers, especially females, explore this issue with women as they are more vulnerable to climate-related risks, a topic that this study could not address due to cultural and religious barriers.

References

Abid, M., Scheffran, J., Schneider, U.A. and Ashfaq, M. (2015), "Farmers' perceptions of and adaptation strategies to climate change and their determinants: the case of Punjab province, Pakistan", *Earth System Dynamics*, Vol. 6 No. 1, pp. 225-243.

Abid, M., Schilling, J., Scheffran, J. and Zulfiqar, F. (2016), "Climate change vulnerability, adaptation and risk perceptions at farm level in Punjab, Pakistan", *Science of the Total Environment*, Vol. 547, pp. 447-460.

Adger, W.N. and Kelly, P.M. (1999), "Social vulnerability to climate change and the architecture of entitlements", *Mitigation and Adaptation Strategies for Global Change*, Vol. 4 No. 34, pp. 253-266.

Adger, W.N., Huq, S., Brown, K., Conway, D. and Hulme, M. (2003), "Adaptation to climate change in the developing world", *Progress in Development Studies*, Vol. 3 No. 3, pp. 179-195.

Ahmad, M., Iqbal, M. and Khan, M.A. (2013), "Climate change, agriculture and food security in pakistan: adaptation options and strategies", *Climate Change Brief*, Pakistan Institute of Development Economics, Islamabad.

Ali, S. (2013), "Groundwater depletion: another crisis looming in Pakistan", available at: http://waterinfo.net.pk/?q=node/356 (accessed 21 September 2015).

Ali, A. and Erenstein, O. (2016), "Assessing farmer use of climate change adaptation practices and impacts on food security and poverty in Pakistan", *Climate Risk Management*, Vol. 16, pp. 183-194.

Asif, M. (2013), "Climatic change, irrigation water crisis and food security in Pakistan", Dissertation, available at: http://urn.kb.se/resolve?urn=urn:nbn:se:uu:diva-211663

Atta-ur-RahmanKhan, A. (2011), "Analysis of flood causes and associated socio-economic damages in the Hindukush Region", *Natural Hazards*, Vol. 59 No. 3, pp. 1239-1260.

Atta-ur-RahmanKhan, A. (2013), "Analysis of 2010-flood causes, nature and magnitude in the Khyber Pakhtunkhwa, Pakistan", *Natural Hazards*, Vol. 66 No. 2, pp. 887-904.

Baig, A. and Amjad, S. (2014), "Impact of climate change on major crops of Pakistan: a forecast for 2020", *Pakistan Business Review*, Vol. 15, pp. 600-617.

Blaikie, P., Cannon, T., Davis, I. and Wisner, B. (2014), *At Risk: Natural Hazards, People's Vulnerability and Disasters*, Routledge.

Bogner, A., Littig, B. and Minz, W. (2009), "Introduction: expert interviews – an introduction to a new methodological debate", *Interviewing Experts*, Palgrave Macmillan, UK, pp. 1-13.

Bryan, E., Deressa, T.T., Gbetibouo, G.A. and Ringler, C. (2009), "Adaptation to climate change in Ethiopia and South Africa: options and constraints", *Environmental Science and Policy*, Vol. 12 No. 4, pp. 413-426.

Bukhari, M. and Sayal, E.A. (2011), "Emerging climate changes and water resource situation in Pakistan", *Pakistan Vision*, Vol. 12 No. 2, pp. 236-254.

Cardona, O.D., Hurtado, J.E. and Chardon, A.C. (2003), *Indicators for Disaster Risk Management. Expert Meeting on Disaster Risk Conceptualization and Indicators Modelling in Universidad Nacional De Colombia*, Manizales, 9–11 July 2003, available at: http://idea.manizales.unal.edu.co/Proyectos Especiales/adminIDEA/CentroDocumentacion/DocDigitals/documents/01%20Conceptual1%20 Framework%20IADBIDEA%20Phase%20I.pdf

Dazé, A. (2011), "Understanding vulnerability to climate change: insights from application of CARE's climate vulnerability and capacity analysis (CVCA) methodology", in C.C.N., E.A. (Ed.), *Care Poverty*, pp. 1-24, available at: www.careclimatechange.org/files/adaptation/CARE_ Understanding_Vulnerability.pdf

Deressa, T.T., Hassan, R.M. and Ringler, C. (2011), "Perception of and adaptation to climate change by farmers in the Nile Basin OF Ethiopia", *The Journal of Agricultural Science*, Vol. 149 No. 1, pp. 23-31.

Deressa, T.T., Hassan, R.M., Ringler, C., Alemu, T. and Yesuf, M. (2009), "Determinants of farmers' choice of adaptation methods to climate change in the Nile Basin of Ethiopia", *Global Environmental Change*, Vol. 19 No. 2, pp. 248-255.

Eakin, H. (2003), *Rural Responses to Climatic Variability and Institutional Change in Central Mexico*, University of California at San Diego, Center for U.S.-Mexican Studies 1004, Center for U.S.-Mexican Studies, UC San Diego.

Field, C.B. (Ed.) (2012), *Managing the Risks of Extreme Events and Disasters to Advance Climate Change Adaptation: Special Report of the Intergovernmental Panel on Climate Change*, Cambridge University Press.

Forrester, D.I., Theiveyanathan, S., Collopy, J.J. and Marcar, N.E. (2010), "Enhanced water use efficiency in a mixed eucalyptus Globulus and Acacia Mearnsii Plantation", *Forest Ecology and Management*, Vol. 259 No. 9, pp. 1761-1770.

Füssel, H.M. (2007), "Adaptation planning for climate change: concepts, assessment approaches, and key lessons", *Sustainability Science*, Vol. 2 No. 2, pp. 265-275.

Greenough, G., McGeehin, M., Bernard, S.M., Trtanj, J., Riad, J. and Engelberg, D. (2001), "The potential impacts of climate variability and change on health impacts of extreme weather events in the United States", *Environmental Health Perspectives*, Vol. 109 No. S2, p. 191.

Hay, J. and Mamura, N. (2010), "The changing nature of extreme weather and climate events: risks to sustainable development", *Geomatics, Natural Hazards and Risk*, Vol. 1 No. 1, pp. 3-18.

Huq, S., Reid, H., Konate, M., Rahman, A., Sokona, Y. and Crick, F. (2004), "Mainstreaming adaptation to climate change in Least Developed Countries (LDCs)", *Climate Policy*, Vol. 4 No. 1, pp. 25-43.

Hussain, S.S. and Mudasser, M. (2007), "Prospects for wheat production under changing climate in mountain areas of Pakistan–an econometric analysis", *Agricultural System*, Vol. 94 No. 2, pp. 494-501.

IUCN (2009), "Climate change; vulnerabilities in agriculture in Pakistan", available at: http://cmsdata. iucn.org/download/pkcc_agrvul.pdf (accessed 18 September 2016).

Jodha, N.S., Singh, N.P. and Bantilan, M.C.S. (2012), "Enhancing farmers' adaptation to climate change in arid and semi-arid agriculture of India: evidences from indigenous practices: developing international public goods from development-oriented projects", Working Paper Series no. 32, International Crops Research Institute for the Semi-Arid Tropics, Patancheru, p. 28.

Kelly, P.M. and Adger, W.N. (2000), "Theory and practice in assessing vulnerability to climate change and facilitating adaptation", *Climate Change*, Vol. 47, pp. 325-352.

Khan, A.N. (2013), "Analysis of 2010-flood causes, nature and magnitude in the Khyber Pakhtunkhwa, Pakistan", *Natural Hazards*, Vol. 66 No. 2, pp. 887-904.

Khan, M.A. (2012), "Agricultural development in Khyber Pakhtunkhwa, Prospects, challenges and policy options", *Pakistaniaat: A Journal of Pakistan Studies*, Vol. 4 No. 1, pp. 49-68.

Khan, R.A. (1994), *Over View of NWFP Agriculture*, Nespak OPCV, Peshawar, pp. 1-15.

Khan, F.A. and Salman, A. (2012), "A simple human vulnerability index to climate change hazards for Pakistan", *International Journal of Disaster Risk Science*, Vol. 3 No. 3, pp. 163-176.

Khan, A.N., Khan, B., Qasim, S. and Khan, S.N. (2013), "Causes, effects and remedies: a case study of rural flooding in District Charsadda, Pakistan", *Journal of Managerial Sciences*, Vol. 7 No. 1, p. 2.

Kreft, S. and Eckstein, D. (2013), *Global Climate Risk Index 2014: Who Suffers Most from Extreme Weather Events? Weather-Related Loss Events in 2012 and 1993 to 2012*, Germanwatch eV, Bonn.

Maheen, H. and Hoban, E. (2017), "Rural women's experience of living and giving birth in relief camps in Pakistan", *PLoS Currents (ed: 1)*, available at: http://doi.org/10.1371/currents. dis.7285361a16eefbeddacc8599f326a1dd

Malik, S. (2012), "Case study: exploring demographic dimensions of flood vulnerability in rural Charsadda, Pakistan", Master thesis: appendix 4, available at: https://assets.publishing. service.gov.uk/media/57a08a65ed915d3cfd000746/Indus_Floods_Research_Appendix_4. pdf

Mendelsohn, R. (2001), in Wallace, E. and Oates, H.F. (Eds), "Global warming and the American Economy", *New Horizons in Environmental Economics*, Edward Elgar Publishing.

Mertz, O., Mbow, C., Reenberg, A. and Diouf, A. (2009), "Farmers' perceptions of climate change and agricultural adaptation strategies in Rural Sahel", *Environmental Management*, Vol. 43 No. 5, pp. 804-816.

Mirza, M.M.Q. (2003), "Climate change and extreme weather events: can developing countries adapt? ", *Climate Policy*, Vol. 3 No. 3, pp. 233-248.

Mueller, V., Gray, C. and Kosec, K. (2014), "Heat stress increases long-term human migration in rural Pakistan", *Nature Climate Change*, Vol. 4 No. 3, pp. 182-185.

Mustafa, D. (1998), "Structural causes of vulnerability to flood hazard in Pakistan", *Economic Geography*, Vol. 74 No. 3, pp. 289-305.

Mustafa, D. (2002), "To each according to his power? Participation, access, and vulnerability in irrigation and flood management in Pakistan", *Environment and Planning D: Society and Space*, Vol. 20 No. 6, pp. 737-752.

Qasim, S., Khan, A.N., Shrestha, R.P. and Qasim, M. (2015), "Risk perception of the people in the flood prone Khyber Pakhtunkhwa province of Pakistan", *International Journal of Disaster Risk Reduction*, Vol. 14, pp. 373-378.

Rafiq, L. and Blaschke, T. (2012), "Disaster risk and vulnerability in Pakistan at a district level", *Geomatics, Natural Hazards and Risk*, Vol. 3 No. 4, pp. 324-341.

Rasul, G., Mahmood, A., Sadiq, A. and Khan, S.I. (2012), "Vulnerability of the Indus Delta to climate change in Pakistan", *Pakistan Journal of Meteorology*, Vol. 8 No. 16, pp. 89-107.

Roncoli, C., Ingram, K. and Kirshen, P. (2002), "Reading the rains: local knowledge and rainfall forecasting among farmers of Burkina Faso", *Society and Natural Resources*, Vol. 15 No. 5, pp. 411-430.

Saif-ur-rehman, S.A. and Shaukat, B. (2013), "The effects of 2010 flood on educational institutions and children schooling in Khyber Pakhtunkhwa: a study of Charsadda and Swat Districts", *International Journal of Environment, Ecology, Family and Urban Studies (IJEEFUS)*, Vol. 3 No. 3.

Shahzada, H., Hussain, E. and Iqbal, J. (2012), "Post flood drinking water quality assessment of Charsadda District using GIT's", *Journal of Himalayan Earth Science*, Vol. 45 No. 2.

Smit, B. and Pilifosova, O. (2003), "From adaptation to adaptive capacity and vulnerability reduction", *Climate Change, Adaptive Capacity and Development*, pp. 9-28.

Tingju, Z., Xie, H., Waqas, A., Ringler, C., Iqbal, M.M., Goheer, M.A. and Sulser, T. (2014), "Climate change and extreme events, impacts on Pakistan's agriculture", *International Food Policy Research Institute (IFPRI) (PSSP Policy Note 002)*.

Ullah, R., Shivakoti, G.P. and Ali, G. (2015), "Factors effecting farmers' risk attitude and risk perceptions: the case of Khyber Pakhtunkhwa, Pakistan", *International Journal of Disaster Risk Reduction*, Vol. 13, pp. 151-157.

Van Aalst, M.K. (2006), "The impacts of climate change on the risk of natural disasters", *Disasters*, Vol. 30 No. 1, pp. 5-18.

Watts, M.J. and Bohle, H.G. (1993), "The space of vulnerability: the causal structure of hunger and famine", *Progress in Human Geography*, Vol. 17 No. 1, pp. 43-67.

Wisner, B. Gaillard, J.C. and Kelman, I. (Eds) (2012), *Handbook of Hazards and Disaster Risk Reduction and Management*, Routledge.

Yousaf, S. and Naveed, S. (2013), "Flood (2010) effects on agriculture, livestock, infrastructure and human health: a case study of Charsadda District", *The Journal of Humanities and Social Sciences*, Vol. 21 No. 1, pp. 81-90.

Further reading

Moser, C.O. (1998), "The asset vulnerability framework: Reassessing urban poverty reduction strategies", *World Development*, Vol. 26 No. 1, pp. 1-19.

About the authors

Wahid Ullah is currently pursuing a PhD at the Department of Human Geography of Hokkaido University, Japan. He is involved in research focusing on vulnerability and adaptation of indigenous communities toward climate change induced natural disasters in Pakistan. His fields of interest are climate change, disaster risk reduction, poverty reduction and livelihood improvement through sustainable agriculture and natural resource management. Previously, he worked for the Center for International Forestry Research on identifying suitable conditions for efficient and sustainable participatory monitoring, reporting and verification systems of carbon emissions so that it can be embedded into a national database. Wahid Ullah is the corresponding author and can be contacted at: waheedullah@live.in

Takaaki Nihei is an Associate Professor at the Department of Human Geography, Graduate School of Letters in Hokkaido University. His research specialty is agricultural and rural geography. The subject relates to food production, land use, tourism and sustainable development. He has conducted fieldwork in Japan, North America and South America.

Muhammad Nafees has been a Professor in the Department of Environmental Sciences, University of Peshawar since 1995. He teaches pollution control technology and environmental impact assessment (EIA) at Masters and PhD levels. His other fields of interest are industrial pollution, solid waste management, cleaner production and fresh water resources.

Rahman Zaman completed Master of Environmental Sciences from the University of Peshawar in 2017. His research expertise includes sustainable agriculture, disaster risk reduction and synthesis of drinking water quality.

Muhammad Ali completed Master of Environmental Sciences from the University of Peshawar in 2017. His thesis was entitled "Quantifying insecticide and herbicide utilization in rural farming practices". He conducted fieldwork in two villages in Charsadda district of Khyber Pakhtunkhwa province for collecting the data needed to complete his study.

12

Farmers' perspectives Impact of climate change on African indigenous vegetable production in Kenya

Winifred Chepkoech, Nancy W. Mungai, Silke Stöber, Hillary K. Bett
and Hermann Lotze-Campen
(Author affiliations can be found at the end of the article)

Abstract

Purpose – Understanding farmers' perceptions of how the climate is changing is vital to anticipating its impacts. Farmers are known to take appropriate steps to adapt only when they perceive change to be taking place. This study aims to analyse how African indigenous vegetable (AIV) farmers perceive climate change in three different agro-climatic zones (ACZs) in Kenya, identify the main differences in historical seasonal and annual rainfall and temperature trends between the zones, discuss differences in farmers' perceptions and historical trends and analyse the impact of these perceived changes and trends on yields, weeds, pests and disease infestation of AIVs.

Design/methodology/approach – Data collection was undertaken in focus group discussions (FGD) ($N = 211$) and during interviews with individual farmers ($N = 269$). The Mann–Kendall test and regression were applied for trend analysis of time series data (1980-2014). Analysis of variance and least significant difference were used to test for differences in mean rainfall data, while a chi-square test examined the association between farmer perceptions and ACZs. Coefficient of variation expressed as a percentage was used to show variability in mean annual and seasonal rainfall between the zones.

Findings – Farmers perceived that higher temperatures, decreased rainfall, late onset and early retreat of rain, erratic rainfall patterns and frequent dry spells were increasing the incidences of droughts and floods. The chi-square results showed a significant relationship between some of these perceptions and ACZs. Meteorological data provided some evidence to support farmers' perceptions of changing rainfall. No trend was detected in mean annual rainfall, but a significant increase was recorded in the semi-humid zone. A decreasing maximum temperature was noted in the semi-humid zone, but otherwise, an overall increase was detected. There were highly significant differences in mean annual rainfall between the zones. Farmers perceived reduced yields and changes in pest infestation and diseases in some AIVs to be prevalent in the dry season. This study's findings provide a basis for local and timely institutional changes, which could certainly help in reducing the adverse effects of climate change.

Keywords Kenya, Farmers, Perceptions, Climate change, African indigenous vegetables, Agro-climatic zones

1. Introduction

Sub-Saharan Africa (SSA) has been identified as one of the regions that are most vulnerable to the impacts of climate change (CC) (Bryan *et al.*, 2013; IPCC, 2014). CC is seen as one of the major factors limiting Africa's efforts to achieve food security because of the continent's dependency on rain-fed agriculture and the low capacity of smallholders to adapt to CC (IPCC, 2014; Phirri *et al.*, 2016). Furthermore, widespread poverty, inequitable land distribution and the declining size of farmland make the region more vulnerable (IPCC, 2014). It is projected that SSA will witness increases in temperature, changes in rainfall intensity and distribution and a rise in incidences of extreme weather events (e.g. droughts and floods), pests, weeds and disease epidemics (FAO, 2015; Connolly-Boutin and Smit, 2015). In Kenya, for example, since the early 1960s, the mean temperature has been increasing and rainfall patterns have become irregular and unpredictable, with the country experiencing an increased number of extreme events such as droughts and floods (GoK, 2013a; Thompson *et al.*, 2015). As the full effects of these changes are yet to be felt, there is evidence in the literature that these changes are already having an impact on agricultural production. For instance, combined infestations of pests and diseases in plants have been reported to result in losses of over 50 per cent of the yield of major crops. Plant diseases are also estimated to cause up to a 20 per cent reduction in the yield of principal food and cash crops worldwide (Gautam *et al.*, 2013; West *et al.*, 2015), while weeds have been found to contribute the greatest potential yield losses of up to 34 per cent compared with those caused by pests and diseases (Oerke, 2005). According to Thornton *et al.* (2014), changes in climate variability and the frequency of extreme events may have a substantial impact on the prevalence and distribution of pests, weeds and crop diseases. Additionally, increases in minimum temperatures have negative impacts on rice yields by up to 10 per cent for each increase in temperature by 1°C in the dry season. Furthermore, increases in maximum temperatures lead to reduced yields of maize by 1.7 per cent for each day above 30°C under drought conditions (Thornton *et al.*, 2013). Research shows that the impact of CC on agricultural production is likely to intensify in future because of an expected further increase in temperature. This increase is predicted to cause a yield decline of 14 per cent (rice), 22 per cent (wheat) and 5 per cent (maize) in SSA. The consequence of this is increased poverty and vulnerability of the people who primarily depend on rain-fed agriculture for their livelihoods (IPCC, 2014; Kabubo-Mariara and Kabara, 2015; Adhikari *et al.*, 2015).

Agricultural production in Kenya contributes about 25.4 per cent of the country's gross domestic product directly and another 27 per cent indirectly through its links with agro-based industries and the service sector (FAO, 2015). In Kenya, the majority of the population lives in rural areas, with smallholder rain-fed agriculture as their key economic activity (Bryan *et al.*, 2013; Barasa *et al.*, 2015b). According to the Nutritional Action Plan 2012-2017, the country is facing the challenges of food and nutrition insecurity. Additionally, the prevalence of micronutrient deficiencies in the population is becoming a matter of concern to the government. The most common include vitamin A deficiency, iron deficiency anaemia, iodine deficiency disorders and zinc deficiency (GoK, 2012). African indigenous vegetables

(AIVs) have been identified as an important component in providing food and nutrition security and compensating for nutrient deficiencies (Habwe *et al.*, 2009; Abukutsa-Onyango, 2010; Chagomoka *et al.*, 2014). Apart from their nutritional importance, they also provide opportunities for generating a higher income and other livelihood gains such as employment opportunities (Prasad and Chakravorty, 2015). AIVs are recognised for their superior provision of micronutrients, particularly iron, zinc, vitamins C and A and protein compared to other widely consumed exotic vegetables such as tomatoes, cabbages and onions (Abukutsa-Onyango, 2010; Kamga *et al.*, 2013; Luoh *et al.*, 2014). This is especially important for a country like Kenya, which is characterised by a rapidly increasing population and high poverty levels and where multiple burdens of malnutrition-persistent hunger and undernutrition are becoming increasingly prevalent (Kenya National Bureau of Statistics-KNBS and Society for International Development-SID, 2013; United Nations Children's Fund (UNICEF), 2013).

There is considerable literature on how farmers perceive CC. Some of the perceptions presented in the literature include intense rainfall; changes in the timing of rainfall; frequent droughts; and changes in temperature, landslides, crop pests, thunderstorms, hailstorms, winds and floods (Babatolu and Akinnubi, 2016; Limantol *et al.*, 2016; Sanogo *et al.*, 2016; Elum *et al.*, 2017; Stöber *et al.*, 2017; Mkonda and He, 2017; Mutunga *et al.*, 2017; Fadina and Barjolle, 2018; Williams *et al.*, 2018). Furthermore, this growing body of literature has also focused on evaluating the impact of CC on major food crops (wheat, maize, rice, millet, sorghum and cassava (Rurinda *et al.*, 2015). Notably, their observations are based on projections and CC crop model simulations, which have concentrated on the impact of CC on cereal crop production, with very little reference to horticultural crops such as vegetables. Some studies based on experiments and crop simulation models have, however, documented the effect of various climatic conditions on vegetable crops. For instance, Luoh *et al.* (2014) found that the fresh edible weight/yield of three AIVs is reduced in water-stressed conditions. Additionally, drought tolerance has also been documented for indigenous vegetables such as amaranth (Van den Heever and Slabbert, 2007). Other studies have focused on monitoring the sensitivity of some exotic vegetables to different climate variables. For example, cabbage and spinach were exposed to water stress and elevated CO_2 conditions in India, and the authors found that heat and water stress reduced the quality of cabbages, while an increased yield was reported for spinach (Jain *et al.*, 2007; Moretti *et al.*, 2010; Choudhary *et al.*, 2015). In other studies, tomatoes were evaluated under conditions of rising temperatures, double CO_2 and water deficiency in Nepal and Macedonia. A yield reduction of 72-84 per cent for tomatoes was found in water-deficient conditions. Decreased ripening and increased productivity by 51 per cent were however reported when CO_2 conditions for tomatoes were doubled (Malla, 2008; Domazetova, 2011).

Nevertheless, some attempts have been made to assess the impact of these changes on specific crops from the perspective of farmers. For instance, farmers have perceived an increase in temperatures, droughts and storms, which has led to problems of reduced yield and increased incidences of pests, diseases and failures in potato and cabbage crops in South Africa (Elum *et al.* (2017). Similarly, Ayyogari *et al.* (2014) noted that the most constraining factors in vegetable cultivation under changing climatic situations include crop failures, a decline in yields, a reduction in quality and increasing pest and disease problems. Furthermore, a general spread of pests and weeds on cropland and reduced yield have been reported by farmers (Mertz *et al.*, 2009; Apata *et al.*, 2009). Even though AIVs are known for being relatively robust to CC, water stress, pests, diseases and weed pressure pose as yet unknown risks to AIV production (Ngugi *et al.*, 2006; Luoh *et al.*, 2014; Stöber *et al.*, 2017). There is a scarcity of evidence-based research information on how CC affects smallholder production of AIVs (Stöber *et al.*, 2017). Consequently, there has always been a tendency among many researchers and practitioners to conclude that AIVs are resistant to CC, with the consequent branding of these vegetables as "hard-core" or "survivor plants". Indigenous crops including AIVs have often been recommended for use in marginal areas for coping with CC and food security because of their ability to adapt to harsh environments with various types of stresses (Abukutsa-Onyango *et al.*, 2010; Capuno *et al.*, 2015). Many also argue that they are resistant to drought and have greater resistance to pests and diseases (Muhanji *et al.*, 2011; Luoh *et al.*, 2014). For instance, slender leaf (*Crotolaria* sp.) has been identified as being particularly hardy during droughts because it quickly establishes its taproot and can survive when rainfall is low (Cernansky, 2015). In contrast, some studies point out that CC-related risks have contributed to the decline in AIV production (Masinde and Stützel, 2005; Ngugi *et al.*, 2006; Muthomi and Musyimi, 2009). However, these broad generalisations are likely to mask local farmers' perceptions of the impact of CC on production.

The production area and value of AIVs rose by 6 and 10 per cent, respectively, in 2014 (HCDA, 2014). This was because of increased awareness of their nutritional and health benefits. In addition, the role of the World Vegetable Center (AVRDC) in enhancing and promoting the AIV value chain cannot be ignored (Palada *et al.*, 2006; Ojiewo *et al.*, 2010; Ochieng *et al.*, 2016). The organisation has been instrumental in genetically enhancing local cultivars and making advances in breeding for resistance/tolerance to biotic/abiotic stresses. It is important to note that although locally adapted seeds have been developed, farmers are more likely to use their own informally produced seeds. In Western Kenya, for example, farmers have been found to produce, distribute and store their own seeds among themselves (Abukutsa-Onyango, 2007). Generally, it has also been documented that less than 10 per cent of the crop seed planted in Africa is purchased from the formal market each year (Rohrbach *et al.*, 2003). This means that farmers may not benefit from the opportunities that could be exploited by using locally adapted modified seed varieties. Additionally, previous studies on AIVs have focused on their diversity, consumption, nutritional value, demand and commercial importance (Habwe *et al.*, 2009; Yang and Keding, 2009; Oluoch *et al.*, 2009; Weinberger and Pichop, 2009; Maundu *et al.*, 2009; Pasquini *et al.*, 2009; Abukutsa-Onyango, 2010; Chagomoka *et al.*, 2014; Gido *et al.*, 2017). The present study recognises the fact that a great deal of research has been carried out to document farmers' perceptions of CC, but very few have provided evidence linking location-specific perceptions of the impact of CC on particular crops such as AIVs. This study therefore contributes new knowledge about local perceptions of CC and its impact on AIV production in light of the uncertainty around future rainfall patterns, very different agro-climatic conditions and the gap in academic literature. Knowledge of this will be important in informing institutional and policy support programmes aimed at promoting specific adaptations in Africa targeted locally and at specific crops. This may enable farmers to exploit the full potential of these AIVs in a constantly changing climate. In connection with this, four distinct questions were addressed in this study:

Q1. How do AIV farmers perceive CC in three different agro-climatic zones (ACZs) in Kenya?

Q2. What are the main differences between the zones in historical rainfall and temperature trends?

Q3. Is there any association between farmers' perceptions and the agro-climatic location?

Q4. What impact do these perceived changes and trends have on yields, weeds, pests and disease infestation of AIVs?

2. Research methodology

2.1 Study areas

The study area comprised three purposively selected ACZs in Kenya (Figure 1). Kenya is divided into seven ACZs by vegetation characteristics, the amount and reliability of rainfall and the land's ecological potential (Bryan *et al.*, 2013). The high to medium potential areas comprise ACZs I (humid), II (sub-humid) and III (semi-humid). Humid and sub-humid zones are well-watered, support arable agriculture and have a high population density, with annual rainfall above 800 mm (MAFAP, 2013). Marginal or low potential areas comprise ACZs IV (semi-humid to semi-arid), V (semi-arid), VI (arid) and VII (very arid) and constitute arid and semi-arid land or rangelands. These areas are generally hot and dry, with low and unpredictable rainfall of less than 600 mm per year (FAO, 2010).

Kakamega County (humid zone I) lies between longitudes 34° and 35° east and latitudes 0° and 1° north of the equator and within altitudes of 1,250-2,000 m above sea level (Barasa *et al.*, 2015a and b). The total area of Kakamega County is 3,020 square kilometres and it has a population of 1,660,651 (GoK, 2013b). Its climate is predominantly hot and wet most of the year, with mean annual rainfall of between 1,800 and 2,000 mm. The mean monthly rainfall trend represents maximums and minimums over the year. The maximums occur in April to June and August to November (GoK, 2013b). Generally, there are two main cropping seasons in most parts of the county that coincide with the "long rains" and "short rains". The "short rains" fall between March and May, while the "long rains" fall between October and December (Kabubo-Mariara and Karanja, 2007). The average temperature in the county is 22.5°C. January and February are generally considered dry months (Barasa *et al.*, 2015a). The county has high temperatures all year round, with slight variations in mean maximum and minimum ranging from 28°C to 32°C and 11°C to 13°C, respectively. The mean annual evaporation is high and ranges from 1,600 mm to 2,100 mm with high humidity (Ngetich, 2013).

Nakuru County (semi-humid zone III) lies within the Great Rift Valley. The county covers an area of 7,235.3 square kilometres and is located between longitudes 35° and 35° east and latitudes 0° and 1° south (GoK, 2013b). The county is between 1,520 and 2,400 m above sea level. The temperature varies between 24°C and 29.6°C. The zone is characterised by a bimodal rainfall pattern, with a high of 1800 mm and a low of 500 mm (Ogeto *et al.*, 2013;).

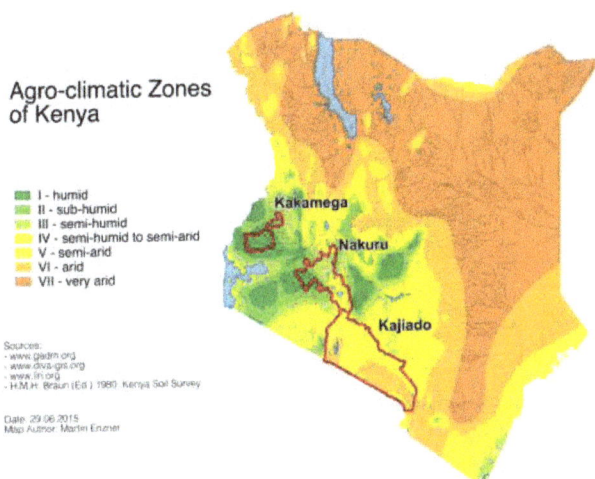

Figure 1. Map of study areas

Kajiado County (semi-arid zone V) has a population of 687,312 and occupies an area of 21,902 square kilometres (GOK, 2009a). It is located between longitudes 360° 5' and 370° 5' east and between latitudes 10° 0' and 30° 0' south. The county has two distinct rainy seasons: a long rainy season from March to May and a short rainy season from October to December (Babadoye *et al.*, 2014; GoK, 2014). The distribution of rainfall between the two seasons changes gradually from east to west across Kajiado County. In eastern Kajiado, more rain falls during the long rains (March-May). The mean annual rainfall ranges from 300 to 800 mm (GOK, 2009b). The main characteristics of the three selected ACZs are presented in Table I.

2.2 Sample size and sampling procedure

A multi-stage purposive sampling procedure was used to select respondents for the study. In the first stage, three counties were selected that reflected the considerable variation in climate across the country. These were Kakamega, Nakuru and Kajiado Counties. In the second stage, a purposive sampling method was used to select two sub-counties with high potential AIV production. In Kakamega, the sub-counties of Butere and Lugari, in Nakuru the sub-counties of Rongai and Bahati and in Kajiado the sub-counties of Kajiado North and Kajiado West were selected. To obtain participants for the focus group discussions (FGDs), purposive sampling was used to identify AIV farmers. Local extension workers were helpful in recruiting participants. In all, 18 FGDs were conducted involving a total of 211 farmers. Eight FGDs (four in each sub-county), six (three in each sub-county) and four (two in each sub-county) were held in Kakamega, Nakuru and Kajiado counties, respectively. FGDs generated information on farmers' perceptions of CC, the effects of climate on AIV production and the sensitivities of AIVs to climate. Each FGD lasted between 1 h and 1 h and 45 min. In the last stage, using the snowball sampling technique, 269 interviews were conducted with smallholder AIV farmers from the three selected ACZs. Interviews were conducted with 100, 107 and 62 AIV farmers in Kakamega, Nakuru and Kajiado Counties, respectively, 93 of whom were male and 176 were female. The interviews with farmers generated information about their perceptions of the changing climate. To understand the effect of CC on AIV production, the influence on yield, pests, diseases and weeds was assessed. This was because of the fact that other than having direct effects on plant productivity, the changes in climate variables also significantly influence productivity through indirect effects mediated by changes in weeds, pests and diseases (Thomson *et al.*, 2010; IPCC, 2014).

The sample summary for the study is given in Table II.

Table I. Main features of the study areas

County	Kakamega (ACZ-II)	Nakuru (ACZ-III)	Kajiado (ACZ-V)
Agro-climatic zone	humid	semi-humid	semi-arid
No. of sub-counties	12	11	5
Rural/urban character	rural	peri-urban	peri-urban
Population	1,660,651	1,603,325	687,312
Poverty status (%)	53	40.1	11.6
Mean temperature max.	29°C	20°C	34°C
Mean temperature min.	18°C	15°C	22°C
Mean precipitation p.a.	2,000 mm	800 mm	300 mm

Source: Stöber *et al.* (2017)

2.3 Data

The study used data collected from farmers combined with climate data to assess the impact of CC on AIV production. Farmers' perceptions of CC were then compared with historical trends from meteorological data. Perception here follows the definition of Ndamani and Watanabe (2015) as the process by which organisms (humans) interpret and organise sensations to produce a meaningful experience of the world. Farmers were asked to describe the changes in CC parameters. These parameters – expressed as changes in mean temperatures, amount of rainfall, the frequency, duration and intensity of dry spells and droughts, the timing, duration and intensity of rain, the start/end of growing seasons, the frequency and intensity of storms and floods – were analysed in this study. Previous studies of such as Mertz *et al.* (2009), Kabubo-Mariara and Kabara (2015), Babatolu and Akinnubi (2016) and Dhanya and Ramachandran (2016) have followed similar approaches.

To examine trends in rainfall and temperature, climate data (rainfall and temperature) from Kenya Meteorological Services (KMS) for the respective counties were analysed. The data obtained from KMS were monthly minimum and maximum temperatures (°C) and rainfall (mm) from 1980 to 2014 for the three ACZs. The data were supplied by three reference weather stations: Kakamega town for the humid zone of Kakamega, Jomo Kenyatta International Airport (JKIA) for the semi-arid zone of Kajiado and Nakuru town for the semi-humid zone of Nakuru. These climate parameters were selected because they have the longest and widest data coverage in the country and are the most common climatic variables considered by many studies in SSA (Ochieng *et al.*, 2016). Furthermore, among small-scale farmers, rainfall is the most important climatic factor critical to their survival, particularly in relation to crop growth (AGRA, 2014). During data screening, the last-observation-carried-forward (LOCF) method was used in cases where numerical values were missing. LOCF is a common statistical approach in the analysis of longitudinal repeated measures data in which some follow-up observations may be missing. In a LOCF analysis, a missing follow-up visit value is replaced by (imputed as) that subject's previously observed value, i.e. the last observation is carried forward. The combination of observed and imputed data is then analysed as though no data were missing (Lachin, 2016). To avoid biased results, the dataset was checked for outliers. In this case, extreme data values were replaced by seasonal or annual average values.

2.4 Data analysis

2.4.1 *Survey data analysis.* Qualitative and quantitative data were collected in parallel and then analysis for integration of the sets of data was undertaken. Finally, the two forms of data were analysed separately and then merged (Fetters *et al.*, 2013). A statistical package for social sciences was used to analyse quantitative data. Percentages and frequencies were used to represent farmers' perceived long-term changes in temperature and rainfall.

Table II. Sample summary

ACZ	County	Sub-county	Farmers interviewed	No. of FGDs	No. of farmers in FGDs
Humid	Kakamega	Butere	50	4	41
		Lugari	50	4	41
Semi-humid	Nakuru	Bahati	54	3	35
		Rongai	53	3	33
Semi-arid	Kajiado	Kajiado North	31	2	14
		Kajiado South	31	2	17
Total			269	18	211

Contingency tables and chi-square tests were performed to investigate whether there was any association between farmers' perception of CC and their location. The hypothesis which stated that there is no association between farmer perception and ACZ was tested at a $p < 0.05$ level of significance.

Qualitative data collected during the FGDs were analysed using qualitative content analysis. The transcription of the audio recordings of the FGDs in the sample areas was translated from Kiswahili into English. The transcriptions were studied repeatedly to develop an analysis structure. A systematic process described by Hsieh and Shannon (2005) involved the development and identification of themes and patterns identified through a series of coding cycles. In accordance with Newing (2011), data from FGDs were illustrated by direct quotations, recounting particularly relevant experiences and the views of smallholder farmers, which are essential for the findings' authenticity. Samples of quotations from the interviews and FGDs depicting the differences and similarities between the changes in climate perceived by farmers from the study sites are presented in tabular form.

2.4.2 Trend analysis.

2.4.2.1 Mann–Kendall test. The Mann–Kendall test is a non-parametric test that does not require the data to be normally distributed and has low sensitivity to abrupt breaks because of inhomogeneous time series (Tabari *et al.*, 2011). The test has been widely recommended by the World Meteorological Organization for public application (Mitchell *et al.*, 1996). Furthermore, studies of Mkonda and He (2017), Karmeshu (2012), Mavromatis and Stathis (2011), Tabari *et al.* (2011), Longobardi and Villani (2009) and many others have valued the use of this test for evaluating trends in climatic data. In this test, each value in the series is compared with others, always in a sequential order (Jaiswal *et al.*, 2015). The trend analysis of long-term climate data was performed to ascertain the changes in climate patterns and analyse the relationship between farmers' perceptions and climatic facts. Addinsoft's XLSTAT 2017 software was used to perform the statistical Mann–Kendall test in this study. The null hypothesis was tested at a 95 per cent confidence level for both temperature and rainfall data for the three ACZs. According to this test, the null hypothesis assumes that there is no trend (the data are independent and randomly ordered) and this is tested against the alternative hypothesis, which assumes that there is a trend (Longobardi and Villani, 2009).

2.4.2.2 Linear regression. Linear regression is a non-parametric test that was used to study trends in rainfall and temperature change data. Regressions are used to estimate temporal trends in rainfall records. The suitability of data for regression analysis and the interpretation of results are taken from case studies that have applied a similar approach (Longobardi and Villani, 2009; Nyatuame *et al.*, 2014). Microsoft Excel was used to conduct linear regression analysis. In the linear regression test, a straight line is fitted to the data and the slope of the line may or may not be significantly different from zero (Jaiswal *et al.*, 2015). In addition, trend lines were plotted for each of the ACZs. The value of R-squared (R^2) or the square of the correlation from the regression analysis was used to show the strength of the correlation and relationship between variables X and Y. The value is a fraction between 0.0 and 1.0. An R^2 value of 1.0 means that the correlation is strong and all points lie on a straight line, while an R^2 value of 0.0 means that there is no correlation or linear relationship between X and Y (Nyatuame *et al.*, 2014).

2.4.2.3 Analysis of variance and least significant difference. One-way analysis of variance (ANOVA) was used to determine whether there were significant differences in the mean annual rainfall between the three climatic zones in the region. Least significant difference (LSD) was used to test the variance between the means.

The hypothesis stated that:

H0.   There is no significant difference in the mean annual rainfall between the climatic zones in Kenya.

H1.   There is a significant difference in the mean annual rainfall between the climatic zones in Kenya.

3. Results and discussion
3.1 Farmers' perceptions of climate change
AIV farmers in Kenya have perceived a notable change in climate in recent years across the study areas (Tables III and IV). The three ACZs differ from each other with respect to ecological and climatic conditions, but also with respect to CC scenarios. Extremes in temperatures have mostly been reported by farmers. In the humid zone, for example, farmers agreed that extremes in temperature prevailed in the area both during the day and at night. For instance, farmers reported that "when it is hot, it is extremely hot and when it is cold, it is extremely cold". In the semi-arid zone, farmers particularly noted that temperatures were very high during the day and very low at night (Table III). Approximately, more than 70 per cent had observed an increase in temperatures in all the zones, while less than 23 per cent had perceived a decrease in temperatures in the past 20 years. These results are in agreement with the findings of numerous studies from Africa, which report that farmers have perceived an increase in temperatures (Babatolu and Akinnubi, 2016; Limantol *et al.*, 2016; Bobadoye *et al.*, 2016; Sanogo *et al.*, 2016; Mutunga *et al.*, 2017; Mkonda and He, 2017; Fadina and Barjolle, 2018).

Farmers in all three zones also reported a change in the amount of rainfall. In the humid zone, farmers' perceptions were that rainfall has increased, whereas in the semi-humid zone and semi-arid zone farmers perceived that the rains had generally decreased over the years and were not sufficient. Erratic rainfall patterns were also reported in all the study areas. In the humid zone (Kakamega), farmers perceived that the rains that used to come regularly during the planting season in previous years have now become more unpredictable. Similar observations were made regarding the onset and cessation of rain in the three zones. In the humid zone, the majority noted that the rain mostly started late and ended early. In the humid zone, farmers furthermore said that "in the 1980s and 90s, rainfall was regular from March to May and July to December, but now the seasons are no longer predictable". Additionally, farmers noted that "long rains are no longer long rains and short rains are no longer short rains, seasons are scattered". These results are in agreement with many studies such as those by Limantol *et al.* (2016), Sanogo *et al.* (2016), Dhanya and Ramachandran (2016), Bobadoye *et al.* (2016), Mutunga *et al.* (2017), Mkonda and He (2017), Fadina and Barjolle (2018) and Williams *et al.* (2018). These authors confirm that farmers perceived less rain, a decreased duration and of rainy seasons and unpredictable and irregular rainfall patterns.

Besides season duration, this study revealed that changes in the intensity of the rainy seasons had also been observed in the study areas. More than half perceived that the intensity of the rainy seasons had increased in the humid and semi-arid zone, while only 34 per cent had perceived this in the semi-humid zone. It is notable that many farmers observed that the rains were unpredictable/unreliable, with clear changes in rain onset. More than half of the farmers noted that rains started late. For example, in the semi-humid zone (Nakuru), the majority of the farmers agreed that "in the past, rain started in March and ended in August, but now it starts in late April and ends in June" (Table III). Other results from this study also revealed that farmers have observed clear changes in rain cessation. There was

Table III. Statements on change in climate parameters in ACZs

Characteristic/ parameter	Temperature	Rainfall	Dry spell	Flood/hailstorm
Humid Zone Amount and frequency	Temperatures have generally increased	Heavy rains Rains mostly come late	Frequent dry spells	More floods and waterlogging problems
Duration/ intensity	When it is cold, it is extremely cold and when it is hot, it is extremely hot	Rainy seasons are short but intense	Dry spell intensity is high	–
Variability/ change	Higher than in the past even during rainy season High during the day and low during the night	Rain is unpredictable Seasons are scattered		–
Semi-humid zone Amount and frequency	Increase in temperatures	Delayed onset of rain and planting Rainfall is no longer predictable	Drought is very common in the area Recent drought in 2014 was severe	Floods have increased
Duration/ intensity	Very high during dry season and very low in wet seasons	Rainfall has declined in the area and has been inadequate	Increase in intensity Longer dry spells	
Variability/ change	High volatile temperature	Unreliable and unpredictable rainfall Large change in seasons		Large amounts of rain cause floods, as in 2012 and 2013
Semi-arid zone Amount and frequency	Increased temperatures Temperatures are high during the night	There is not enough rain/rainfall has decreased in recent years	Dry spells have increased Since 2000, there have been more droughts	
Duration/ intensity	Extreme temperatures	Rainy seasons are short Rain started in mid-March and ended in August, but now it ends in June	Prolonged and intense dry spell Droughts last up to one year and farming is temporarily abandoned	–
Variability/ change	Temperatures are very high, even during rainy season	Rainfall is erratic, unreliable and onset is delayed	In 2014 and 2015 the dry spell was harsh and of high intensity Unpredictable droughts	–

an overall agreement by more than 60 per cent of the farmers that the rains ended early, while less than 26 per cent in each of the zones perceived that they ended later. Similar observations have been reported in various studies (Apata *et al.*, 2009; Gandure *et al.*, 2012; Ochenje *et al.*, 2016; Fadina and Barjolle, 2018), which confirm that farmers perceive a delayed and decreased amount of rainfall and an early cessation of it. For example, in research on farmers' awareness of climate variability in Kenya, farmers experience excessive downpours and the rainfall has become unreliable, appearing out of season

(Barasa *et al.*, 2015b). Similarly, in a study in Tanzania, farmers declared that rainfall onset has changed because they used to plant crops in October/November, but the season has now shifted to December/January (Lema and Majule, 2009). Besides changes in the start and end of the rainy seasons, the findings from the semi-humid and semi-arid zones in the present study agree with those of other studies who report farmers' observations of a general decrease in the amount of rainfall (Limantol *et al.*, 2016; Mutunga *et al.*, 2017). In line with these findings, Mugalavai *et al.* (2008) argued that early onset translates into early cessation, while for the short rains, early onset translates into a longer growing season.

The interviewed farmers notably observed an increase in the frequency, duration and intensity of dry spells. More than half of the farmers in the semi-humid and semi-arid zones perceived that the frequency of dry spells had increased, while 36 per cent from the humid zone did so. Over half the farmers in the humid and semi-arid zones perceived that the duration of dry spells had increased, while 30 per cent in the semi-humid zone did so. Evidence of the perceptions of farmers in Zambia and Benin agrees with these findings. The authors reported that farmers were experiencing prolonged and frequent dry spells (Nyanga *et al.*, 2011; Sanogo *et al.*, 2016; Fadina and Barjolle, 2018).

Overall, more than half of the interviewed farmers had also experienced droughts (Table IV). Approximately 30 per cent of them perceived that the frequency of droughts

Table IV. Percentage distribution and correlation between farmers' perceptions and ACZs

Perceptions	Kakamega (humid) n =	Name of ACZ Nakuru (semi-humid) n =	Kajiado (semi-arid) n =	Total N = 269	χ^2 (*p*-value)
Rain onset early	17	7	45	19.7	38.72****
Late	76	90	52	75.8	(0.000)
No change	7	3	3	4.5	
Rain cessation early	63	93	72	77.3	36.31 ****
Late	23	5	26	16.4	(0.000)
No change	14	2	2	6.3	
Duration of rain season					11.18 **
Increased	23	11	23	18.2	(0.025)
Decreased	65	84	71	74	
No change	12	5	6	7.8	
Temperature					0.421
Increased	79	81	77	79.3	(0.810)
Decreased	21	19	23	20.7	
No change	0	0	0	0	
Dry spell frequency					18.19 ****
Increased	36	57	53	48.3	(0.001)
Decreased	37	13	26	24.9	
No change	27	30	21	26.8	
Drought frequency					
Increased	30	55	34	41.8	26.05 ****
Decreased	53	13	46	34.1	(0.000)
No change	17	32	20	24.2	
Drought duration					
Increased	40	65	0	50.5	16.01 ****
Decreased	58	30	59	46.7	(0.003)
No change	2	5	41	2.7	

Note: **** and ** indicate a statistical significance at $p < 0.01$ and $p < 0.05$ respectively

had increased, while the remainder perceived that it had decreased. In the semi-arid zone (Kajiado), farmers agreed that the most recent drought occurred in 2014 and was very severe. The Kajiado County integrated development plan 2013-2017 agrees that Kajiado County is characterised by cyclical and prolonged droughts, which confirms the findings of this study (GoK, 2014). Furthermore, these findings are a confirmation of the perceptions of climate documented for farmers in the Sahel and Benin. According to those studies, farmers perceive less rain, longer dry spells and a reduced number of rainy seasons (Mertz et al., 2009; Fadina and Barjolle, 2018). Similarly, in recent years, increased incidences of extreme droughts have also been reported in studies such as those by Sanogo et al. (2016), Mkonda and He (2017).

The interviewed farmers also confirmed that hailstorms were common in the humid zone. According to the farmers, the number of storms had increased over time, resulting in flash floods that wash away their crops. They mentioned that "short rains turn into hailstorms that come with strong winds and destroy our crops, including vegetables" (Table III). Barasa et al. (2015b) agreed that more farmers perceive an increase in the frequency of violent storms in Kakamega. In the humid zone, farmers stated that they had experienced frequent floods caused by heavy rains, leading to waterlogging problems. In the semi-humid region, only 20 per cent of farmers had experienced flooding. In the semi-arid zone, they stated that sometimes the rainfall is very high and causes floods. For instance, farmers mentioned that rainfall in 2012 and 2013 resulted in floods because it was too heavy. Similarly, hail stones, flooding and frequent droughts have also been reported by farmers in countries such as Ethiopia, Zambia, Senegal and Nigeria (Mertz et al., 2009; Apata et al., 2009; Mengistu, 2011; Nyanga et al., 2011). The authors reported that farmers are experiencing increased frequency of floods, excess rainfall, violent rains and hailstorms. Globally, research shows that farmers' perceptions of the climate will inform their response to CC (Abid et al., 2015; Fadina and Barjolle, 2018). It is therefore important to understand how farmers perceive climate in order to support effective context specific adaptations. Chi-square test results to determine the association between farmers' perceptions of CC for each ACZ are given in Table IV. The hypothesis stated that there is no relationship between farmers' perceptions and ACZ (HO). Except for the temperature and duration of the rainy season, the results showed a highly significant association with the ACZ at a $p < 0.05$ level of significance. This means that AIV farmers perceived different climatic parameters depending on the ACZ in which they live.

3.2 Results of the long-term average annual rainfall trends

A trend is a significant change over time exhibited by a random variable, detectable by statistical parametric and non-parametric procedures (Longobardi and Villani, 2009). Climate trends were analysed for both temperature and rainfall. Figure 2 shows graphs for 12-month precipitation accumulation observations for each of the three ACZs. A fitting of trend lines showed that there is increasing rainfall for Kakamega (humid) and Nakuru (semi-humid). The slope of the trend line was not very large in magnitude for Kakamega and Nakuru, but it was positive. Meanwhile, decreasing rainfall for Kajiado (semi-arid zone) over the 30-year period was noted. The slope of the trend line was not very large in magnitude for Kajiado, but it was negative. Based on the above results, it is of immense importance to discuss the potential impact on agricultural production if increasing and decreasing rainfall trends continue in these ACZs in future. Research has shown that excess rainfall, for instance, could lead to soil saturation, as well as to runoff, and soil erosion problems (Frunhoff et al., 2007). However, it is important to note that a decrease in rainfall in future

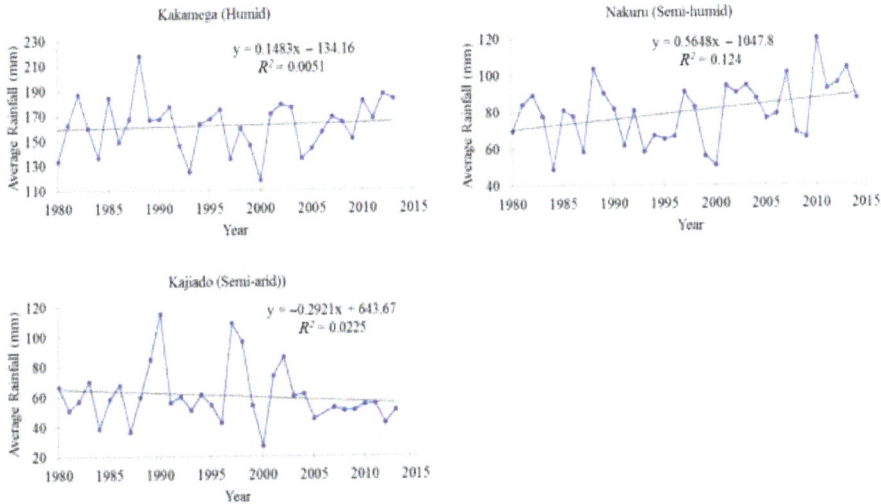

Figure 2. Mann–Kendall graphs showing average annual rainfall trends

could have repercussions on the sustainability of surface water resources and groundwater recharge (Green *et al.*, 2011).

To establish trends in average annual rainfall, the Mann–Kendall test was applied to climate data. The results in Table V were obtained for the three ACZs. If the *p*-value is less than the significance level α (alpha) = 0.05, *H0* is rejected. Rejecting *H0* indicates that there is a trend in the time series, while accepting *H0* indicates no trend is detected. On rejecting *H0*, the result is said to be statistically significant. For this test, *H0* was accepted for the three ACZs. Linear regression results, however, showed that the trend was only significant in the semi-humid zone (Table V).

3.3 Results of seasonal rainfall trends

In this study, only two seasons useful for crop production under a rain-fed system were considered. Normally, there are two cropping seasons in Kenya that coincide with bimodal rainfall regimes in which long rains fall between March and May (MAM) and the short rains between October and December (OND) [Famine and Early Warning System Network (FEWS NET, 2013)]. The Mann–Kendall trend line plotting for seasonal rainfall for the three ACZs is shown in Figure 3.

The analysis of the seasonal rainfall variations is important because of farmers' dependence on rain-fed vegetable production. The delays in rain onset, early cessation of rain and inadequate rainfall reported by farmers could mean that the seasons are getting shorter. An analysis of climate data showed that seasonal MAM mean rainfall was

Table V. Mann–Kendall test and regression analysis of mean annual rainfall data for the ACZs in Kenya

| ACZs | Mann–Kendall test | | | | | Regression analysis | | |
	MK statistic (S)	Kendall's Tau	Var (S)	*p*-value	Test interpretation	Equation	R^2	*p*-value (slope)
Humid	65.0	0.12	4550.3	0.343	No trend	Y = 0.148 + 159.31	0.005	0.86
Semi-humid	137.0	0.23	4958.3	0.053	No trend	Y = 0.565x + 69.87	0.124	0.053**
Semi-arid	−76.0	−0.14	4165.3	0.245	No trend	Y = -0.433x + 69.25	0.036	0.36

Note: **Shows statistical significance at $p < 0.05$

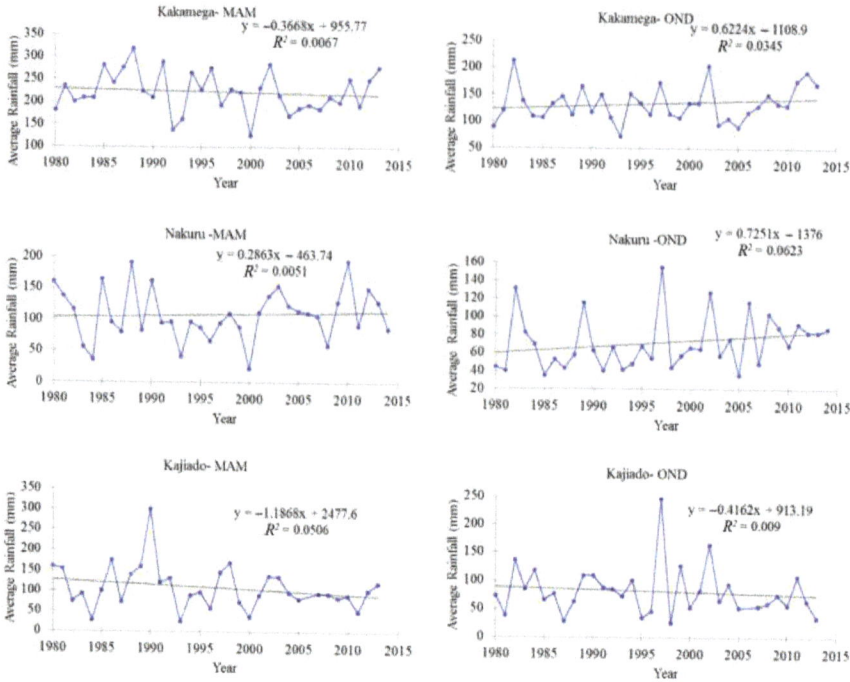

Figure 3. Mann–Kendall graphs showing average annual seasonal rainfall trends

decreasing for the humid and semi-arid zones; however, it was increasing in the semi-humid zone. The OND mean rainfall was increasing for the humid and semi-humid zone but decreasing in the semi-arid zone. However, the Mann–Kendall test was statistically significant only for OND rains in the semi-humid zone. Otherwise there was no trend in either the MAM and OND rains in all three zones. However, the MK test result showed no trend in MAM and OND in the humid and semi-arid zones, and in MAM in the semi-humid zone since *H0* was accepted. Although the MAM and OND rainfall was increasing for the humid and semi-humid zones, and decreasing for the semi-arid zone, the regression analysis results showed that the change was not significant in the three zones (Table VI). Differences in mean annual rainfall between the three ACZs were tested using ANOVA (Table VII). Highly significant differences in mean annual rainfall were detected ($F = 247.7944$,

Table VI. Mann-Kendall and Regression Statistics Results for the ACZs in Kenya

ACZs	Mann–Kendall test			Regression analysis		
	MK statistic (S)	p-value	Test interpretation	Regression equation	R^2	p-value (slope)
Humid						
MAM	−21.00	0.767	No trend	Y = -0.3668x+229.8	0.01	0.548
OND	77.00	0.260	No trend	Y = 0.6224x+122.85	0.04	0.293
Semi-humid						
MAM	41.000	0.570	No trend	Y = 0.2863x+102.84	0.01	0.741
OND	155.000	0.029	Trend detected	Y = 0.7251x+58.98	0.06	0.169
Semi-arid						
MAM	−80.000	0.221	No trend	Y = -1.2195x+129.11	0.05	0.189
OND	−66.00	0.314	No trend	Y = -0.7931x+102.98	0.02	0.565

$P = 1.79E$-39). The hypothesis that there is no significant difference in the mean annual rainfall between the climatic zones in Kenya over the 30-year period was therefore not accepted (HO). To determine which of the zones actually differed in terms of mean annual rainfall, multiple mean comparisons were performed using LSD. The results in Table VIII indicated that the mean difference in annual rainfall between the zones was highly significant for all the ACZs.

3.3.1 Results of the coefficient of variation in seasonal rainfall for the selected agro-climatic zones. This study also investigated variability through annual and seasonal meteorological time series. A coefficient of variation (CV) of 30 per cent or more is considered high (Thornton *et al.*, 2014). Rainfall in the two growing seasons was highly variable, with a CV ranging from 31 to 105 per cent (Table IX). This was shown in all three zones except in May and October (29 per cent) in the humid zone. These climate data confirmed farmers' perception that the rainfall in the study areas was highly variable. According to Mamba *et al.* (2015), an analysis of farmers' perception of CC in Swaziland shows that rainfall variability presents a great challenge to farmers. It makes rainfall highly

Table VII. Analysis of variance of mean annual rainfall between ACZs in Kenya

Source of variation	Sum of squares	df	MS	F	*p*-value	*F* crit
Between groups	191329.7	2	96164.85	247.7944	1.79E-39****	3.09
Within groups	38808.32	100	388.08			
Total	231138	102				

Table VIII. LSD Multiple comparisons of mean annual rainfall in the ACZs in Kenya from 1980 to 2014

(I) Zone	(J) Zones	Mean difference (I-J)	Std. Error	Sig.	95% confidence interval Lower bound	Upper bound
Semi-arid	humid	−99.6*	4.8	****	−109.1	−90.1
	Semi-humid	−17.7*	4.7	****	−27.1	−8.3
Humid	Semi-arid	99.6*	4.8	****	90.1	109,.1
	Semi-humid	81.9*	4.7	****	72.5	91.3
Semi-humid	Semi-arid	17.71*	4.7	****	8.3	27.1
	Humid	−81.9*	4.7	****	−91.3	−72.5

Note: ****Shows significance at $p < 0.001$ level

Table IX. MAM and OND seasonal variability of rainfall

	Mean			SD			CV (%)		
	Humid	Semi-humid	Semi-arid	Humid	Semi-humid	Semi-arid	Humid	Semi-humid	Semi-arid
March	163	66	74	77	50	54	47	75	73
April	259	134	132	81	68	69	31	51	52
May	248	124	120	73	74	116	29	60	96
October	163	85	54	47	40	57	29	47	106
November	150	86	115	66	42	70	44	49	61
December	89	45	80	62	39	56	69	87	70

Notes: SD = standard deviation; CV = coefficient of variation

unpredictable and tends to confuse farmers. Increased rainfall variability will affect agricultural growth and economic development. CC and increased climate variability, through their impact on food production, will have a negative impact on the prevalence of undernutrition, increasing severe stunting by 62 per cent in southern Asia and 55 per cent in eastern and southern Africa by the 2050s (Lloyd *et al.*, 2011).

3.4 Results of temperature trends in the selected agro-climatic zones of Kenya

Figure 4 shows graphs for the 12-month average maximum and minimum temperatures for each of the ACZs of Kakamega (humid), Nakuru (semi-humid) and Kajiado (semi-arid). The results showed an increase in both the maximum and minimum temperatures by 0.02°C and 0.08°C in the humid and semi-arid zones, respectively. The climate data agreed with farmers' perceptions in the two zones. Even though the farmers in the semi-humid zone perceived increased temperatures, the climate data showed a decreasing trend in the annual maximum temperature.

For the temperature data, the Mann–Kendall test was statistically significant for all three ACZs, except the maximum and minimum temperatures in the semi-humid and semi-arid zones respectively (Table X). In this case, *HO* was rejected, thereby implying that there was a trend in the data for the two zones. With temperature showing an increasing trend for the three ACZs over the 30-year time period, it is essential to understand how this may affect AIV production if such a trend continues. It also confirms recent findings by Bobadoye *et al.* (2016) and Mulinya *et al.* (2016).

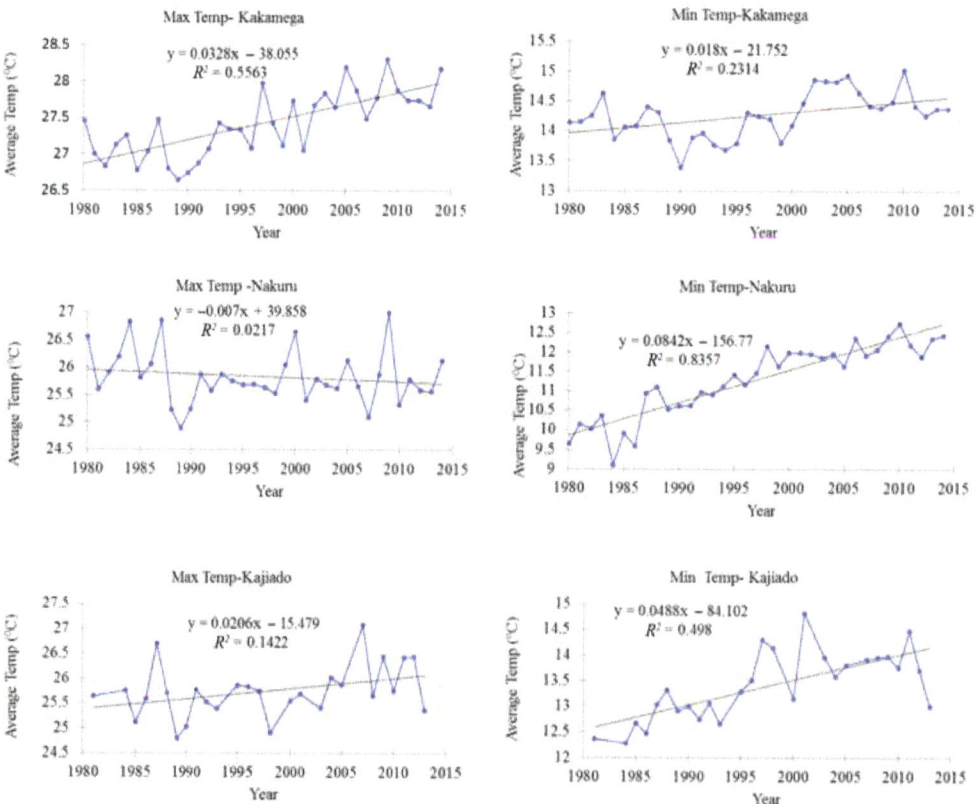

Figure 4. Mann–Kendall graphs for annual maximum and minimum mean temperatures

3.5 Farmers' perceptions of the effect of changing climate on African indigenous vegetable production

The most common AIVs produced across the study areas included cowpea *(Vigna unguiculata)*, jute mallow, African nightshade *(Solanum villosum)*, pumpkin leaves *(Cucurbita moschata)*, amaranth *(Amaranthus* sp.) and spider plant *(Cleome gynandra)*. The least produced included Ethiopian kale and African spinach and slender leaf *(Crotolaria* sp.). It is worth noting that the impact of CC on AIV production is both positive and negative. The findings of this study clearly revealed that AIV farmers in the study areas had noted that the climate is indeed changing. Farmers' perceptions discussed in this paper show that the humid, semi-humid and semi-arid zones of Kenya already face a range of CCs that include more frequent, heavy or inadequate rainfall events, erratic rainfall and unpredictable onset/retreat of rain and change in the duration of dry spells characterised by increasing temperatures. The negative effects on these CCs described by AIV farmers that are predicted to have implications on AIV production cannot be ignored. The two major parameters of CC that have far-reaching implications on agriculture in general, and horticulture in particular, are more erratic and inadequate rainfall patterns and unpredictable high temperatures (Datta, 2013). According to Thornton *et al.* (2013), CC is already affecting the amounts, distribution and intensity of rainfall in many places in the tropics. This has direct effects on the timing and duration of crop-growing seasons, with related effects on plant growth. Research has shown that farming activities rely on favourable climate conditions and are at risk from a changing climate (Porter *et al.*, 2014). Changes in yields, weeds, pests and diseases of AIVs in two different conditions, the rainy season and the dry season, are discussed in this study.

Table XI: Changes in the pests and diseases of AIVs because of CC.

An understanding of the impact of CC conditions on the distribution and intensity of plant pests and diseases is important in informing decisions that will lead to better ways of managing them in a changing climate. Perceived changes in the severity and frequency of attacks by pests and diseases varied with the seasons and also with the type of AIVs. There was uniform agreement among farmers that AIVs, for instance, cowpea, were affected by some pests, which included aphids and whiteflies *(Bemisia tabaci)*, with increased incidences of disease such as rust *(Uromyces anthyllidis)* during the dry spells, while in the wet season farmers noticed that cowpea was affected by a number of pests and diseases.

Table X. Mann–Kendall and regression statistic results for annual maximum and minimum temperatures for ACZs

| ACZs | Mann–Kendall | | | | Regression analysis | | |
	MK (S)	p-value	Sig.	Test interpretation	Equation	R^2	p-value (slope)
Kakamega (Humid)							
Max	312.0	0.0001	****	trend	Y = 0.0032x + 26.83	0.56	0.0007 ****
Min	182.0	0.010	***	trend	Y = 0.0394 + 27.08	0.23	0.0657
Nakuru (Semi-humid)							
Max	−73.0	0.306		no trend	Y = -0.007 + 25.96	0.022	0.020 ***
Min	454.0	0.0001	**	trend	Y = 0.04x + 9.9.78	0.84	1.15E-13 ****
Kajiado (Semi-arid)							
Max	83.0	0.087		no trend	Y = 0.0239x + 25.41	0.12	0.5423
Min	184.0	0.000	****	trend	Y = 0.598x + 12.564	0.51	0.000 ****

Notes: ****, *** and ** indicate significance at $p < 0.001$, $p < 0.01$ and $p < 0.05$ respectively

Table XI. Comparison of the changes in sensitivities of common AIVs to changes in rainfall

AIVs	Changes in sensitivities of AIVs because of too much rain			Changes in sensitivities of AIVs because of dry spell/drought/no rain		
	Yield	Pests and diseases	Weeds	Yield	Pests and diseases	Weeds
Cowpea	Heavy unpredictable downpours reduce yield and quality (turns yellow)	Changes observed – many pests. High severity of black spot on the leaves	Increased weeds	No changes	Many pests and diseases have occurred	Weeds remain low (no changes)
Nightshade	No change in yield (yield remains high)	Changes observed – high severity of blight, Frequent pests such as spider mite Frequent moderate infestation of diseases such as bacterial wilt and powdery mildew	Emergence of new weeds that farmers do not recognize	Moderate yield reduction Tough leaves	Sudden dry spells and drought cause moderate to high severity of pests such as spider mite and white flies observed Frequent high severity of diseases such as bacterial wilt and powdery mildew	no observed changes
Spider plant	Not affected by too much rain if germination has occurred	Changes occur but few pests and moderate severity of diseases have been observed		Moderate yield if dry spell/drought is not prolonged Tough leaves	Few pests and moderate severity of diseases have been observed	No observed changes in weeds
Amaranth	Not affected No change reported	No changes on pests and diseases		Yield change to moderate	No changes	No observed changes in weeds
Pumpkin leaves	No change in yield	No changes observed		Reduced yield	No changes observed	No observed changes in weeds

Source: Focus group discussion, 2015

Increased infestation of pests and diseases was also reported in nightshade during the dry season. Pests reported in this period included red spider mites (*Tetranychus urticae*), which cause yellowing of leaves, while aphids (Aphis spp.) affect quality because they cause leaves to curl and shrink. Importantly, farmers agreed that attacks by aphids had become more common in recent years. Thus, continued droughts and dry spells coupled with high temperatures caused by CC are expected to alter the manifestation of pests and diseases for some AIVs. In agreement, Abang *et al.* (2014), noted that aphids are the most recurrent insect pests in vegetable crops. Findings from this study also revealed that increased wilting of nightshade caused my nematodes has been recurrent in recent years. During the rainy season, some pests, for example, spider mites, are reported with a moderate recurrence of diseases such as bacterial wilt (*Ralstonium solanacearum)* and powdery mildew (*Erysiphe polygoni, Laveillula taurica*), which affect the yield and quality of nightshade. Furthermore, farmers have observed frequent and severe attacks by bacterial blight (*Clavibacter michiganense pv. michiganense)* in African nightshade in recent years because of too much rain. A moderate to high severity of pests such as spider mites and whiteflies in African nightshade is reported during dry spells and droughts. Farmers noted that changes in attacks by these pests in spider plants were higher during dry season as opposed to the rainy season. In agreement, Stöber *et al.* (2017) noted that a lack of rainfall and extreme heat cause wilting, attract pests and reduce yields. Overall, a high attack of blight, mildew and wilt are the most serious diseases in AIV production noted by farmers in the recent past. This could be because high moisture caused by frequent and extreme rainfall events creates favourable environments for diseases, destroying fields and causing nutrient leaching (Stöber *et al.*, 2017). Datta (2013) confirmed that more extreme weather may make certain diseases (such as rust and powdery mildews) more sporadic and encourage those that develop quickly in warm conditions.

Farmers also agreed that no clear changes in pests and diseases for amaranth and pumpkin leaves had been detected either during the dry or the wet seasons. Abang *et al.* (2014) confirmed that insect pest attacks are more numerous in the dry season and diseases are more problematic in the rainy season. Crop resistance and pest dynamics are altered by CC, creating more complex challenges for the management of pests and disease (Keatinge *et al.*, 2010). Rain, heavy dews, warm temperatures and dry climates have been identified as causes of increased pests and diseases (Abang *et al.*, 2014). Research has shown that pests and diseases are important constraints to vegetable production in the tropics. Plant pests and diseases are one of the important factors that have a direct impact on global agricultural productivity, and CC will further aggravate the situation. Further changes in temperature could directly affect the spread of infectious diseases and their survival between seasons (Gautam *et al.*, 2013). Overall, CC is predicted to have a direct impact on the occurrence and severity of diseases in crops, which will have a serious impact on food security (West *et al.*, 2015; Gautam *et al.*, 2013). Similarly, Gregory *et al.* (2009) noted that CC is already altering the distribution and intensity of weeds and plant pests and diseases.

3.5.1 Changes in weeds in African indigenous vegetable production because of climate change. Besides altering yields, pests and disease infestation, CCs also influence the manifestation of weeds in vegetable production. Weeds are undesirable, unattractive or troublesome plants, not intentionally grown, especially when they grow where they are not wanted, such as in gardens or agricultural fields (Prakash *et al.*, 2015). For all the common AIVs grown by the farmers in the study areas, weeds grew and spread more quickly in the wet season than in the dry season. High weed growth during the rainy season has led to competition and more pest and disease incidence because of CC for some AIVs (Table X). One exception to this was pumpkins which, according to AIV farmers, do not allow weeds to grow, as they have large leaves that cover the ground entirely (Tables XI). The emergence of new persistent weeds was also reported by farmers in semi-humid and semi-arid zones. However, they stated that they were not knowledgeable about the names of the weeds. Although labour intensive, farmers in this case are forced to regularly weed their gardens. According to the farmers, this is necessary, as most AIVs only do well in weed-free plots and gardens. Research has also found that crops and weeds alike respond to elevated CO^2 concentrations or an increase in temperature in the atmosphere. As the weeds are found in the same agricultural fields in which crops grow, any change in weed behaviour will directly or indirectly depend on the relative advantage or disadvantage crops have because of CC (Prakash *et al.*, 2015). Weeds affect crop cultivation in many ways by reducing crop yield, stealing water, light, space and soil nutrients, producing substances that are toxic to crop plants, often serving as hosts for many crop diseases, and also providing shelter for insects and disease pests during their off-season (Prakash *et al.*, 2015).

3.5.2 Changes in the yield of African indigenous vegetables because of climate change. All the farmers agreed that there had been changes in terms of vegetable yield because of the combined effects of climate conditions (Table XI). It was apparent that rainfall and temperature changes had already posed some challenges to AIV production in Kenya.

Cowpea performs better with little rain. Farmers agreed that it germinates well with little water and the yield is good. However, farmers expressed concerns about unexpected heavy downpours and hailstones, which result in cowpea crops being lost. There was overall agreement that cowpeas are more sensitive to heavy rains. They turn yellow and are often completely destroyed. However, other vegetables, such as the spider plant, African nightshade, pumpkin leaves and amaranth, record high yields even in the heavy rainy season. In particular, farmers noted that the spider plant requires a lot of water and the yield is high during the wet seasons. However, the frequent prolonged dry spells and droughts experienced in the recent past have been the main challenge faced by farmers. They agreed that the spider plant is stunted during the dry season, and turns yellow and flowers early, therefore producing low yields. Farmers agreed that timing in the planting of AIVs such as spider plants was very important as heavy rains limit their germination. This means that the frequent and increased intensity of dry spells and droughts reported by farmers was a hindrance in their production. Farmers also mentioned that pumpkin leaves have thick succulent stems that store water to be used by the plant during dry spells. Overall, farmers agreed that they experienced yield reduction for all AIVs during the dry season but could generate a fair amount of money at the same time. In the humid zone, farmers experienced frequent hailstones that completely destroyed their vegetables. AIV farmers were aware of the type of vegetables that thrive under certain climatic conditions. Amaranth, jute mallow and pumpkin leaves were identified as most resistant to changes in yield reduction, pests and diseases. According to Stöber *et al.* (2017), AIVs thrive better in rainy seasons. This implies that as a matter of priority, to respond to these threats caused by CC, farmers need to combine adaptation strategies that address water, pests and disease problems in production. According to Thornton *et al.* (2013), changes in agricultural inputs and the way farmers use them may be able to more than offset projected yield declines by using adaptation options. These strategies may include, but are not limited to, a combination of more drought and pest-tolerant varieties of AIVs, water harvesting, irrigation, mulching and intercropping.

The most tolerant AIVs grown across all the ACZs were pumpkin leaves, wild amaranth, slender leaf, jute mallow and Indian spinach. In general, farmers agreed that AIVs register low sensitivities to climate variables compared to other vegetables. Luoh *et al.* (2014) confirmed that some indigenous vegetables are considered to have stronger water-stress adaptation mechanisms than common, commercially available vegetables. Additionally,

Stöber *et al.* (2017) confirmed that many AIVs tolerate a wide spectrum of climate variability and are therefore considered less sensitive to climate variations. Despite this, farmers agreed that a high reduction in yields was observed whenever there was a prolonged dry spell or drought. Besides yield reduction, high temperatures also caused yellowing of leaves for some vegetables such as spider plant and African nightshade. For most AIVs, the leaves become tough during droughts. In the FGDs the majority of the farmers mentioned that they temporarily abandon AIV farming whenever there is a prolonged drought/dry spell because of a lack of water for irrigation. Additionally, theft of vegetables, which mostly occurs at night, was also reported by some farmers in the humid and semi-humid zones. This was attributed to the scarcity of the vegetables during the season. According to the farmers, AIVs are tastier during the dry season and they can be sold for a good price in markets. There is no doubt that AIV farmers in Kenya are more concerned now with the rising temperatures, heavy and or unpredictable rainfall patterns, prolonged droughts, frequent and prolonged dry spells, together with increased pests and disease problems. Even though the impact is not severe for AIVs, the findings of this study are in agreement with those of Ayyogari *et al.* (2014) and Mattos *et al.* (2014) who found that vegetables are generally sensitive to environmental extremes. Therefore, high temperatures and limited soil moisture are the major causes of low yields since they greatly affect several physiological and biochemical processes. Within the scope of this study, it was not possible to ascertain the extent of yield losses caused by the climatic conditions.

4. Conclusions and recommendations

Most of the research that has been carried out on AIVs has ignored and downplayed the range of potential effects on production caused by CC variables. AIV farmers are aware that their area is getting warmer and drier. Increased temperatures, an increased frequency of droughts and dry spells, late onset, early cessation and decreased rainfall, are among the major climate hazards reported. Climate data revealed no trends in the mean annual rainfall in the selected ACZs in Kenya, but the mean annual rainfall in the semi-humid zone is increasing significantly. The main constraining factors identified in AIV production were increased temperatures, erratic rainfall patterns and inadequate rainfall caused by frequent dry spells and droughts. Farmers were aware of the moderate effect of climate on AIVs. The current volatility in rainfall patterns has made it difficult for farmers to plan their cropping calendar to suit the changes. Therefore, CC adaptation programmes need to provide accurate projected weather patterns to farmers. This study also revealed that farmers' perceptions were significantly associated with their ACZs, while there are highly significant differences in the mean rainfall of the three ACZs. Severity of pests and disease are more prevalent in the dry season than in the wet season. Too much rain and frequent dry spell conditions caused by CC alter the infestation of pests and manifestation of diseases as well as growth of some new weeds. This has resulted in lower yields for AIVs. Changes in the manifestation of pests, diseases and weeds in some AIVs could imply that farmers may be pushed into applying pesticides and herbicides on their vegetables. There is therefore a need to make farmers aware of the need to use pesticides and herbicides sustainably, including integrated pest management practices, to reduce environmental damage. The changes in the severity of pests and diseases in spider plant and African nightshade calls for a development of more resistant cultivars of these AIVs. This could be designed within programmes aimed at ensuring that farmers benefit from opportunities offered by engaging in AIV production. To achieve this, it is important to ensure that farmers gain access to good quality and affordable resistant seed varieties. Farmers' perceptions and observed trends should be communicated to policy makers at local, national and international levels. This will provide a pathway for effective planning of context-specific adaptation options and climate communication programmes. However, further research is recommended to ascertain the quantity of yield losses because of pests, disease and CC.

References

Abang, A.F., Kouamé, C.M., Abang, M., Hanna, R. and Fotso, A.K. (2014), "Assessing vegetable farmer knowledge of diseases and insect pests of vegetable and management practices under tropical conditions", *International Journal of Vegetable Science*, Vol. 20 No. 3, pp. 240-253, doi: 10.1080/19315260.2013.800625.

Abid, M., Scheffran, J., Schneider, U.A. and Ashfaq, M. (2015), "Farmers' perceptions of and adaptation strategies to climate change and their determinants: the case of Punjab province, Pakistan", *Earth System Dynamics*, Vol. 6 No. 1, pp. 225-243, doi: 10.5194/esd-6-225-2015.

Abukutsa-Onyango, M. (2007), "Seed production and support systems for African leafy vegetables in three communities in Western Kenya", *African Journal of Food, Agriculture, Nutrition and Development*, Vol. 7.

Abukutsa-Onyango, M.O. (2010), *African Indigenous Vegetables in Kenya: Strategic Repositioning in the Horticultural Sector*, Inaugural Lecture, Jomo Kenyatta University of Agriculture and Technology, Nairobi.

Adhikari, U., Nejadhashemi, A.P. and Woznicki, S.A. (2015), "Climate change and Eastern Africa: a review of impact on major crops", *Food and Energy Security*, Vol. 4 No. 2, pp. 110-132, doi: 10.1002/fes3.61.

AGRA (2014), *Africa Agriculture Status Report: Climate Change and Smallholder Agriculture in Sub-Saharan Africa*, Nairobi.

Apata, T.G., Samuel, K.D. and Adeola, A.O. (2009), "Analysis of climate change perceptions and adaptation among arable food crop farmers in South Western Nigeria", *Contributed Paper Presented at 23rd Conference of International Association of Agricultural Economists*, Beijing, p. 16-22.

Ayyogari, K., Sidhya, P. and Pandit, M.K. (2014), "Impact of climate change on vegetable cultivation – a review", *International Journal of Agriculture, Environment and Biotechnology*, Vol. 7 No. 1, pp. 145, doi: 10.5958/j.2230-732X.7.1.020.

Babatolu, J.S. and Akinnubi, R.T. (2016), "Smallholder farmers' perception of climate change and variability impact and their adaptation strategies in the upper and lower Niger river basin development authority areas, Nigeria", *Journal of Petroleum & Environmental Biotechnology*, Vol. 7 No. 3, pp. 279, doi: 10.4172/2157-7463.1000279.

Barasa, B.M.O., Oteng'i, S.B.B. and Wakhungu, J.W. (2015b), "Farmer awareness of climate variability in Kakamega county, Kenya", *The International Journal of Science & Technoledge*, Vol. 3 No. 7.

Barasa, M.O.B., O'tengi, S.B. and Wakhungu, J.W. (2015a), "Farmer adaptation strategies to the impacts of climate variability in Kakamega county, Kenya", *Researchjournali's Journal of Agriculture*, Vol. 2 No. 5.

Babadoye, A.O., Ogara, W.O., Ouma, G.O. and Onono, J.O. (2014), "Comparative analysis of rainfall trends in different sub counties in Kajiado County Kenya", *International Journal of Innovative Research Studies*, Vol. 13 No. 2, ISSN 2319-9725.

Bobadoye, A.O., Ogara, W.O., Ouma, G.O. and Onono, J.O. (2016), "Pastoralist perceptions on climate change and variability in Kajiado in relation to meteorology evidence", *Academic Journal of Interdisciplinary Studies*, Vol. 5 No. 1, doi: 10.5901/ajis.2016.v5n1p37.

Bryan, E., Ringler, C., Okoba, B., Roncoli, C., Silvestri, S. and Herrero, M. (2013), "Adapting agriculture to climate change in Kenya: household strategies and determinants", *Journal of Environmental Management*, Vol. 114, pp. 26-35, doi: 10.1016/j.jenvman.2012.10.036.

Capuno, O.B., Gonzaga, Z.C., Dimabuyu, H.B. and Rom, J.C. (2015), "Indigenous vegetables for coping with climate change and food security", *Acta Horticulturae*, No. 1102, pp. 171-178, doi: 10.17660/ActaHortic.2015.1102.21.

Cernansky, R. (2015), "The rise of Africa's super vegetables", *Nature*, Vol. 522 No. 7555, pp. 146-148, doi: 10.1038/522146.

Chagomoka, T., Afari-Sefa, V. and Pitoro, R. (2014), "Value chain analysis of traditional vegetables from Malawi and Mozambique", *International Food and Agribusiness Management Review*, Vol. 17 No. 4.

Choudhary, S.K., Singh, R.N., Upadhyay, P.K., Singh, R.K., Choudhary, H.R. and Vijay, P. (2015), "Effect of vegetable intercrops and planting pattern of maize on growth, yield and economics of winter maize (zea mays L.) in Eastern Uttar Pradesh", *Environment & Ecology*, Vol. 32 No. 1, pp. 101-105.

Connolly-Boutin, L. and Smit, B. (2015), "Climate change, food security, and livelihoods in Sub-Saharan Africa", *Regional Environmental Change*, Vol. 16 No. 2, pp. 385-399, doi: 10.1007/s10113-015-0761-x.

Datta, S. (2013), "Impact of climate change in Indian horticulture a review", *International Journal of Science, Environment and Technology*, Vol. 2 No. 4, pp. 661-667.

Dhanya, P. and Ramachandran, A. (2016), "Farmers' perceptions of climate change and the proposed agriculture adaptation strategies in a semi-arid region of South India", *Journal of Integrative Environmental Sciences*, Vol. 13 No. 1, pp. 1-18, doi: 10.1080/1943815X.2015.1062031.

Domazetova, V.L. (2011), "The impact of climate change on food production in the former Yugoslav Republic of Macedonia", Ivanyi, Z. (Ed.), *The Impacts of Climate Change on Food Production in the Western Balkan Region*, available at: http://documents.rec.org/topic-areas/Impacts-climage-change-food-production.pdf (accessed 27 April 2018).

Elum, Z.A., Modise, D.M. and Marr, A. (2017), "Farmer's perception of climate change and responsive strategies in three selected provinces of South Africa", *Climate Risk Management*, Vol. 16, pp. 246-257, doi: 10.1016/j.crm.2016.11.001.

Fadina, A.M.R. and Barjolle, D. (2018), "Farmers' adaptation strategies to climate change and their implications in the Zou department of South Benin", *Environments*, Vol. 5 No. 1, doi: 10.3390/environments5010015.

FAO (2010), *Analysis of Climate Change and Variability Risks in the Smallholder Sector Case Studies of the Laikipia and Narok Districts Representing Major Agro-Ecological Zones in Kenya*, Rome.

FAO (2015), *Climate Change and Food Systems: Global Assessments and Implications for Food Security and Trade*, Food Agriculture Organization of the United Nations (FAO), Rome.

Fetters, M.D., Curry, L.A. and Creswell, J.W. (2013), "Achieving integration in MixedMethods designs–principles and practices", *Health Services Research*, Vol. 48 No. 6, pp. 2134-2156, doi: 10.1111/1475-6773.12117.

FEWS NET (2013), "Kenya food security brief. United States agency for international development (UASID) famine early warning systems network (FEWS NET)", www.fews.net/sites/default/files/documents/reports/Kenya_FoodSecurity_In_Brief_2013_final_0.pdf (accessed 20 July 2017).

Frunhoff, P., Carthy, J., Melillo, J., Moser, S. and Wuebbles, D. (2007), "Confronting climate change in the US northeast – science, impacts, and solutions, 26", available at: www.climatechoices.org/assets/documents/climatechoices/confronting-climatechange-in-the-u-s-northeast.pdf (accessed 26 December 2017).

Gandure, S., Walker, S. and Botha, J.J. (2012), "Farmers' perceptions of adaptation to climate change and water in a South African rural community", Environment Development, available at: http://dx.doi.org/10.1016/j.endev.2012.11.004.

Gautam, H.R., Bhardwaj, M.L. and Rohitashw, K. (2013), "Climate change and its impact on plant diseases", *Current Science*, Vol. 105 No. 12.

Gido, E.O., Ayuya, O.I., Owuor, G. and Bokelmann, W. (2017), "Consumption intensity of leafy African indigenous vegetables: towards enhancing nutritional security in rural and urban dwellers in Kenya", *Agricultural and Food Economics*, Vol. 5 No. 1, doi: 10.1186/s40100-017-0082-0.

GoK (2009a), *The National Population Census Report*, Kenya National Bureau of Statistics, available at: http://cdkn.org/wp-content/uploads/2012/04/National-

GoK (2009b), *Kajiado County Development Plan 2008-2012. Kenya Vision 2030. Towards a Globally Competitive and Prosperous Kenya*, Ministry of Planning and Development, Nairobi.

GoK (2012), "National nutrition action plan 2012-2017", available at: https://scalingupnutrition.org/wp-content/uploads/2013/10/Kenya-National-Nutrition-Action-Plan-2012-2017-final.pdf (accessed 8 March 2018).

GoK (2013a), "National climate change action plan 2013-2017", Executive Summary, Nairobi, available at: https://cdkn.org/wp-content/uploads/2013/03/Kenya-National-Climate-Change-Action-Plan.pdf (accessed 10 February 2017).

GoK (2013b), *Kenya County Fact Sheet Commission on Revenue Allocation*, 2ne ed., Nairobi.

GoK (2014), "Kajiado county integrated development plan 2013-2017", available at: https://cog.go.ke/images/stories/CIDPs/Kajiado.pdf (accessed 10 March 2017).

Green, T., Taniguchi, M., Kooi, H., Gurdak, J., Allen, D., Hiscock, K., Treidel, H. and Aureli, A. (2011), "Beneath the surface of global change: impacts of climate change on groundwater", *Journal of Hydrology*, Vol. 405 Nos 3/4, pp. 32-560.

Gregory, P.J., Johnson, S.N., Newton, A.C. and Ingram, J.S.I. (2009), "Integrating pests and pathogens into the climate change/food security debate", *Journal of Experimental Botany*, Vol. 60 No. 10, pp. 2827-2838, available at: https://doi.org/10.1093/jxb/erp080

Habwe, F.O., Walingo, M.K., Abukutsa-Onyango, M.O. and Oluoch, M.O. (2009), "Iron content of the formulated East African indigenous vegetable recipes", *African Journal of Food Science*, Vol. 3 No. 12, pp. 393-397.

HCDA (2014), *Horticulture Validated Report 2014*, USAID, available at: www.agricultureauthority.go.ke/wp-content/uploads/2016/05/Horticulture-Validated-Report-2014-Final-copy.pdf (accessed 10 March 2018).

Hsieh, H.-F. and Shannon, S.E. (2005), Three approaches to qualitative content analysis.

IPCC (2014), "Climate change 2014 impacts, adaptation, and vulnerability. part b: regional aspects", *Contribution of Working Group II to the Fifth Assessment Report of the Intergovernmental Panel on Climate Change*, Cambridge University Press, Cambridge, New York, NY, p. 688.

Jain, V., Pal, M., Raj, A. and Khetarpal, S. (2007), "Photosynthesis and nutrient composition of spinach and fenugreek grown under elevated carbon dioxide concentration", *Biologia Plantarum*, Vol. 51 No. 3, pp. 559-562.

Jaiswal, R.K., Lohani, A.K. and Tiwari, H.L. (2015), "Statistical analysis for change detection and trend assessment in climatological parameters", *Environmental Processes*, Vol. 2 No. 4, pp. 729-749, doi: 10.1007/s40710-015-0105-3.

Kabubo-Mariara, J. and Kabara, M. (2015), "Climate change and food security in Kenya", Environment for Development. Discussion Paper Series March 2015 E f D D P 15-0, available at: www.efdinitiative.org/sites/default/files/publications/efd-dp-15-05.pdf (available 8 February 2018).

Kabubo-Mariara, J. and Karanja, F.K. (2007), "The economic impact of climate change on Kenyan crop agriculture: a Ricardian approach", The World Bank Development Research Group Sustainable Rural and Urban Development Team August 2007 Policy Research Working Paper 4334.

Kamga, R.T., Kouamé, C., Atangana, A.R., Chagomoka, T. and Ndango, R. (2013), "Nutritional evaluation of five African indigenous vegetables", *Journal of Horticultural Research*, Vol. 21 No. 1, doi: 10.2478/johr-2013-0014.

Karmeshu, N. (2012) "Trend detection in annual temperature & precipitation using the mann kendall test – a case study to assess climate change on select states in the Northeastern United States", Master of Environmental Studies Capstone Projects, p. 47, available at: http://repository.upenn.edu/mes_capstones/47

Keatinge, J.D.H., Waliyar, F., Jamnadas, R.H., Moustafa, A., Andrade, M., Drechsel, P. and Luther, K. (2010), "Relearning old lessons for the future of food–by bread alone no longer: diversifying diets with fruit and vegetables", *Crop Science*, Vol. 50 No. 1, pp. 51-62, doi: 10.2135/cropsci2009.09.0528.

Kenya National Bureau of Statistics-KNBS and Society for International Development-SID (2013), *Exploring Kenya's Inequality: Pooling Apart or Pooling Together?*, Kenya National Bureau of Statistics (KNBS) and Society for International Development (SID) Nairobi.

Lachin, J.M. (2016), "Fallacies of last observation carried forward analyses", *Clinical Trials (London, England)*, Vol. 13 No. 2, p. 161-168.

Lema, M.A. and Majule, A.E. (2009), "Impacts of climate change, variability and adaptation strategies on agriculture in semi-arid areas of Tanzania: the case of Manyoni district in Singida region, Tanzania", *African Journal of Environmental Science and Technology*, Vol. 3 No. 8, pp. 206-218, doi: 10.5897/AJEST09.099.

Limantol, A.M., Keith, B.E., Azabre, B.A. and Lennartz, B. (2016), "Farmers' perception and adaptation practice to climate variability and change: a case study of the vea catchment in Ghana", *Springerplus*, Vol. 5 No. 1, pp. 830, doi: 10.1186/s40064-016-2433-9.

Lloyd, S.J., Kovats, R.S. and Chalabi, Z. (2011), "Climate change, crop yields, and undernutrition: development of a model to quantify the impact of climate scenarios on child undernutrition", *Environmental Health Perspectives*, Vol. 119 No. 12, pp. 1817-1823, doi: 10.1289/ehp.1003311, ISSN 0091-6765.

Longobardi, A. and Villani, P. (2009), "Trend analysis of annual and seasonal rainfall time series in the Mediterranean area", *International Journal of Climatology*, doi: 10.1002/joc.2001.

Luoh, J.W., Begg, C.B., Symonds, R.C., Ledesma, D. and Yang, R.-Y. (2014), "Nutritional yield of African indigenous vegetables in water-deficient and water-sufficient conditions", *Food and Nutrition Sciences*, Vol. 5 No. 9, pp. 812-822, doi: 10.4236/fns.2014.59091.

MAFAP (2013), "Review of food and agricultural policies in Kenya 2005-2011", *MAFAP Country Report Series*, Rome.

Malla, G. (2008), "Climate change and its impact on Nepalese agriculture", *Journal of Agriculture and Environment*, Vol. 9.

Mamba, S.F., Salam, A. and Peter, G. (2015), "Farmers' perception of climate change a case study in Swaziland", *Journal of Food Security*, Vol. 3 No. 2, pp. 47-61.

Masinde, P.W. and Stützel, H. (2005), "Plant growth, water relations, and transpiration of spiderplant [gynandropsis gynandra (L.) briq.] under water-limited conditions", *Journal of American Society Horticultural Science*, Vol. 130 No. 3, pp. 469-477.

Mattos, M.L., Celso, L.M., Sumira, J., Sargent, S.A., Eduardo, C.P., Lima, P. and Mariana, R.F. (2014), Climate Changes and Potential Impacts on Quality of Fruit and Vegetable Crops In book: Emerging Technologies and Management of Crop tree Tolerance publisher: Academic press edited: Parvaiz Ahmad and Salema Rasool, doi: 10.1016/B978-0-12-800876-8.00019-9.

Maundu, P., Achigan-Dako, E. and Morimoto, Y. (2009), "Biodiversity of African vegetables", in Shackleton, C.M., Pasquini, M. and Drescher, A.W. (Eds), *African Indigenous Vegetables in Urban Agriculture*, Earthscan, London, Sterling, VA, pp. 105-144.

Mavromatis, T. and Stathis, D. (2011), "Response of the water balance in Greece to temperature and precipitation trends", *Theoretical and Applied Climatology*, Vol. 104 Nos 1/2, pp. 13-24, doi: 10.1007/s00704-010-0320-9.

Mengistu, D.K. (2011), "Farmers' perception and knowledge on climate change and their coping strategies to the related hazards: case study from Adiha, Central Tigray, Ethiopia", *Agricultural Sciences*, Vol. 2 No. 2, pp. 138-145, doi: 10.4236/as.2011.22020.

Mertz, O., Mbow, C., Reenberg, A. and Diouf, A. (2009), "Farmers' perceptions of climate change and agricultural adaptation strategies in rural Sahel", *Environmental Management*, Vol. 43 No. 5, pp. 804-816.

Mitchell, J.M., Dzezerdzeeskii, B., Flohn, H., Hofmeyer, W.L., Lamb, H.H., Rao, K.N. and Wallen, C.C. (1996), *Climatic Change*, WMO Technical Note 79, WMO No. 195.TP-100, Geneva.

Mkonda, M. and He, X. (2017), "Are rainfall and temperature really changing? farmer's perceptions, meteorological data, and policy implications in the Tanzanian semi-arid zone", *Sustainability*, Vol. 9 No. 8, pp. 128-136, doi: 10.3390/su9081412ology 151.

Moretti, C.L., Mattos, L.M., Calbo, A.G. and Sargent, S.A. (2010), "Climate changes and potential impacts on postharvest quality of fruit and vegetable crops: a review", *Food Research International*, Vol. 43 No. 7, pp. 1824-1832, available at: http://dx.doi.org/10.1016/j.foodres.2009.10.013

Mugalavai, E.M., Kipkorir, E.C., Raes, D. and Rao, M.S. (2008), "Analysis of rainfall onset, cessation and length of growing season for Western Kenya", *Agricultural and Forest Meteorology*, Vol. 148 No. 6/7, pp. 1123-1135.

Muhanji, G., Roothaert, R.L., Chris, W. and Mwangi, S. (2011), "African indigenous vegetable enterprises and market access for small-scale farmers in East Africa", *International Journal of Agricultural Sustainability*, Vol. 9 No. 1, pp. 194-202, doi: 10.3763/ijas.2010.0561.

Mulinya, C., Ang'awa, F. and Tonui, K. (2016), "Constraints faced by small scale farmers in adapting to climate change in Kakamega county", *IOSR Journal of Humanities and Social Science (IOSR-JHSS)*, Vol. 21 No. 10, pp. 8-18, doi: 10.9790/0837-2110110818.

Muthomi, J. and Musyimi, D.M. (2009), "Growth and responses of African nightshades (solanum scabrum MILL) seedlings to water deficit", *ARPN Journal of Agricultural and Biological Science*, Vol. 4 No. 5.

Mutunga, E.J., Ndungu, C. and Muendo, P. (2017), "Smallholder farmers perceptions and adaptations to climate change and variability in Kitui county, Kenya", *Journal of Earth Science & Climatic Change*, Vol. 8 No. 3, doi: 10.4172/2157-7617.1000389.

Ndamani, F. and Watanabe, T. (2015), "Farmers' perceptions about adaptation practices to climate change and barriers to adaptation: a micro-level study in Ghana", *Water*, Vol. 7 No. 12, pp. 4593-4604, doi: 10.3390/w7094593.

Newing, H. (2011), *Conducting Research in Conservation. A Social Science Perspective*, Routledge, London.

Ngetich, J.K. (2013), "Planning and development of Kakamega county in Kenya: challenges and opportunities", *Research Journal in Organizational Psychology & Educational Studies*, Vol. 2 No. 3, pp. 111-118, Emerging Academy Resources.

Ngugi, I.K., Gitau, R. and Nyoro, J.K. (2006), *Access to High Value Markets by Smallholder Farmers of African Indigenous Vegetables in Kenya*, London.

Nyanga, P., Johnsen, F., Aune, J. and Kahinda, T. (2011), "Smallholder farmers' perceptions of climate change and conservation agriculture: evidence from Zambia", *Journal of Sustainable Development*, Vol. 4 No. 4, pp. 73-85, available at: http://dx.doi.org/10.5539/jsd.v4n4p73

Nyatuame, M., Owusu-Gyimah, V. and Ampiaw, F. (2014), "Statistical analysis of rainfall trend for Volta region in Ghana", *International Journal of Atmospheric Sciences*, Vol. 2014, available at: http://dx.doi.org/10.1155/2014/203245

Ochenje, I.M., Ritho, C.N., Guthiga, P.M. and Mbatia, O.L.E. (2016), "Assessment of farmers' perception to the effects of climate change on water resources at farm level: the case of Kakamega county, Kenya", Invited poster presented at the 5th International Conference of the African Association of Agricultural Economists, *Addis Ababa*.

Ochieng, J., Kirimi, L. and Mathenge, M. (2016), "Effects of climate variability and change on agricultural production: the case of small scale farmers in Kenya", *NJAS – Wageningen Journal of Life Sciences*, Vol. 77, pp. 71-78, doi: 10.1016/j.njas.2016.03.005.

Oerke, E.C. (2005), "Crop losses to pests", *The Journal of Agricultural Science*, Vol. 144 No. 1, doi: 10.1017/s0021859605005708.

Ogeto, R.M., Cheruiyot, E., Mshenga, P. and Onyari, C.N. (2013), "Sorghum production for food security: a socio-economic analysis of sorghum production in Nakuru county", *Kenya. African Journal of Agricultural Research*, Vol. 8 No. 47, pp. 6055-6067.

Ojiewo, C., Tenkouano, A., Oluoch, M. and Yang, R. (2010), "The role of AVRDC–the world vegetable center in vegetable value chains", *African Journal of Horticultural Science*, Vol. 3, pp. 1-23.

Oluoch, M.O., Germain, N., Pichop, D.S., Mary, O., Abukutsa-Onyango, M.D. and Shackleton, C.M. (2009), "Production and harvesting systems for African indigenous vegetables", in Shackelton, C.M., Pasquini, M. and Drescher, A.W. (Eds), *African Indigenous Vegetables in Urban Agriculture*, Earthscan, London, Sterling, VA, pp. 105-144.

Palada, M.C., Kalb, T.J. and Lumpkin, T.A. (2006), "The role of AVRDC–the world vegetable center in enhancing and promoting vegetable production in the tropics", *Horticultural Science*, Vol. 41 No. 3.

Pasquini, M.W., Assogba-Komlan, F., Vorster, I., Shackleton, C.M. and Abukutsa-Onyango, M.O. (2009), "The production of African indigenous vegetables in urban and peri-urban agriculture: a comparative analysis of case studies from Benin, Kenya and South Africa", in Shackleton, C.M., Pasquini, M. and Drescher, A.W. (Eds), *African Indigenous Vegetables in Urban Agriculture*, Earthscan, London, Sterling, VA, pp. 105-144.

Phirri, G.K., Egeru, A. and Ekwamu, A. (2016), "Climate change and agriculture nexus in Sub-Saharan Africa: the agonizing reality for smallholder", *International Journal of Current Pharmaceutical Review and Research*, Vol. 8 No. 2.

Porter, J.R.L.X., Challinor, A.J., Cochrane, K., Howden, S.M., Iqbal, M.M., Lobell, D.B. and Travasso, M.I. (2014), "Food security and food production systems", Field, C.B., Barros, V.R., Dokken, D.J., Mach, K.J., Mastrandrea, M.D., Bilir, T.E., Chatterjee, M., Ebi, K.L., Estrada, Y.O., Genova, R.C., Girma, B., Kissel, E.S., Levy, A.N., MacCracken, S., Mastrandrea, P.R. and White, L.L. (Eds), *Climate Change 2014: Impacts, Adaptation, and Vulnerability. Part a: Global and Sectoral Aspects. Contribution of Working Group II to the Fifth Assessment Report of the Intergovernmental Panel on Climate Change*, Cambridge University Press Cambridge, United Kingdom and New York, NY, pp. 485-533.

Prakash, A.R.J., Mukherjee, A.K., Berliner, J., Pokhare, S.S., Adak, T., Munda, S. and Shashank, P.R. (2015), Climate Change: Impact on Crop Pests, Cuttack-753 006, Odisha, India, Applied Zoologists Research Association (AZRA) Central Rice Research Institute.

Prasad, B.V.G. and Chakravorty, S. (2015), "Effects of climate change on vegetable cultivation – a review", *Nature Environment and Pollution Technology*, Vol. 14 No. 4, pp. 923-929, Qualitative Health Research, 1277-1288, Vol. 15 No. 9.

Rohrbach, D.D., Minde, I.J. and Howard, J. (2003), "Looking beyond national boundaries: regional harmonization of seed policies, laws and regulations", *Food Policy*, Vol. 28 No. 4, pp. 317-333, doi: 10.1016/j.foodpol.2003.08.005.

Rurinda, J., van Wijk, M.T., Mapfumo, P., Descheemaeker, K., Supit, I. and Giller, K.E. (2015), "Climate change and maize yield in Southern Africa: what can farm management do?", *Global Change Biology*, Vol. 21 No. 12, pp. 4588-4601, doi: 10.1111/gcb.13061.

Sanogo, K., Sanogo, S. and Ba, A. (2016), "Farmers' perception and adaption to land use change and climate variability in fina reserve, Mali", *Turkish Journal of Agriculture–Food Science and Technology*, Vol. 4 No. 4, doi: 10.24925/turjaf.v4i4.291-297.544.

Stöber, S., Chepkoech, W., Neubert, S., Kurgat, B., Bett, H. and Lotze Campen, H. (2017), "Adaptation pathways for African indigenous vegetables' value chains", in Leal Filho, W., Belay, S., Kalangu, J., Menas, W., Munishi, P. and Musiyiwa, K. (Eds), *Climate Change Adaptation in Africa, Climate Change Management*, pp. 413-433.

Tabari, H., Marofi, S., Aeini, A., Talaee, P.H. and Mohammadi, K. (2011), "Trend analysis of reference evapotranspiration in the Western half of Iran", *Agricultural and Forest Meteorology*, Vol. 151 No. 2, pp. 128-136, doi: 10.1016/j.agrformet.2010.09.009.

Thompson, K., Kruszewska, I. and Tirado, R. (2015), Building environmental resilience: a snapshot of farmers adapting to climate change in Kenya, *Greenpeace Research Laboratories Technical Report: 04-Commissioned by: Greenpeace Africa*.

Thomson, L.J., Macfadyen, S. and Hoffmann, A.A. (2010), "Predicting the effects of climate change on natural enemies of agricultural pests", *Biological Control*, Vol. 52 No. 3, pp. 296-306, doi: 10.1016/j.biocontrol.2009.01.022.

Thornton, P.K., Ericksen, P.J., Herrero, M. and Challinor, A.J. (2014), "Climate variability and vulnerability to climate change: a review", *Global Change Biology*, Vol. 20 No. 11, pp. 3313-3328, doi: 10.1111/gcb.12581.

Thornton, P.K., Lipper, L., Baas, S., Cattaneo, A., Chesterman, S., Cochrane, K., de Young, C., Ericksen, P., van Etten, J., de Clerck, F., Douthwaite, B., DuVal, A., Fadda, C., Garnett, T., Gerber, P.J., Howden, M., Mann, W., McCarthy, N., Sessa, R., Vermeulen, S. and Vervoort, J. (2013), "How does climate change alter agricultural strategies 'to support food security?'" *Background paper for the conference 'Food Security Futures:' Research Priorities for the 21st Century, Dublin*.

United Nations Children's Fund (UNICEF) (2013), *Improving Child Nutrition. The Achievable Imperative for Global Progress*, UNICEF, New York, NY, available at: www.unicef.org/publications/index. html, (accessed 12 June 2015), ISBN: 978-92-806-4686-3.

Van den Heever, E. and Slabbert, M.M. (2007), *Proceedings of the 1st International conference on indigenous vegetables and legumes prospectus for fighting poverty, hunger and malnutrition, Acta Horticulturae*, pp. 752-281.

Weinberger, K. and Pichop, G.N. (2009), "Marketing of African indigenous vegetables along urban and peri-urban supply chains in Sub-Saharan Africa", in Shackelton, C.M., Pasquini, M., Drescher, A. W. (Eds), *African Indigenous Vegetables in Urban Agriculture*, Earthscan, London, Sterling, VA, PP. 105-144.

West, J.S., Fitt, B.D.L., Townsend, J.A., Stevens, M., Edwards, S.G., Turner, J.A. and Edmonds, J. (2015), Impact of climate change on diseases on sustainable crop systems: CLIMDIS, Project Number 539. HCGA.

Yang, R.Y. and Keding, G.B. (2009), "Nutritional contributions of important African indigenous vegetables", in Shackelton, C.M., Pasquini, M. and Drescher, A.W. (Eds), *African Indigenous Vegetables in Urban Agriculture*, Earthscan, London, Sterling, VA, pp. 105-144.

Further reading

Abukutsa, O.M.O. (2003), "Unexploited potential of indigenous African vegetables in Western Kenya", *Maseno Journal of Education Arts and Science*, Vol. 4 No. 1, pp. 103-122.

Akponikpe, P., Johnston, P. and Agbossou, E.K. (2010), "Farmers' perceptions of climate change and adaptation strategies in Sub-Sahara West Africa", 2nd International Conference on Climate, Sustainability and Development in Arid Regions, *Fartaleza-Ceara*.

AVRDC (2004), *AVRDC Medium-Term Plan: 2004–2006*, AVRDC – the World Vegetable Center, Shanhua.

Bebber, D.P., Holmes, T. and Gurr, S.J. (2014), "The global spread of crop pests and pathogens", *Global Ecology and Biogeography*, Vol. 23 No. 12, pp. 1398-1407, doi: 10.1111/geb.12214.

Chabala, M.L., Kuntashula, E. and Kaluba, P. (2013), "Characterization of temporal changes in rainfall, temperature, flooding hazard and dry spells over Zambia", *Universal Journal of Agricultural Research*, Vol. 1 No. 4, pp. 134-144, doi: 10.13189/ujar.2013.010403.

Juana, J.S., Kahaka, Z. and Okurut, F.N. (2013), "Farmers' perceptions and adaptations to climate change in Sub-Sahara Africa: a synthesis of empirical studies and implications for public policy in African agriculture", *Journal of Agricultural Science*, Vol. 5 No. 4, p. 121, doi: 10.5539/jas. v5n4p121.

Nzau, M. (2013), *Mainstreaming Climate Change Resilience into Development Planning in Kenya*, IIED country report, IIED, London.

Nzeadibe, T.C., Egbule, C.L., Chukwuone, N. and Agu, V. (2011), "Farmers' perceptions of climate change governance and adaptation constraints in Niger delta region of Nigeria", African Technology Policy Network, Research Paper No. 7.

Ochieng, M.A. and Koske, J. (2013), "The level of climate change awareness and perception among primary school teachers in Kisumu municipality, Kenya", *International Journal of Humanities and Social Science*, Vol. 3 No. 21, p. 174.

Ojwang, G.O., Agatsiva, J. and Situma, C. (2010), *Analysis of Climate Change and Variability Risks in the Smallholder Sector Case Studies of the Laikipia and Narok Districts Representing Major Agro-Ecological Zones in Kenya*, Rome.

Okonya, J.S., Syndiku, K. and Kroschel, J. (2013), "Farmers' perception of and coping strategies to climate change: evidence from six agro-ecological zones of Uganda", *Journal of Agricultural Science, Canadian Centre for Science and Education*, Vol. 5 No. 8, p. 252.

Author affiliations

Winifred Chepkoech, Centre for Rural Development (SLE), Albrecht Daniel Thaer-Institute of Agricultural and Horticultural Sciences, Humboldt University of Berlin, Berlin, Germany

Nancy W. Mungai, Department of Crops, Horticulture and Soils, Egerton University, Egerton, Kenya

Silke Stöber, Centre for Rural Development (SLE), Albrecht Daniel Thaer-Institute of Agricultural and Horticultural Sciences, Humboldt University of Berlin, Berlin, Germany

Hillary K. Bett, Department of Agricultural Economics and Agribusiness Management, Egerton University, Egerton, Kenya, and

Hermann Lotze-Campen, Potsdam Institute for Climate Impact Research (PIK), Potsdam, Germany and Faculty of Life Sciences, Department of Sustainable Land Use and Climate Change, Humboldt University of Berlin, Berlin, Germany

Corresponding author

Winifred Chepkoech can be contacted at: sangwinfred@gmail.com

PERMISSIONS

All chapters in this book were first published in IJCCSM, by Emerald Publishing Limited; hereby published with permission under the Creative Commons Attribution License or equivalent. Every chapter published in this book has been scrutinized by our experts. Their significance has been extensively debated. The topics covered herein carry significant findings which will fuel the growth of the discipline. They may even be implemented as practical applications or may be referred to as a beginning point for another development.

The contributors of this book come from diverse backgrounds, making this book a truly international effort. This book will bring forth new frontiers with its revolutionizing research information and detailed analysis of the nascent developments around the world.

We would like to thank all the contributing authors for lending their expertise to make the book truly unique. They have played a crucial role in the development of this book. Without their invaluable contributions this book wouldn't have been possible. They have made vital efforts to compile up to date information on the varied aspects of this subject to make this book a valuable addition to the collection of many professionals and students.

This book was conceptualized with the vision of imparting up-to-date information and advanced data in this field. To ensure the same, a matchless editorial board was set up. Every individual on the board went through rigorous rounds of assessment to prove their worth. After which they invested a large part of their time researching and compiling the most relevant data for our readers.

The editorial board has been involved in producing this book since its inception. They have spent rigorous hours researching and exploring the diverse topics which have resulted in the successful publishing of this book. They have passed on their knowledge of decades through this book. To expedite this challenging task, the publisher supported the team at every step. A small team of assistant editors was also appointed to further simplify the editing procedure and attain best results for the readers.

Apart from the editorial board, the designing team has also invested a significant amount of their time in understanding the subject and creating the most relevant covers. They scrutinized every image to scout for the most suitable representation of the subject and create an appropriate cover for the book.

The publishing team has been an ardent support to the editorial, designing and production team. Their endless efforts to recruit the best for this project, has resulted in the accomplishment of this book. They are a veteran in the field of academics and their pool of knowledge is as vast as their experience in printing. Their expertise and guidance has proved useful at every step. Their uncompromising quality standards have made this book an exceptional effort. Their encouragement from time to time has been an inspiration for everyone.

The publisher and the editorial board hope that this book will prove to be a valuable piece of knowledge for researchers, students, practitioners and scholars across the globe.

LIST OF CONTRIBUTORS

Ahmad Rajabi and Zahra Babakhani
Department of Water Engineering, College of Agriculture, Kermanshah Branch, Islamic Azad University, Kermanshah, Iran

William M. Fonta
West African Science Service Centre on Climate Change and Adapted Land Use, Ouagadougou, Burkina Faso

Abbi M. Kedir
Management School, University of Sheffield, Sheffield, UK

Aymar Y. Bossa, Karen M. Greenough and Bamba M. Sylla
West Africa Science Service Center on Climate Change and Adapted Land Use, Ouagadougou, Burkina Faso

Elias T. Ayuk
United Nations University Institute for Natural Resources in Africa (UNU-INRA), University of Ghana, Accra, Ghana

Xiao-jun Wang
State Key Laboratory of Hydrology-Water Resources and Hydraulic Engineering, Nanjing Hydraulic Research Institute, Nanjing, China and Research Center for Climate Change, Ministry of Water Resources, Nanjing, China

Jian-yun Zhang
State Key Laboratory of Hydrology-Water Resources and Hydraulic Engineering, Nanjing Hydraulic Research Institute, Nanjing, China and Research Center for Climate Change, Ministry of Water Resources, Nanjing, China

Shamsuddin Shahid
Faculty of Civil Engineering, Universiti Teknologi Malaysia (UTM), Johor Bahru, Malaysia Lang Yu, China Institute of Water Resources and Hydropower Research, Beijing, China

Chen Xie
Yellow River Conservancy Commission, Zhengzhou, China Bing-xuan Wang, Hohai University, Nanjing, China

Xu Zhang
State Key Laboratory of Hydrology-Water Resources and Hydraulic Engineering, Nanjing Hydraulic Research Institute, Nanjing, China and Research Center for Climate Change, Ministry of Water Resources, Nanjing, China

Arega Shumetie
Haramaya University, Dire Dawa, Ethiopia, and Makerere University, Kampala, Uganda

Molla Alemayehu Yismaw
Radboud University, Nijmegen, The Netherlands

Francis Wasswa Nsubuga
Geography, Geoinformatics and Meteorology, University of Pretoria, Pretoria, South Africa

Hannes Rautenbach
South African Weather Service, Pretoria, South Africa and School of Health Systems and Public Health, University of Pretoria, South Africa

Stuart Capstick
School of Psychology and Tyndall Centre for Climate Change Research, Cardiff University, Cardiff, UK

Sarah Hemstock
Secretariat of the Pacific Community (SPC), Suva Regional Office, Suva, Fiji

Ruci Senikula
University of the South Pacific, Suva, Fiji

Guillaume Rohat
Institute for Environmental Sciences, University of Geneva, Geneva, Switzerland and Faculty of Geo-Information Science and Earth Observation, University of Twente, Enschede, The Netherlands

Stéphane Goyette
Institute for Environmental Sciences, University of Geneva, Geneva, Switzerland

Johannes Flacke
Faculty of Geo-Information Science and Earth Observation, University of Twente, Enschede, The Netherlands

Clara Inés Pardo Martínez
Universidad del Rosario, Bogotá, Colombia and Colombian Observatory of Science and Technology, Bogotá, Colombia

William H. Alfonso P.
Universidad del Rosario, Bogotá, Colombia

Aideen Maria Foley
Department of Geography, Birkbeck University of London, London, UK

Hilary Bambrick
School of Public Health and Social Work, Queensland University of Technology, Queensland, Australia

Wahid Ullah and Takaaki Nihei
Department of Regional Geography, Graduate School of Letters, Hokkaido University, Sapporo, Japan

Muhammad Nafees, Rahman Zaman and Muhammad Ali
Department of Environmental Sciences, University of Peshawar, Peshawar, Pakistan

Winifred Chepkoech
Centre for Rural Development (SLE), Albrecht Daniel Thaer-Institute of Agricultural and Horticultural Sciences, Humboldt University of Berlin, Berlin, Germany

Nancy W. Mungai
Department of Crops, Horticulture and Soils, Egerton University, Egerton, Kenya

Silke Stöber
Centre for Rural Development (SLE), Albrecht Daniel Thaer-Institute of Agricultural and Horticultural Sciences, Humboldt University of Berlin, Berlin, Germany

Hillary K. Bett
Department of Agricultural Economics and Agribusiness Management, Egerton University, Egerton, Kenya

Hermann Lotze-Campen
Potsdam Institute for Climate Impact Research (PIK), Potsdam, Germany and Faculty of Life Sciences, Department of Sustainable Land Use and Climate Change, Humboldt University of Berlin, Berlin, Germany

Index

www.ingramcontent.com/pod-product-compliance
Lightning Source LLC
Chambersburg PA
CBHW061949190326
41458CB00009B/2825